Reclaiming
the
Sacred

Healing Our Relationships
with Ourselves
and the World

Jeff Golden

Unless otherwise cited, all poems from the Penguin publication *Love Poems from God: Twelve Sacred Voices from the East and West* by Daniel Ladinsky, copyright 2002 and used with permission.

"Where is the Door to the Tavern?" and "Now Is the Time" from *The Gift: Poems of Hafiz* by Daniel Ladinsky, copyright 1999 and used with permission.

Robin Wall Kimmerer, excerpts from *Braiding Sweetgrass: Indigenous Wisdom, Scientific Knowledge and the Teachings of Plants*. Copyright © 2013, 2015 by Robin Wall Kimmerer. Reprinted with the permission of The Permissions Company, LLC on behalf of Milkweed Editions. www.milkweed.org.

Plant Intelligence and the Imaginal Realm: Into the Dreaming of Earth by Stephen Harrod Buhner published by Inner Traditions International and Bear & Company, ©2014. All rights reserved. Reprinted with permission of publisher. www.Innertraditions.com

"Love After Love" from *Sea Grapes* by Derek Walcott. Copyright © 1976 by Derek Walcott. Reprinted by permission of Farrar, Straus and Giroux. All Rights Reserved.

"The Guest House" from *The Illuminated Rumi* by Jalal Al-Din Rumi, translated by Coleman Barks, copyright © 1997 by Coleman Barks and Michael Green. Used by permission of Broadway Books, an imprint of Random House, a division of Penguin Random House LLC. All rights reserved

ISBN: 9798986725420 (pbk.) — ISBN: 9798986725475 (e-book)

www.reclaimingthesacred.net

All proceeds from this book are being donated to nonprofits doing critical work related to the themes of this book. See www.reclaimingthesacrd.net/proceeeds for more information.

Contents

Introduction

WE IN THE United States live amidst material wealth unimaginable to most people throughout human history, and far beyond the reach of most people throughout the world today. Yet we also live amidst an immense poverty. Surveys reveal that, on average, we are less happy, more depressed, and lonelier than ever. We consume two-thirds of the world's antidepressants, drug overdoses are the number one cause of death for those of us under age fifty, and each year one in twenty-five of us seriously considers suicide.[1]

This is not a coincidence, or a contradiction. This material wealth and poverty of spirit are intimately related.

Research shows that money and possessions do very little for our happiness or well-being—far less than most Americans imagine. To be clear, money *can* boost our happiness, but only up until we're able to meet our basic needs at a very simple level. After that, additional money does almost nothing for us.

Research also shows that materialism is "toxic" for happiness, that the more importance we place on money and possessions, the more strained our relationships tend to be, the lower our sense of self-worth, and the more fleeting our happiness. Add to that, the pursuit of money and possessions often crowds out the things that really *do* matter, like spending time with friends and family, being physically active, and having time to do the things we find enjoyable and just relax.

In many ways, the tremendous materialism and material wealth of this society are, in fact, *evidence* of the deeper poverty we live in. Research shows that one of the major reasons money is so important to so many of us, despite doing so little to actually benefit us, is that we use it to try to make up for a lack of some of the things that really *do* matter: we tend to buy more and give more significance to money and work when we feel unhappy, stressed, or lonely, when we feel inadequate, scared, or rejected, or when we feel a lack of purpose in our lives.

Working more and buying things are ways we try to numb these feelings of lack, or to distract ourselves from them. Ultimately, they are ways we try to fill the holes in ourselves and our lives.

---

[1] All facts and quotes in the introduction are cited in the main body of the book. Endnotes are also available online, for easier reading and clickable links:

www.reclaimingthesacred.net/endnotes

In other words, the fact that we in the US care so much about money and possessions, and have collectively accumulated such staggering material wealth—these are indicators that something is very wrong. They speak to how diminished our lives are in very important respects, and how disconnected we have become from the fundamental wealth of ourselves and the world.

But we don't just suffer this poverty ourselves. Our addiction to consumption causes tremendous harm to others as well. We create waste that will be hazardous to life for millions of years; we discard so much trash that soon there will be more plastic than fish in the oceans; we relegate huge numbers of people to poverty, even to slavery—there are more slaves in the world today than ever before, many of them making products for the American market. In the US alone, we condemn ten billion animals each year to spend their lives locked in factory farms.

We are, meanwhile, destroying the very fabric of life on the planet. Humans have killed half the population of wild animals in the last forty years, we have destroyed or degraded 85 percent of the world's forests, and we have transformed more than 70 percent of the earth's land surface.

Overshadowing all of this though, is what we're doing to the climate. There is nothing in human history—not world wars or plagues or famines—that can compare with the amount of violence and destruction we are unleashing through climate change. Droughts, wildfires, the spread of disease, the collapse of food supplies, an estimated one billion climate refugees—these are just a few of the consequences we're facing.

Between 40 and 80 percent of all species may not survive to see the next century.

We are one of those species.

Americans are only 4 percent of the world population, yet we have caused as much as one-third of global warming. We—Americans—have to make significant changes. That includes things like shifting to renewable energy and efficient technologies. It also includes ending the $20 billion in subsidies we give to oil companies every year, and banning new fossil fuel projects. But we have to go deeper than that.

We have to address the root causes of this violence, the root causes of our addiction to consumption. Otherwise we as a society will do what we've always done, steamroll right over these changes and keep consuming more and more.

For those of us who have access to the things that have to be scaled back—which is many of us in the US but very few of us worldwide—these changes can evoke a sense of sacrifice or loss. But what we have before us is actually an opportunity to live richer and more joyful lives.

As we've surrounded ourselves with more and more possessions, we've grown further from the sacredness of the world, and the sacredness of ourselves. As we've elevated economic growth and production and consumption to the highest measures of success and purpose, we've closed ourselves off from so much of the joy and wonder and belonging that are inherent in us and the world.

We have an opportunity to reorient our lives toward the things that really *do* matter. An opportunity to slow down, to be present, to hear our deeper callings. An opportunity to shift our attention from "the things we *don't* have" or "the things we have to give up," to **so much that we *do* have.** From food and shelter, to our breath, the sky, movement, life.

We have an opportunity to reweave ourselves back into the human community and the family of all living beings, the family of the land and trees, the otters and grasses—to live with them in relationships of respect and wonder.

We have an opportunity to reclaim ourselves and this world as sacred.

It is a profound blessing that so many of the changes our current crisis requires are precisely what best serve us. So much of what we must surrender actually impoverishes us—even those of us who have supposedly benefitted the most from it. And so much of what we must embrace actually nourishes happiness, abundance, and belonging.

Indeed, the extensive research that has been done in regards to the sciences of happiness, abundance, and belonging offer a way forward out of this system and its deceptions and poverty, and into entirely different ways of thinking and feeling and living. **They are like golden threads we can follow home to the heart of ourselves and the world.**

That is the journey of this book: to follow these golden threads of happiness, abundance, and belonging, **to return home***.*

GUIDING US IN this journey will be literally thousands of experts—Nobel and Pulitzer Prize-winning psychologists and economists, Indigenous people, saints and poets, cosmologists and activists, and a few fictional characters thrown in for good measure. Their work and insights flow through every page, yielding the roughly 2,500 references at the end of the book.

I also offer to you, woven into this journey, the terrain of my own mind and heart. Only occasionally does my own life make an explicit appearance, but it is present in the vision I bring to these pages.

I am here as a father, and also as a six-year-old boy cracked open by the death of my own father. I am here as a husband of sixteen years, and as a man cracked open again with the end of that marriage. I am a child become a teenager, marked by periodic bursts of violence in the form of being bullied. And I am a teenager grown into an adult, carrying deep insecurities around masculinity and sex and being a very sensitive person. And I am that sensitive person, nourished by significant time spent with the stars and trees and rivers and animals, and by living and traveling widely throughout the world, and by just doing a lot of deep personal work.

Also woven into this book are my more explicitly professional experiences. The most obvious of these, in terms of what would appear on a resume, are my two decades of directing several nonprofits, and cofounding and living in two intentional communities dedicated to sustainability and justice. Nonprofits and intentional communities are both notoriously challenging, and I definitely experienced a lot of difficulty and heartbreak during those times, along with much success and joy.

Also present in this book, though, are many experiences hidden behind those broad strokes, each with its own mix of accomplishments and shortfalls. For example, about fifteen years ago, while heading one of those nonprofits, I directed the creation of one of the greenest certified buildings in the US. Then, when nobody in the local area could even tell me which Native Americans' homeland the building was located in, I researched that history, and I went to visit members of the Stockbridge-Munsee Band of Mohican Indians in Wisconsin, and I undertook projects to help raise awareness back in New York about them and their history. Then, when it came time for the nonprofit to sell the building, we directed the proceeds to the Mohican Language and Culture Committee as a way of supporting them and acknowledging their rightful stake in the land we had benefitted from and bought and sold.

To offer another example, in 2015, when corrections officers murdered a man, Sam Harrell, who was incarcerated in a prison mere minutes from my home (a young Black man with bipolar disorder, incarcerated for selling drugs), I dropped everything to help organize the local community and work with his family to seek justice for him, and to try to help prevent anything like that from happening again. We led protests, including a brief hunger strike by members of his family, we got national media coverage, and we met with more than twenty state representatives. We were successful in raising awareness about the need for prison reform, and we contributed to the passage of restrictions on the use of solitary confinement in New York State. But we had to live with the heartache and rage that nothing else at the prison changed, and not one of the officers involved in the murder was ever held accountable.

Some of my other professional experiences that inform my life and the journey of this book are: six years I spent launching an online education program for students to collaborate with Nobel Peace Prize laureates and others around the world to learn about current events and culture; five years where I participated in powerful dialogue work on "Race, Class, Gender and Power," led by the incredibly skilled and visionary people at Be Present, Inc.; five years I gave to teaching Latin American literature in the Spanish bilingual program at Mission and Leadership High Schools in San Francisco; and a year working with community organizers in a shantytown in Venezuela (Barrio Bolivar) and evaluating environmental education programs throughout the country.

I've been researching and writing this book for twelve years, I've been leading courses on these topics for over five years, most recently at Vassar College, and I've been writing and teaching about them in other settings for over thirty years.

Humbly, and with a full heart, I offer myself (and our thousands of guides) and this book to you.

OVER THE COURSE of seven parts, two codas, and three interludes, this book will lay before you the golden threads of happiness, abundance, and belonging, inviting you to follow them.

The science of happiness—the vast body of research that has been conducted by psychologists, economists, and sociologists, their countless studies, surveys, and peer reviews—**this science will illuminate for you the most critical factors that contribute to human well-being and how to apply them in your own life**. As mentioned earlier, this science will definitely invite you to spend time with family and friends, to be physically active, and to take time to relax. But more than that, it will take you soaring through galaxies and diving into the foundations of the universe, seeking to stir in you wonder and delight. It will sit with you in stillness before a cherry blossom, a blade of grass, your own breath, seeking to nourish presence and gratitude. It will walk with you gently into your own heart, helping you heal wounds, touch emotions, and listen to your deepest callings.

The science of abundance—and the evolutionary biologists, Indigenous writers, activists, and others who will join our team of guides—will take you behind the scenes of this careening, materialistic train we are on. **This science will make it clear just how limited the role is of money and possessions in our well-being**, it will tell you concretely how much money really is enough to maximize your well-being, and it will lay bare the ways that money so often hooks us, despite not doing much for our happiness. *And then…* the science of abundance will take you by the hand and jump off this train with you. It will lie with you looking up at the sky as the train disappears into the distance, as the sounds of the wind and birds and crickets emerge in its place. **This science will reveal the vast abundance that exists in the world around you and the world within you.**

Then, when you are ready, **the science of belonging**—along with the physicists, brain researchers, cosmologists, saints, and poets who will add their voices—will climb with you to the tops of the pillars of the scientific revolution, and then dive with you into the vast realm of possibility that lies beyond. It will affirm that the forests and rivers and bison and salamanders, are so much more than resources, that they are each a face of creation and deserve our reverence. And it will affirm that *you* are also a face of creation, that you are as central to the unfolding of the universe as any supernova or waterfall or gravity, and that you also deserve reverence. **It will read love notes to you from the world, and it will remind you that you are yourself a love note to the world.**

YOU.

You are sacred.

When this book refers to *Reclaiming the Sacred*, the most important thing it is referring to is *you*.

By "sacred," I do not mean something that is considered sacred within a particular religious tradition or institution. I mean it in the sense of those things that are of the most profound meaning and importance, worthy of being held with the utmost awe and respect.

Author Arne Garborg once wrote, "To love a person is to learn the song that is in their heart and to sing it to them when they have forgotten." **That is the purpose of this book, to sing to you parts of your song, your beautiful, vibrant song.** It is to sing to you of your immense wondrousness and inherent belonging. It is to sing to you of the sacredness of you and the world. It is to help you connect with the joy and purpose that blossom when we live in these truths. It is to help you forge the kind of life and the kind of world that we all deserve.

Jeff Golden
June, 2022

Part I

The
Abundance
Point

Chapter 1:

Money Just Can't Buy
What it Used to

IN MANY WAYS, Americans today live in an entirely different world from the one we knew just a century ago. As late as the 1940s most Americans lived in conditions that few today would consider acceptable.

A third of homes didn't have indoor toilets or running water, more than half lacked central heating, and almost none had air-conditioning.[1] There was one car for every five people, a phone was still a luxury for most, and a 1945 poll found that most Americans didn't even know what a "television" was.[2] Food, penicillin, and many manufactured goods were in short supply due to World War II.[3]

We've come a long way since then:

- Running water is nearly universal.[4]
- Most homes not only have air-conditioning and central heating, but also a fridge, washer, dryer, and dishwasher, as well as a TV, and an average of *twenty-five* consumer electronic devices.[5]
- New houses have more than doubled in size, even while the average household has shrunk by more than one person.[6]
- Two-thirds of homes have a garage or carport, and there are fifty million more registered vehicles than licensed drivers.[7]
- Not only do the vast majority of Americans have a phone, but two-thirds of us have a smartphone, which 47 percent of us report we "couldn't last a day without."[8]

Our per capita income, adjusted for inflation, has increased by more than 250 percent since the 1940s.[9] Most of those gains have gone to the wealthiest of us, but median income, a statistic that better reflects the reality of most Americans, has still increased by nearly 60 percent.[10]

Fifteen of every one hundred of us in the US still live in poverty, and our lives are shaped by the higher rates of violence, abuse, neglect, and certain mental health issues that come with that.[11] For some of us, living in poverty involves hunger or even homelessness. Yet even those of us living in poverty are generally far better off *in material terms* than the average person was in the 1940s. Most Americans in poverty today not only have the "basics" that 1940s Americans did *not* have—running water, heat, air-conditioning, and a washer/dryer—but also generally have a car, two TVs, cable or satellite, several electronic devices, and more living space than the average European.[12]

What has the impact of these monumental material gains been on our happiness? On a scale of one to ten, in the 1940s we scored 7.5. As of 2008 we scored 7.2.[13] That's right; we are *less* happy. In fact, **our rates of happiness have been in a pretty constant decline since the 1940s, even during periods of great economic growth.**[14] On top of that:

- Our depression rates have increased tenfold.[15]
- One in fifteen of us abuses alcohol.[16]
- We consume two-thirds of the world's antidepressants—more than any other prescribed drug.[17]
- Drug overdoses are the number one cause of death for those under age fifty.[18]
- The number of us with no close friends has quadrupled since 1990, and more than half of us feel that there is not a single person who knows us well.[19]
- We have the fourteenth highest murder rate in the world, eighteen times higher than any Western European country, and we commit suicide at about twice the rate we kill each other, up 33 percent in roughly the last twenty years. *One in twenty-five of us seriously considers suicide every year.*[20]
- The average child experiences more anxiety on a regular basis than the level of anxiety that in the 1950s caused a child to be referred for psychiatric treatment, and more than one in twenty teens today is on some kind of psychiatric medication.[21]

In his landmark book *The Loss of Happiness in Market Democracies*, political psychologist Robert E. Lane offers a visual representation of one aspect of this: he charts our declining happiness against our increasing incomes from 1940 to 1990 (see next page)

As the title of Lane's book suggests, we're not alone in this general trend. Levels of happiness have remained virtually unchanged in most industrialized countries since the 1940s, with some edging upward slightly, but many declining. This is perhaps most striking in the case of Japan, whose per capita income following World War II increased five-fold in just thirty years, doubling what we have accomplished in the US, and in half the time. Yet their happiness has remained the same.[22]

On the other hand, there are many countries—Nigeria, Mexico, and Vietnam, for example—whose average income is only a fraction of ours, yet whose happiness is greater.[23]

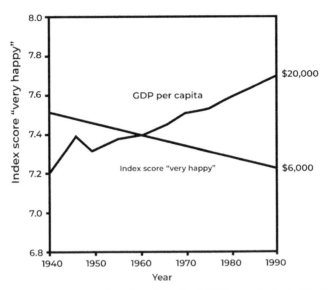

Figure 1. Index of people reporting being "very happy" and GDP per capita in the US, 1946–1990.[24]

Likewise, the Irish are happier than the Germans, who are twice as financially wealthy, the Taiwanese are as happy as the Japanese, who are three times as wealthy, and so on.[25]

How is this possible? How is it that we are swimming in this vast material wealth, and yet we are less happy than we were in the relative poverty of the 1940s? Central to understanding this, and to seizing a profound opportunity to live more joyful and sustainable lives, is *the abundance point.*

AT THE HEART of the abundance point is the fact that money simply has a very limited impact on our happiness. To be clear, money *does* buy happiness—a good deal of it, in fact, for many of us worldwide. Study after study confirms that if we do not have enough money to meet our basic needs such as food, clothing, and shelter, then, on average, every bit of money we get until we can meet those needs *does* boost our happiness.[26] It's just that after that the relationship between money and happiness drops off rapidly.

The abundance point (my term) is the point where that drop-off happens, which is basically the point where we are able to meet our basic needs. **Up until the abundance point, there is a consistent link between money and happiness. Beyond the abundance point, not so much.** More money does continue to buy more happiness, just very little of it, and its impact diminishes rapidly until, **at *the saturation point* (again, my term), additional money does *nothing* for our happiness.**[27]

This means you could give a small amount of money to someone who lives *below* the abundance point and they will usually be happier for it. Give someone who lives *above* the abundance point some money, even a lot of it, and they will also tend to be happier, but only very slightly. Hand someone who lives at or beyond the *saturation* point a lot of money, even vast amounts of it, and (after an initial rush) and it will generally have no effect; they will tend to be just as happy as they were before.

This helps explain why people with incomes over $10 million annually tend to be only slightly happier than the average person. And why 37 percent of the wealthiest Americans are actually *less* happy than the average American.[28]

Even for those of us who live below the abundance point, we need to be careful not to exaggerate the importance of money to happiness. For a long time, as one gets further and further below the abundance point, people still tend to be happy. It is only when we reach the point of the most extreme and persistent poverty (what we could call *the deprivation point*) that that flips and people tend to be unhappy. Even then the relationship is weak enough that, for example, most people in Malawi are fairly happy, even though three-fourths of them live on the equivalent of $1.25 a day or less.[29]

Interestingly, the amount of money we *spend* seems to impact our happiness even *less* than the amount of money we have or receive in income. With the exception of a few specific and limited kinds of spending, like donating to charity and spending on social activities, research suggests that for people living above the abundance point, *there is zero connection between spending and happiness.*[30]

Put all of this together and researchers have concluded that **our income accounts for only 2 to 4 percent of our happiness.**[31]

THE RELATIVELY LIMITED ability of money to impact our happiness is underscored by just how much the reverse is true, that is, by how much power our happiness actually has over our financial situation. It's not that being happy brings us more money (though research suggests there is something to that as well, because we tend to be more productive, healthier, and more creative when we are happy, and we are more likely to be hired and promoted[32]); it's rather that *our happiness largely determines our financial outlook.*

You would think that the more money a person has, the more satisfied they would be with their financial situation, right? And that's true to a degree. But how happy someone is, in general, is a *four times* better predictor of how satisfied they are with their financial situation than their income.[33] In other words, **if we're generally happy, then we tend to be happy with our income, whatever it is. If we're not happy, then no matter how much money we make, we generally aren't happy with it.**

Perhaps it's not surprising then that we Americans are not only slightly less happy than we were in the 1940s, but we are also less happy with our financial situations, despite our stunning material gains.[34]

THERE IS A saying, "Enough is as good as a feast." This captures the spirit behind the name *the abundance point*. Once we have enough money to meet our basic needs, having more money is like having more food set before us when we're already satisfied—we can continue to enjoy it for a while, but nothing like when we were hungry, and at some point all the food in the world could be set before us to no effect. **Once our needs are met, we have truly crossed over into a realm of abundance, where, in terms of happiness anyway, there is very little difference between what we actually have and having everything in the world.**

It's not just that once we have enough money to meet our basic needs, more money doesn't do much for us; it's that *at that point we already generally have all the money we need to maximize our happiness*, and pursuing more money actually becomes an *obstacle* to realizing that potential. This is because there are many things that have a greater impact on our happiness than money, and pursuing more money often means sacrificing those other things.

We will be exploring these critical happiness factors in parts II and III, "Charting the Path of a Joyful Life" and "Happiness and the Journey Within," but consider for now this one example. Political economist Stefano Bartolini has calculated that "a person with no friends or social relations with neighbors would have to earn $320,000 more each year" just to enjoy the same level of happiness as someone who does.[35] Most of us could dedicate our entire lives to making money and couldn't ever make the $320,000 a year needed to compensate for having a very limited social life. Or we could much more easily, and for a fraction of the time, simply commit to making some friends and meeting our neighbors, and tend to some of the other major happiness factors on top of that, and be just as happy, or happier.

And, as striking as this statistic is, it actually gives money far too much credit. It does not take into account the very fact we're exploring here about money, that it buys less and less happiness until, at the saturation point, it can buy no more. As we will soon see, researchers have been able to put specific amounts to the abundance and saturation points, and $320,000 is *vastly* beyond both of them. This means we could actually keep heaping money on the heads of people with no social relations forever and (on average) they would *never* be as happy as people who do have friends and interactions with their neighbors.

So even though, yes, more money does generally boost our happiness up until the saturation point, **pursuing more money often actually means being *less* happy than we could be otherwise.**[36] In fact, this is true *even for those of us who do not have enough money to meet all our basic needs*, those of us who live *below* the abundance point—these other factors are *still* often much more significant than money.[37]

ALL OF THIS offers at least a cursory answer to our question: "How is it that we're swimming in this vast material wealth yet we're less happy than we were in the 1940s?" But it raises another important one: "Why is money so important to so many of us and our society if it has so little effect on our happiness?" We need to have some understanding of this if we are going to be able to take seriously just how little money does for our happiness, and to resist whatever the hooks are that catch so many of us, and cause so many of us to choose the pursuit of money over the things that really do matter.

The answer involves biology, culture, sex, and more, which will be the focus of part IV, "Dueling with Death, Destiny, and Other Minor Demons." In short, though, it has a lot to do with the same happiness factors we will explore in parts II and III. Specifically, money is able to hook us largely by rushing in and (to a certain degree) making up for an absence of those things that really *do* have a significant impact on our happiness. Money can't actually replace them, but it can help distract us and somewhat numb the discomfort and pain of those absences.

In part V, "Living the Great Lie," we will explore the vast consequences of our making money such a priority and neglecting those things that really do matter. We'll also meet with some youth activists, indigenous leaders, formerly enslaved people, and the heads of revolutionary movements, as we plot our exit strategies from this system of money-and-possessions-first.

Then in part VI, "Coming Home," we will go beyond all of this as we consider some final critical questions, "What is the alternative?" and "Where do we go from here?" **We will walk right up to the edge of what this system can see and imagine, and we will leap into the immense and dazzling possibility beyond.** We will explore entirely different ways of thinking and feeling about ourselves and the world, ways that are rooted in abundance, belonging, and love.

But before we get ahead of ourselves, there is a bit more for us to explore about the relationship between money and happiness. And central to that is what psychologists refer to as the *set point*. Understanding the set point will provide critical insight into how the abundance point works, while also introducing us to *the single greatest factor affecting our happiness*.

Chapter 2:

Twins, Lotteries, and Human Resilience: Welcome to the Set Point

I N 1998, FIVE decades into our country's astonishing material and financial transformation described in the last chapter, and five decades into our slow-motion decline in happiness, the newly elected head of the American Psychological Association, Martin Seligman, chose "positive psychology" as the theme of his presidency.[38]

The genesis story of positive psychology goes something like this: In the beginning there was psychology. Psychology was dedicated to studying the whole range of mental disorders that afflict humans. But then some psychologists decided to turn that on its head and focus instead on what leads to human flourishing. Thus was born a new field of study, one that focuses its attention on all things related to happiness and well-being instead of illness.[39]

Happiness has been a central concern for humans as long as we have been around— we are naturally drawn to those things that make us happy and seek to avoid those that make us *un*happy.[40] We have thousands of years of texts of people writing about happiness, tens of thousands if you consider the earliest known cave paintings of horses and reindeer to be reflections on the good things in life.[41]

But it's only in the past several decades that we've begun to rigorously test our ideas about happiness. Research on the subject has blossomed during that time, as Seligman both acknowledged and spurred on by making it the theme of his presidency. Today there's a whole industry dedicated to positive psychology, involving journals, conferences, books, podcasts, personal coaches, tote bags, and more. The field is rife with disagreements and there are still huge gaps in what we know, but it has also made huge strides and has achieved consensus in many areas.

One of the most significant of these points of consensus came to light in 1996, three years before positive psychology actually got its name. A landmark study by psychologists David Lykken and Auke Tellegen found that for identical twins the number one predictor of how happy they were wasn't their income, their level of education, or whether they were

David Lykken and Auke Tellegen found that for identical twins the number one predictor of how happy they were wasn't their income, their level of education, or whether they were married. Rather, it was how happy their twin was. This was not just the number one predictor, it was *by far* the number one predictor.

The happiness of *non*identical twins varied significantly, but for identical twins the effect was just as strong for those who'd been raised apart as it was for those who'd been raised together, meaning the very similar levels of happiness weren't the result of their having had the same upbringing.[5] There was clearly something genetic going on.

Thanks to that study and others like it, psychologists have established that roughly 50 percent of our happiness is genetically based.[6] Over time, independent of much of what happens in our lives, we tend to settle back in around a level of happiness that's natural for each of us—what's referred to as our *set point*.

On one level this seems kind of obvious: some of us are just naturally happier than others. But on another level it kind of defies how most of us think about happiness: **no matter what we do or what happens in our lives, we tend to return to a level of happiness that is natural for us.**

This is testimony to the power of human "adaptation."[7] For example, that super cool flat-screen TV might have been a thrill the first week you had it, but you get used to it and soon it becomes your new normal. The same goes for negative things: over time they generally lose their edge as we get used to them, whether it's a breakup or a broken bone or the loss of a job.

At the same time, each new experience becomes a new reference point for us, and raises or lowers the bar for other experiences. We could call this the "first-day-back-at-work-after-vacation" effect, or the "first-day-back-at-work-after-camping-in-the-rain-for-days-with-previously-really-good-friends" effect.

Adaptation has its pros and cons. On the downside, it means that contrary to what your friendly local car dealer, realtor, or phone salesperson might have you believe, the thrill of buying a new car, house, or phone wears off pretty quickly. A famous study by psychologist Philip Brickman involving big lottery winners found it took only a few months for the thrill to fade and for the winners to generally return to whatever level of happiness they were at before they won.[8] On the upside, we also tend to rebound pretty quickly when bad things happen. In another part of that same study, Brickman found that it took people who were partially paralyzed in an accident roughly the same amount of time to return to their previous level of happiness as it did the lottery winners.[9]

Psychologist Daniel Gilbert refers to this as our "psychological immune system."[10] As we will see in chapter four, our circumstances *can* have a significant and lasting impact on our happiness, and the effects of negative events tend to last longer than positive ones. The effects of *really* negative events can last *much* longer. However, as psychologist Sonya Lyubomisrky notes, we usually adapt to most significant life events "such as being accepted into graduate school, becoming an uncle, experiencing the death of a close friend, having financial problems, and getting promoted," within three to six months.[11] Psychologist

David Meyers notes, "**Within a matter of weeks, one's current mood is more affected by the day's events—an argument with one's spouse, a failure at work, a rewarding call or a gratifying letter from a dear friend or child**" than by seemingly more significant life events.[12]

WHILE THERE IS definitely variation from person to person, our set points tend to be on the happier side of the spectrum. The majority of us are happy regardless of nationality, marital status, education, physical handicaps, and, yes, income. In other words, **humans are generally hardwired to be happy**. Which makes sense—it's a pretty big evolutionary benefit for us to *want* to be alive and do the things it takes to stay alive.[13]

Now, depending on who you read, our set points aren't really set—they can and do change over time. According to others, they're not really points but ranges. Still others write about each of us having multiple set points.[14] Nonetheless, forgiving positive psychologists these minor discrepancies that are still being worked out, we owe them a debt of gratitude for their insight into this most significant of happiness factors—the set point. This insight also goes a long way to helping us understand why money has so little impact on our happiness: not only does money have to compete with a number of other far more powerful factors, but thanks to our set points, 50 percent of our happiness is already off the table.

This will be very helpful when we consider the actual dollar amounts behind the abundance and saturation points in the next chapter, because those numbers are much lower than most people expect.

Chapter 3

How Much is Enough?
Nailing Down
The Abundance Point

I F YOU WANT to know how much is really enough for people to be their happiest selves, you would be advised not to ask them.

Richard Easterlin, credited as the first economist to officially study happiness, found that in 1978 a sample of people in their thirties felt they needed an average of 4.3 items on a list of "big-ticket consumer goods" in order to live a "good life." The list of twenty-four items included a home, a car, and a TV, as well as a vacation home, a swimming pool, and travel abroad. At the time, those people had an average of 2.5 of these items.

Sixteen years later the same people owned an average of 3.2 items, meaning they were well on their way to that good life. Except that by then the average number of items they felt they needed had shifted to 5.4. So they were actually further from the "good life" than they were before, even though they had more of the things on the list.[1]

Further studies by Easterlin and others have confirmed that a similar thing happens with money: **the more we get, the more we feel we need.**[2] Sociologist Lee Rainwater found that over a thirty-six-year period, the amount of money we report that we need to just "get along" increased in exact proportion to our incomes.[3] Economist Juliet Schor likewise documented that between 1987 and 1996, when asked how much money we think it would take to fulfill our dreams, the average answer jumped from less than $70,000 to $90,000 (adjusted for inflation).[4]

Certainly there must be a point, though, where we *do* feel we have enough, right? At least if we're *really* rich? Alas, even then a sense of "enough" seems to be elusive. Schor found that 25 percent of households with incomes that in 2021 would be at least $174,000 felt they didn't have enough money to buy everything they "really need." An additional 19 percent felt they spent nearly all their money on just "the basic necessities of life."[5]

To be clear, someone with that much income ranks in the top 5 percent of incomes in the US, and the top one-tenth of 1 percent worldwide.[6] So **nearly half of us in this richest fraction of humanity feel that we spend almost all our money on basic necessities, or that we don't even have enough to do that.**

Psychologist Michael Norton surveyed the richer clients of a big investment bank on these topics. ("The poor people in the survey were millionaires.") As these people accumulated more and more money, they were no happier for it. And when he asked them how much they would need to be happier, "All of them said they needed two to three times more than they had."[7]

In the words of Jeff Yeager, an author on simple living, people "convince themselves that they need something when in fact they really don't, or could choose a less costly alternative. … We *need* shelter, but we *want* it in the form of a seven-thousand-square-foot home with a swimming pool."[8]

That may be a bit extreme for most of us, but the idea is entirely accurate. Remember, in the 1940s almost nobody had air conditioning or a phone, not to mention indoor plumbing, and most Americans didn't even know what a television was. As of 2010, the last time we were surveyed, 55 percent of us considered air conditioning a "necessity," 62 percent said the same thing about a phone, and 47 percent of us said the same thing about a TV.[9]

It's helpful to not only compare Americans today with 1940s Americans, but to compare us with more contemporary people in other parts of the world as well. A 2006 Pew survey asked Americans how many of the items presented to them on a list they felt were necessities. For people with incomes of $30,000 a year or less, most people said six to nine items. For people with incomes of $100,000 or more, the answer was in the range of ten to fourteen.[10] By comparison, in a similar survey conducted in Vietnam around the same time, *not a single item* on the US survey made it onto most people's lists. Only 21 percent thought a TV was a necessity, coming in behind a radio at 37 percent. Many people, 77 percent, thought that having a home constructed of stone was a necessity, while 23 percent thought that bamboo, straw, or mud were fully adequate. The top-ranked items were a bicycle, a wooden rice chest, and a thick blanket, all of which made 98 percent of people's lists. (Note, that means two out of every hundred people thought they could still get by fine without those.)[11]

CLEARLY, OUR SUBJECTIVE sense of "enough" isn't the most reliable measure. In fact, it is extremely *un*reliable, constantly leaping ahead of us at every turn. In his book *Stumbling on Happiness,* Daniel Gilbert points out that inherent in this continual ratcheting up of our expectations is a continual disregard for our own lived experience. We somehow think that *this* time is going to be different, that *this* time when we get that additional money or product or experience it is going to really make a difference. Even though it doesn't. Over and over.[12]

One of the important reasons for this is that we tend to remember and anticipate the most dramatic aspects of an experience and disregard the less dramatic but far more substantive aspects. Gilbert cites a study in which college students were asked to predict how they would feel a few days after an upcoming football game against a rival college if their team won. Students consistently focused on how they would feel immediately after the win ("The clock will hit zero, we'll storm the field, everyone will cheer …") but they didn't think much beyond that ("And then I'll go home and study for my final exams").[13]

One of the central points Gilbert makes in his book is that we are much better off if we consider people's reports of how happy something has *actually* made them rather than trying to imagine that possibility for ourselves.[14] So let's return to our original question, but rather than looking at how much money people *think* they need to be happy, let's consider what the research reveals about our lived reality.

A range of leading economists, political scientists, and psychologists have all chimed in on this, and the consensus is that **the abundance point lies somewhere around $10,000 per person per year.** Looking at over sixty countries worldwide including the US, up until that point our income makes a consistent difference in our happiness, but after that there are "virtually no increases" in happiness, or "only small increases."[15] (When considering other countries, researchers use "purchasing power parity." That means it's not ten thousand actual dollars but the equivalent in the local currency. That is, whatever amount of money it would take in the local currency to buy what you can generally buy for $10,000 in the US.)

That number is by no means exact; it is a very broad generalization, which we will get to in a moment.[16] But first, let's consider what it means that "there are virtually no increases" in happiness after $10,000. A study by Ed Diener, arguably the most renowned figure in positive psychology, gives an idea. It was found that, in 2021 dollars, to get the same boost in happiness we get by going from an income of $3,000 per person to $4,000, would then require an increase from $4,000 to $9,000, which would then require an increase from $9,000 to $20,000.[17]

The following chart, reproduced and updated from one published by political scientist Ronald Inglehart, can help us visualize this. It shows the percentage of people who are happy and generally satisfied with their lives in sixty-four countries, and the average income per person in those countries. Up until around $9,000 there is great variation in levels of happiness, but after that things take a sharp curve, with most people being happy, but additional income correlating with relatively minor gains in happiness. (The US, for example, has the highest income of the countries considered, yet the percentage of people who report being happy is only a few points higher than it is in Colombia, with about 80

13

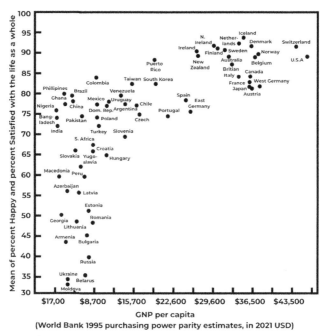

Figure 2. Happiness and GNP per capita by country.[19]

percent less income.) Inglehart notes that, beyond roughly $15,000 per person (in 2021 dollars) "there is practically no relationship" between income and happiness.[18]

Then at some point that "practically no relationship" turns into *absolutely* no relationship. Psychologist Daniel Kahneman and economist Angus Deaton, both Nobel Prize winners, found that **the saturation point is around $30,000 per person** (or $75,000 per household, with an average of 2.6 people). After that amount, additional income doesn't do *anything* for us in the day-to-day happiness department. Technically it still provides a slight boost in our overall "satisfaction with life," because it can feed a sense of pride or accomplishment, but even that is so marginal that, as political scientist Andrew Quinn has noted, we are left with "a rather academic argument that's interesting to discuss but not all that relevant in terms of the average person's life."[20]

AGAIN, THAT ABUNDANCE point amount of $10,000 is a very broad generalization. It has limited relevance for most of us as any kind of one-size-fits-all amount for how much we can or should get by on. But it is nonetheless a very important figure. **There is a lot of truth and perspective to it that are indeed immediately relevant to all of us, and that can be extremely helpful.**

It's important to remember, first of all, that it is $10,000 *per person*, so for the average household of 2.6 people that's $26,000. Also, that number is an *average*, so it naturally

14

varies for each of us depending on a number of factors, such as the number of people living together (it is significantly cheaper per person when more people are living together), age (health care costs increase with age), and the different costs of housing and food around the country.

Nonetheless, that number is actually consistent with the federal government's own findings, which indicate that it takes $13,200 for an individual in the US to cover the basic necessities, or $6,800 per person for a family of four. (The midpoint between those two numbers is exactly $10,000.)[21]

We're not talking about how much money it would take to sustain each of us in roughly the lives we're living right now, but pared down. This is more of a start from scratch, if you needed to let go of everything, "How little could you really get by on if you had to?"

Even then, for many of us in the US, $10,000 per person might still seem low. And depending on our personal situations and where we live, for many of us it really *is* low. And for others it is high. More importantly, though, we already know that our sense of "need" shifts constantly depending on how much we and the people around us already have, always leaping out ahead of us. And we in the US live in one of the most materially wealthy countries, not only in the world, but in all of human history. So, **for those of us in the US, our sense of what we "need" is only naturally, and massively, skewed**.

Psychologist Lara Aknin found that if you ask Americans to predict how happy other people are based solely on their income, we are pretty accurate when it comes to people with incomes over $22,000 per person, but we significantly underestimate how happy people are with incomes below that. *This is despite the fact that many of us participating in these surveys ourselves have incomes below $22,000 and report high levels of happiness.*[22]

For a little perspective, consider that **the average income for those happier-than-us Americans in the 1940s was $11,140 in 2021 dollars**.[23] And even that is deceptive because so many goods and services that we take for granted today didn't even exist then, or were in very limited supply, regardless of income.

Or consider that **for 50 percent of the people on the planet $10,000 is more than they make in a *decade*** (again, adjusted for purchasing power parity). For another 30 percent it takes three years to earn that.[24]

To be clear, those people, *80 percent of the planet*, are not doomed to lives devoid of happiness. On the contrary, people who have minimal possessions, and who may not even be able to always meet their basic needs, still tend to be happy. That's true whether it's a nomad in Kenya or a slumdweller in Calcutta—*or pretty much all of our ancestors for all of time*. In fact, in a 2013 study, Ed Diener documented that both Kenyan nomads and Calcutta slumdwellers are generally happy. (The nomads were actually happier than a sample of nurses and college students in the US.)[25]

Remember, humans are genetically inclined to be happy, and this is generally true regardless of our income. It is true for people in the lowest income bracket in the US, and it is true of people in a vast majority of countries.[26]

15

There is, however, the level of income I mentioned earlier, the *deprivation point*, below which our inability to meet our basic needs—plus the many challenges that often accompany extreme poverty, such as disruption of family and social networks, displacement, and violence—does tend to overwhelm our natural inclination to be happy. One study by Ed Diener and Shigehiro Oishi suggests that the deprivation point might be around $3,000 per person per year.[27]

Even extreme poverty, though, isn't destiny when it comes to happiness. Those slumdwellers in Calcutta, for example, who reported being slightly happy overall, live on about $2,200 a year.[28] Three-fourths of the people in Malawi live on $450 a year or less, yet one study found that the people there tend to score solidly in the positive range.[29]

Diener speaks eloquently to this:

*Clearly, members of these communities are living in extremely adverse conditions. They suffer from poor health and sanitation, live in crowded conditions, and occupy dwellings of poor quality. Examples of the negative memories reported were "I did not eat yesterday," "I had to have an operation," and "A relative died." In fact, of the seventy-three respondents who completed the memory measure, twenty mentioned poor health and ten mentioned a friend or relative dying within the past year. How, then, can they be happy? **The very fact that we ask this question is indicative of our heavy prejudice against poverty and our stereotypes of the poor. Perhaps we should be asking why we assume they are miserable.**[30]*

It's important for us not to romanticize the lives of extremely poor people; persistent deep poverty does fairly consistently overwhelm people's natural inclination to be happy. And **it would be cruel and unhelpful to ignore the ways that a lack of money and financial stress can profoundly impact our lives even when we do not live in persistent deep poverty—the struggles and uncertainty, what it means to go hungry, to lose our home, to not be able to give our children the things they deserve.** The impact of money on our happiness can surge well above that average of 2 to 4 percent for each of us at certain times, and for some of us a lot of the time.

But it is also important to not exaggerate the impact money has, even at amounts far below $10,000 a year, or to project a bleakness onto the lives of all poor people, the majority of whom are generally happy all the way down to $3,000 a year, and who in some places are *still* generally happy even with incomes a fraction of that.

It is not only insulting; it is simply false to imagine that people with little money or possessions cannot and do not often lead abundantly joyful lives. **Or that any of us can't. Or that happiness is a luxury instead of a natural way of being, largely independent of our material wealth.**

ALMOST NONE OF us in the US report having "a lot of money." This is true regardless of how much money we have.[31] Yet by almost any objective measure, most of us actually have vast amounts of money and tremendous buying power:

- If your annual per-person income is above $1,225 after taxes, then it's higher than that of half the people in the world (adjusted for purchasing power parity). If it's above $12,000, you are in the top 10 percent. Above $34,000 and you are a one percenter on the global stage. (**One percenters collectively control more than half the wealth in the entire world.**)[32]
- Seventy percent of Americans live above the abundance point ($10,000 per person), meaning we already have all we need financially to maximize our happiness, *on average*.[33]
- And a third of Americans live beyond the saturation point ($30,000 per person), meaning we have more money coming in than does *anything* for our happiness anymore (again, on average).[34]

This does not take away from the fact that many of us do struggle financially, sometimes terribly, or the fact that depending on our circumstances and the local cost of living, these numbers are low for some of us.

But the power of the abundance point and these other numbers doesn't lie in their exactness, it lies in the general *rightness* of them. It lies in the vital and vibrant truths behind them regardless of what the exact numbers might be for each of us:

- That money usually plays only a tiny role in our happiness, and we generally need very little of it to be happy.
- That many of us have vastly more money than we need to be as happy as we can be, yet we often completely lose sight of that fact.
- That we often place far too much importance on money, and in so doing we undermine our own well-being, and that of the people close to us.

The abundance point and these statistics do not exist as a rebuke, as something to feel bad about or hold over ourselves if we are living beyond them. They are a beautiful invitation, a gift, a prayer. *The abundance point frees us, in many ways, to make the profound shift from a world of scarcity to a world of abundance just by shifting our perceptions.*

It reminds us that when we are able to meet our needs, and perhaps have a little to spare beyond that, we are as blessed as if we had all the material things in the world laid before us, for as little good as they would do for us happiness-wise.

And it reminds us that even when we do *not* have enough to meet our basic needs, there is often *still* considerable room for us to lead joyful lives.

So it is for you and me. So it is for the nomad in Kenya, and the slumdweller in Calcutta, and for twins and lottery winners and people confined to wheelchairs. So it has been for all the human beings who have come before us, and so it will be for all those who come after us—blessings on us all!

And it is to recognize that, ultimately, there is a vast richness and potential for joy in our lives that lies in a realm beyond anything that money can generally touch or do for us.

It is to that realm that we now turn.

Part II

Charting the Path of a Joyful Life

Chapter 4

The *Hap* in Our Happiness: The Role of Circumstances

There are certain queer times and occasions in this strange mixed affair we
call life when a [person] takes this whole universe for a vast practical joke,
though the wit thereof [they] but dimly discern, and more than suspect that
the joke is at nobody's expense but [their] own.[1]

—Ishmael
(from Herman Melville's *Moby Dick*)

THE CLOSEST WORD the ancient Greeks had for happiness was *eudaimonia*, which literally means "good demon or spirit." This is very telling—the ancient Greeks believed happiness was a matter of chance, whether it was a whim of the gods or the luck of the draw as to which spirit would guide each of us.[2]

The word *happiness* has similar origins. There are clues to the meaning of the word's root, *hap*, in words like mishap, haphazard, and happenstance. In languages as far-ranging as Swahili, Chinese, Farsi, Japanese, and Arabic, at least one of the ways people say "happy" relates to blessings or luck. And so it is for happiness and the root *hap*, which does indeed mean luck or chance.[3]

Given that about half our happiness is driven by our set points, there is something very true in this way that our ancestors spoke of happiness in these many languages. And we're not out of the woods yet. Before we turn to the factors that *are* largely within our control, we need to acknowledge one more that is not. I am referring to our circumstances, that is, the things that *hap*-pen around us and to us.

Based on the work of Sonja Lyubomirsky, psychologist and author of *The How of Happiness*, **circumstances largely beyond our control account for some 4 to 8 percent of our happiness.**[4] (The 2-4 percent of our happiness related to income and material wealth falls partly under this heading, since income and wealth are definitely to some degree outside our control, as we will explore in chapter twenty-seven.) This means that, on average, our circumstances play an important but relatively minor role in our happiness.

We should be careful, though, when we deal with averages. In this case we are talking about both an average over the course of an individual's life and an average among many people. While circumstances may generally play a small role in our happiness, for most of us there are also times when circumstances blast open our existence and crush our hearts or raise us up and change our lives, affecting our happiness far beyond 2 to 4 percent for a while. And for some of us the impact of our circumstances is long-lasting, sometimes our entire lives.

That famous study we considered earlier about lottery winners and accident victims offers an example of this. While the most striking finding was that the impact of those experiences on people's happiness was usually short-lived, it was also true that most people's happiness did clearly spike or plummet initially. It was also true that for those accident victims who became paralyzed to the point they could no longer work, their drop in happiness tended to be significant and sustained. Likewise, lottery winners who were poor before winning tended to be a little happier even after the initial spike passed.[5]

There are experiences that can be simply devastating, causing long-term and significant declines in happiness, such as losing a spouse or child, or suffering physical or sexual abuse.[6] (Somatic psychologist Pavini Moray writes, "Whoever came up with that axiom, 'You'll never be given more than you can bear' was a masochist. And a liar."[7])

A 2010 study in the *Journal of Personality and Social Psychology* also notes that we don't need to suffer a singular profound loss for our circumstances to have a sustained impact. Multiple stressors in a person's life (such as illness, injury, financial stress, and living in dangerous housing) tend to have a cumulative effect that goes beyond the impact of each of them alone.[8]

The oversized role that circumstances can have in our lives is particularly relevant to certain populations. People who are marginalized in the US, such as Black, Indigenous, and other people of color, as well as women, queer people, and poor people, are frequently subjected to negative, threatening, and degrading situations that have been documented to create stress and undermine psychological well-being. To different degrees they also have a greater risk of homelessness, hunger, abuse, and violent crime, and suffer higher rates of arrest, longer prison sentences, and greater infant mortality.[9]

So it is important that we not exaggerate the importance of our circumstances, or how much changing our circumstances can boost our happiness. On average they really do only count for 4-8 percent of our happiness. But it is also important to not ignore how important our circumstances can truly be—the far more significant role they play in some of our lives long term, and the more significant role they play in most of our lives for shorter periods of time.

SPEAKING OF SOCIAL identities (race, gender, class, sexual orientation, and so on), before we go any further in our exploration of the psychology of happiness, **it is important to note that the vast majority of psychological research is done with an extremely unrepresentative population, one that is wildly different from most humans**. As many as 96 percent of participants in psychology studies are white, highly educated, and from wealthy, industrialized, democratic societies. Yet these people represent only 12 percent of the world's population. Even when the studies are conducted in nonindustrialized countries, the population is often overwhelmingly urban, educated, and relatively wealthy. On top of that, *a full 67 percent of participants are college students,* because that is the population easily at hand for many of the academics pursuing this research.[10]

The findings of these studies are often treated as universal truths, when, in fact, they often reflect the tenuous truths of only a fraction of humanity. They can be relevant, accurate, and helpful to all of us, even those of us who do not fit the demographic that was studied, but we should keep in mind that every one of these studies reflects to some degree the biases of the participants *and the people who designed and interpreted them*—who tend to be a pretty homogeneous and unique bunch.

I have tried to draw from research that is expansive in its sampling, and I have left out many studies that were one-offs conducted on small numbers of people. Nonetheless, I invite you to actively consider where your own lived experience seems to differ from the research findings, and to allow that to inform how you understand these lessons and apply them to your life.

THE SET POINT has been referred to as "one of the most deflating concepts" in positive psychology, because it takes so much of our happiness out of our hands.[11] Add to that the impact of circumstances beyond our control, and the picture gets grimmer still.

Yet, as Ed Diener writes, "Set-point is not destiny … people's long-term levels of happiness *can* and *do* change."[12] It's true that, on average, between our set points and circumstances beyond our control, "hap" (chance) has got some 55 percent of our happiness locked up, and that circumstances can absolutely have an effect that goes beyond this. However, that still leaves roughly 45 percent of our happiness to some degree within our control. That doesn't mean we can increase our happiness by 45 percent. But with 45 percent of our happiness open to our influence, Martin Seligman, the former president of the American Psychological Association, suggests **we can boost our happiness by 10 to 15 percent overall. And researchers have documented increases *much* greater than that.**[13]

With this in mind, let us leave off now with the "good demons" of the Greeks and the "luck" of the English and so many others, and draw our inspiration instead from the Welsh, whose word for happiness doesn't relate to luck at all, but to *wisdom*. Let us turn to the important role of our intentional choices, and to the factors that we *do* largely have some control over.[14]

Chapter 5

Spouses and Kids and Sex, Oh My! The Role of Kith and Kin

Admit something:
Everyone you see, you say to them, "Love me."

Of course you do not do this out loud, otherwise
someone would call the cops on you.

Still, though, think about this,
this great pull in us to connect.[1]

—Hafiz

T HE RESEARCH IS clear: when it comes to food, shelter, and safety, people who have them are generally happier than people who don't.[2] But not always.

In 2005, psychologists Ed Diener and Robert Biswas-Diener interviewed homeless people in California, Oregon, and Calcutta, India. The participants consistently reported below-average life satisfaction, However, the life satisfaction of the homeless people in Calcutta was still in the positive range, and it was higher than that of the American homeless. This was despite the fact that the Indians had "poorer access to food, clean water, medical care, opportunities for employment, and adequate shelter than their counterparts in the United States."[3]

What could account for this? The researchers noted that what the homeless people in California missed most of all, more than "physical things such as good housing," was "close and trusting friendships." Meanwhile, the people in Calcutta were better able to maintain ties with family and friends, and reported "significantly greater satisfaction with both their overall social lives and their families."[4]

What this and other research underscores is how significant our social lives are to our happiness.[5] In fact, some of the leading figures in positive psychology have suggested that the quality of our relationships and how much time we spend with family and friends is **the single most reliable predictor for our happiness**.[6] That is true for introverts as well as extroverts.[7]

- The number and quality of our friendships have as much of an impact on our risk of death as smoking and drinking, and they have more of an impact than physical activity and obesity.[8]
- Someone who doesn't have anyone to "call in times of trouble" would need, on average, a 150 percent boost in income to be as happy as someone who does.[9]
- Solitary confinement has been found to cause lasting psychological damage after even a few days, and the United Nations considers it torture if it lasts more than fifteen days.[10] (Many people incarcerated in the US spend months, years, even decades in solitary.[11])
- Spending money on experiences is one of the few ways that money *does* tend to boost happiness, largely because they tend to have a social element to them.[12]

Political psychologist Robert E. Lane notes that the decline in the amount of time spent with family, community, and friends in the US between 1975 and 2000 "is almost as dramatic as the increase in disposable income." He argues that this has been central to "the decline in happiness in the United States and the rise in depression in most advanced economies."

David Brooks, in his *New York Times* column titled "The Great Affluence Fallacy," notes that "people in wealthy countries suffer depression by as much as eight times the rate as people in poor countries." He likewise attributes this to our loss of community.[13]

Not all relationships are equal, though, and they don't all automatically translate into greater happiness.

Marriage

Roughly 70 percent of adult Americans have been married at some point in their lives.[14] Roughly 40 percent of marriages end in divorce, but roughly 70 percent of people who have been divorced get married again.[15] Obviously being married is no guarantee of "happily ever after," but it doesn't seem to do a lot for our happiness in the not-ever-after either.

What *does* pretty consistently increase our happiness is *getting* married. It's been calculated that getting married is the "happiness equivalent" of quadrupling our annual income.[16] Adaptation has its way with married couples, though, as it does with all of us, and that boost tends to wear off after about two years.[17]

What is also clear is that people who are married are indeed generally happier. Economists David Blanchflower and Andrew Oswald have estimated that being married is the equivalent of receiving a $100,000 bonus every year.[18] What is *not* clear, however, is that it's the *being* married that makes us happier. This is the difference between correlation (people who are married are generally happier) and causation (being married *contributes to* people being happier). Even when we have hundreds of studies showing a correlation between one thing and another (as is the case with marriage and happiness), it doesn't mean the one *causes* the other.[19] A number of studies and some heavy hitters in the positive psychology world, including Sonja Lyubomirsky and Ed Deiner, believe that happy people are simply more likely to get married and stay married.[20]

Very importantly, **even the common *correlation* between marriage and happiness varies from person to person. Being *unhappily* married is a very dependable predictor of "depressive symptoms," while many of us are absolutely happy being single.**[21]

If nothing else, married people do seem to have sex more often than unmarried people...[22]

Sex

Yes, those of us who are having sex are generally happier than those who aren't, and the more we're having it, the happier we generally are (presumably within reason). Another study by Blanchflower and Oswald suggests that a jump from having sex once a month to once a week impacts our happiness about the same as a $50,000 raise.[23]

However, researchers have found themselves confounded again by the question of correlation and causation. They have simply not been able to determine whether sex makes us happier or happier people tend to have more sex.

Leaving behind for a moment the already uncertain terrain of correlation, and embracing pure and reckless speculation, I invite you to consider … bonobos. For those of you not familiar with bonobos, they tie with chimpanzees as our closest genetic relatives, with their DNA 98.7 percent identical to ours.[24] Their sexual behavior is, however, vastly different from that of any humans I've encountered. They have sex every which way, including while hanging upside down and holding each other in the air. They have oral sex and masturbate, they "penis fence," "rump rub," and more. They also have sex about every ninety minutes (though it usually only lasts about fifteen seconds.)[25]

They use sex to welcome strangers and show affection for friends and family. They use sex to avoid tension, they use it to ease tension, and they use it to reconcile after tension.[26] We're talking very low-tension primates here. I haven't seen any research on how happy bonobos are, but *there hasn't been a single documented case of one bonobo killing another.*[27] They'll scream and throw things at each other and even bite on occasion, but they won't kill each other—unlike the two species most closely related to them: chimpanzees and, ahem, humans.[28] Draw whatever reckless conclusions you will.

24

Either way, sex educator Cory Silverberg wisely emphasizes that all of us have different relationships with sex. **"For some people sex is great, for some people it's terrible, for some people it means nothing." None of us should feel any pressure to either have sex or feel a certain way about it or even care about it at all.**[29]

Kids

Steve Wien, a pastor and father of three children, has written:

> *If you are a parent of small children, you know that there are moments of spectacular delight, and you can't believe you get to be around these little people. But let me be the one who says the following things out loud:*

- *You are not a terrible parent if you yell at your kids sometimes. You have little dictators living in your house. If someone else talked to you like that, they'd be put in prison.*
- *You are not a terrible parent if you can't figure out how to calmly give them appropriate consequences in real time for every single act of terrorism that they so creatively devise.*
- *You are not a terrible parent if you'd rather be at work.*
- *You are not a terrible parent if you just can't wait for them to go to bed.*[30]

Some 70 percent of adult Americans have had at least one child, and the vast majority (90 percent) of those of us between eighteen and forty who don't have any children report a desire to have at least one.[31] Most of us who have kids report being happy about it, and over a third of us rate having children or grandchildren as our single greatest source of joy.[32]

At the same time, though, one prominent study found that when it comes to in-the-moment happiness, caring for our children scores pretty low, tied with doing housework and just ahead of commuting.[33] Another study collected over 1,500 hours of video footage of families going about their normal daily routines. The goal was to simply document homes with two working parents and multiple kids, but the result was what one person involved in the project called "The very purest form of birth control ever devised."[34]

Welcome to what is known as "the parenthood paradox": we generally want kids and value them, but they don't generally make us any happier. The blissful moments and the deeper sense of meaning and connection that often come with having children tend to get steamrolled by the relentless changing of diapers, cleaning up, arguing over vegetables, homework, bedtime, and so on.[35] Meanwhile, all of that time we spend taking care of our kids is also time we're *not* doing many of the things we *do* enjoy. Parents tend to sleep less, spend less time with friends and spouses, have sex less, and have more strained relationships.[36]

Jennifer Senior, author of *All Joy and No Fun: The Paradox of Modern Parenthood*, puts it this way, "While children deepen your emotional life, they shrink your outer world to the size of a teacup, at least for a while." She cites a psychologist who puts it more bluntly: "They're a huge source of joy, but they turn every other source of joy to s***."[37]

In the end, on average, **having kids leaves us no happier than not having them, perhaps even a little *less* happy.**[38] Some reports suggest that each additional child undermines our happiness a little more.[39]

Ed Diener reminds us that this is again an average, that for some of us parenthood definitely does boost our happiness, while for some of us it definitely undermines it.[40] It is important to note, also, that our experience of parenthood is also very much dependent on where we live. **Parents tend to be happier in Europe, where there is more support for parenting**, including a year of paid maternity leave and subsidized child care.[41]

Friends

The bird, a nest.
The spider, a web.
[Humans], a friend.[42]

—William Blake

In his book *The Loss of Happiness in Market Democracies*, Robert E. Lane asks the very question we posed earlier: "If money does not contribute to the happiness of those above the poverty level, what does?" The main answer, he concludes, is "companionship."[43]

You may recall from chapter one that **a person with no friends or social relations with neighbors could never get enough money to make up for it, happiness-wise.** Psychologist David M. G. Lewis notes, "Across cultures, friendship is reliably linked to the experience of positive emotions such as happiness."[44] There are numerous reasons for this: there is often a pleasure in simply being with friends and doing things with them, friends give us a sense of belonging and affirmation, and they can be a support to us in challenging times.[45] Unlike family, we get to choose our friends. And, unlike spouses, there is often greater flexibility around which friends we spend time with and making new friends based on our changing needs and interests.

Quality generally counts more than quantity. Friendships that involve deeper understanding and caring with someone we can confide in and turn to in times of need, those are the ones that support our happiness the most.[46] Having many friends on social media does tend to promote feelings of connectedness, familiarity, and self-worth, yet it has no measurable impact on our happiness.[47]

Still, economist John F. Helliwell has calculated that doubling our number of friends has roughly the same impact on our happiness as a 50 percent increase in income.[48] And one study found that in communities where people move a lot, spreading our attention to a larger number of friends may serve us better than nurturing a few deep relationships with people who might very well move away.[49]

THE BENEFITS OF relationships are the strongest and last the longest when we take an active approach to them.[50] Psychiatrist Robert Waldinger is the director of a study on health and happiness that has spanned more than eighty years, one of the world's longest studies on adult life. Over those many years, social connections have continually shown up as vitally important. Waldinger was asked how the study has impacted him personally. His answer was simple, yet potentially offers some guidance to the rest of us. Along with meditating daily (which we will come back to), he has made relationships a higher priority in his life. "It's easy to get isolated, to get caught up in work and not remember, 'Oh, I haven't seen these friends in a long time.' So I try to pay more attention to my relationships than I used to."[51]

Barbara DeAngelis, a relationship consultant, suggests that "marriage is not a noun; it's a verb. It isn't something you get; it's something you do. It's the way you love your partner every day."[52] It's an inspiring sentiment, and one we can apply not just to spouses but to all of the people in our lives, from family and lovers to children and friends.

Chapter 6

More Sleep, Less Cow: Physical Health and Happiness

PHYSICAL HEALTH CUTS pretty close to the heart of basic human needs, so it makes sense that there's a clear connection between it and happiness. That said, the effect is actually rather limited—*on average*. Remember the study about people who became paralyzed, where the ones who were *severely* debilitated often suffered a significant decline in happiness, but those who weren't so severely debilitated tended to adjust fairly quickly? That's true of health issues in general.

Having a severe health problem or multiple problems often *does* have a negative effect on happiness, and we should be sensitive to how much a significant condition, serious illness, or a lot of pain can impact us and others. Still, we tend to adapt fairly quickly to issues that aren't too serious.[1] This bears out at a societal level too: once a country achieves a basic degree of health and longevity for most of its people, increasing them further doesn't generally result in greater happiness.[2]

If you're skeptical of this (and you're healthy), you're not alone. Healthy people consistently underestimate the well-being of people with disabilities and serious illnesses.[3] That said, **health still has a more significant impact on happiness than money, making the pursuit of good health generally a much better bet**.[4] Caring for our health can also be pretty easy, and it can actually save us money.

Brent Bauer, director of the Mayo Clinic's Complementary and Integrative Medicine Program, believes a commitment to "the four pillars of health"—exercise, diet, stress reduction, and spirituality—can prevent 90 percent of health-care problems in the US.[5]

Exercise consistently boosts our happiness and helps protect against depression and anxiety.[6] For people who are already depressed, one study found that exercise is a more effective treatment than drugs ("and the cost profile is very favorable"), while another study found it to be as effective as therapy.[7] You can reduce your risk of death by half with just thirty minutes of moderate activity—any activity that simply raises your heart rate—five days a week.[8]

As for **diet**, the Harvard School of Public Health recommends that roughly half of what we eat be fruits and vegetables, leaning heavily toward vegetables, plus one-quarter whole grains and one-quarter protein.[9] A study of over twelve thousand people found that "increased fruit and vegetable consumption was predictive of increased happiness, life satisfaction, and well-being."[10]

Fruits and vegetables are also often cheaper than meat and dairy products, not to mention the money saved by being in better health.[11] Americans tend to eat about twice as much protein as is recommended.[12] Eating a lot of meat, dairy products, and eggs can raise cholesterol levels, increasing the risk of heart disease, and eating too much dairy can actually undermine bone strength and increase the risk of some cancers.[13]

One prominent report concluded that 30 to 40 percent of health-care expenditures in the US "go to help address issues that are closely tied to the excess consumption of sugar."[14] Eating a lot of junk food and processed food has also been found to significantly increase rates of depression.[15]

IF YOU FEEL somewhat daunted by physical activity and eating healthy, this next one's for you. A study by economist Daniel Kahneman found that **the quality of our sleep has as big an effect on our happiness the following day as our personality and temperament**.[16]

Wellness consultant Marie-Josée Shaar underscores the importance of sleep by pointing out the behavior of children. They're so cute and playful—until they're not. When they get tired they become restless and irritable. Shaar argues that being tired doesn't affect us any less as adults; we're just better at masking it.[17]

In 1910, before TV and electric lighting, people in the US slept an average of nine hours a night.[18] In some parts of the world people traditionally slept eight hours a night in two spans, with two or three hours of general relaxing in between, for a total of ten or eleven hours of sleep and rest.[19] The consensus seems to be that most adults should sleep seven to nine hours a night, while a small percentage of us need more or less.[20]

One study found that people who didn't sleep for twenty-four hours functioned on the same level as someone who is legally drunk. *The same was true for people who slept six hours a night for two weeks.* These people recognized they were tired, but insisted they had adjusted to that amount of sleep, when in fact "their performance had tanked." People who slept seven hours a night in that study fared better, but their performance still suffered.[21]

When our average drops below seven hours a night, we're also three times more likely to get sick, and we run higher risks of obesity, diabetes, cardiovascular disease, and stroke.[22] Oxford neuroscientist Robert Foster suggests that "over 30 percent of the medical problems that doctors are faced with stem directly or indirectly from sleep."[23]

THE RELATIONSHIP BETWEEN good health and happiness actually flows both ways. Multiple studies have documented that not only does good health contribute to happiness, happiness contributes to good health. **People who are happier generally live longer—as much as** *four to ten years longer* by some estimates. Part of this is that people who are happy tend to take better care of themselves (like get more sleep and exercise), but it also seems that, in the same way that chronic stress directly impacts the body and undermines our health, happiness also directly boosts our health., though in ways that are not yet clear.[24]

Relationships also play a powerful role in our health. Close relationships help to delay mental and physical decline, and several studies have found that **for older people, their "level of satisfaction with their relationships at age fifty was a better predictor of physical health than their cholesterol levels."**[25]

Chapter 7

A "Fool's Life" or
a "Labor of Love"?
Work and Happiness

To allow oneself to be carried away by a multitude of conflicting concerns,
to surrender to too many demands,
to commit oneself to too many projects,
to want to help everyone in everything,
is to succumb to the violence of our times.[1]

—Thomas Merton

IN 2013, GRETCHEN Reynolds, writing the "Phys Ed" column for The New York Times, received the following inquiry:

How do I balance an extra hour of sleep against 40 minutes of exercise? When should I give in to the desire for more sleep and when should I push myself to exercise?

Reynolds replied that it's unclear which is more important, and that, in fact, the two are related. "A good night's sleep, consisting of at least seven hours of slumber, results in better and more prolonged exercise sessions later that day … while fewer hours of sleep frequently lead to reduced motivation to exercise." Likewise, "exercise can improve the quality of sleep."

One expert tried to avoid the trade-off by suggesting biking to work, running at lunchtime, or doing intense but brief, effective bouts of exercise. Another leaned toward getting more sleep, suggesting that the person could aim to go to bed fifteen to thirty minutes earlier on work nights.[2]

Hidden behind these tortured attempts to squeeze a little exercise or more sleep into this person's life was a sort of tacit agreement that the things this person is doing throughout the rest of the day are sacrosanct, not to be touched or questioned, even at the expense of exercising or getting a little more sleep. It's as if we all understand, without having to mention it, that the tremendous busyness in this person's life is important, normal, or just unavoidable.

Time affluence is "the feeling that one has sufficient time to pursue activities that are personally meaningful, to reflect, to engage in leisure," and it is "a consistent predictor of well-being," even among people who indicate that they prefer to be busy. Yet many of us are far more familiar with a sense of time *poverty*, the feeling that we are "constantly stressed, rushed, overworked, behind."[3]

In this chapter and the next we'll look at two of the ways we Americans spend a significant portion of our lives that have a mixed or even negative impact on our happiness, and can therefore potentially be important sources of more time in our lives.

Most [people], even in this comparatively free country … are so occupied with … the coarse labors of life that its finer fruits cannot be plucked by them. … It is a fool's life, as they will find when they get to the end of it, if not before.[4]

—Henry David Thoreau

WE AMERICANS SPEND more time over the course of our lives working than doing any other single activity except sleeping.[5] Those of us who are employed work an average of thirty-five hours a week. *That average includes those of us who work part-time.* It also includes vacation time and holidays (meaning we actually work more than thirty-five hours a week on the weeks we work), and it doesn't count time spent commuting.[6] Full-time employees average forty-seven hours a week.[7]

Work is not only a means of making money; it can also provide structure to our days; it can engage us in absorbing and creative activities; it can give us a sense of identity, purpose, and status; it can be a place we socialize; and it can involve physical activity.[8] Studies indicate that roughly a third of us in any given occupation experience our work as a kind of "calling"—the paycheck is important, but we also take a lot of pride in our work, are perhaps even passionate about it.[9]

At the same time, though, less than half of us who have jobs report that we are satisfied with them.[10] A full 70 percent of us say our workdays "seem like they will never end," and 20 percent of us report that we dislike work so much we actively sabotage our employers.[11]

I did my own informal survey, tracking the bumper stickers I see on people's cars, and I came up with the following:

I'd tell you to go to hell, but I work there and
don't want to see you every day
I used to have a life but my job ate it

I love my job (shoot me now)

Wake me when it's Friday

Do all jobs suck or just mine?

I always arrive late at the office,
but I make up for it by leaving early

A bad day fishing is better than a good day working

Take this job and shove it!

It wasn't the most scientific survey, and there are probably some bumper stickers out there about how much people love their jobs, but I didn't see any. There are also probably a good number of us who both like and dislike our jobs in different measures. Still, the research is clear: **most of us in the US experience "significantly higher well-being" on nonwork days, and generally the less we work, the happier we are.** People who work forty-hour weeks are significantly happier than those working more than forty hours, and people working twenty-hour weeks are significantly happier than forty-hour people.[12]

There is an important exception to the "less is more" of work. Being unemployed (when we did not choose it) consistently undermines our happiness.[13] Interestingly, this doesn't seem to be so much about the lost income as it is about the blow to our self-esteem, a lost sense of purpose, and a lack of control of our lives.[14] Being unemployed has less of an effect if we live in a region with high unemployment because we tend not to judge ourselves or be judged by others as much.[15] And we are generally happier after we retire from working because it doesn't carry the same stigma as being unemployed.[16]

Still, we should be careful not to exaggerate the impact of unemployment, considering that **a one-hour commute to work impacts our happiness as much as being unemployed.**[17] The average person commutes fifty-four minutes each workday. One study ranked commuting as our most unpleasant daily experience, and another calculated that it would take a raise of more than 30 percent to make up for the happiness we lose commuting.[18]

Short of being unemployed, though, the less-is-more rule seems to apply all the way down to just a few hours of working. A study in the UK followed five thousand people during the coronavirus pandemic, a time of great change in the workplace for many, and found that the most likely to be unhappy were those who had lost their jobs. Not far behind them, though, were those who were employed full-time. The happiest were those who were working part-time, with **those working one day a week being the happiest of all.**[19]

Another study involving seventy thousand people documented that when unemployed people found a job their happiness and other aspects of their mental health recovered, but "**it only took eight hours of work per week**" (about 1.5 hours a day) to achieve those gains, "**with no sign of extra benefits with more time on the job**." *That was true independent of a person's household income*, as well as marital status, age, and health.[20]

In other words, while there may be many reasons that we need or want to work more hours, and many reasons that some of us truly enjoy working more, generally speaking, we are happiest if we can do it part-time, ideally about eight hours a week.

Perhaps the most powerful way in which we conspire against ourselves is the simple fact that we have jobs.[21]

—Curtis White

IN 1950 AMERICANS worked 25 percent fewer hours than they did in 1900. President Kennedy noted in 1965, "We had a 58-hour week, a 48-hour week, a 40-hour week. As machines take more and more of the jobs of [people], we're going to find the work week reduced."[22] Around that same time, a prominent futurist, Herman Kahn, predicted that by the year 2000, Americans would have thirteen weeks of vacation and a four-day workweek.[23]

If that's not what you see around you, you can't blame it on our gains in productivity, which have actually exceeded Kahn's predictions. The "number of weekly hours needed to produce the 1950 worker's output declined by almost *one hour per year* until the mid-1970s, and has been declining by about half an hour per year since then." In other words, speaking very roughly, we would've only needed to work about one hour a week in 2020 to achieve the same production as someone working forty hours in 1950. Or we could have accepted our 1975 level of production and only needed to work thirteen-hour weeks.[24]

But that's not what we've done. The people vested with political and economic power—and all of us to a certain degree—have consistently made decisions that drove us in the other direction. We have prioritized channeling most of those gains in productivity into massive increases in wealth for a very small number of us while generally *increasing* our work hours. As charted by the Center for American Progress and Timothy Noah, author of *The Great Divergence*, between 1980 and 2005, despite massive gains in productivity, the two-thirds of us with the lowest incomes saw our median income *fall* as much as 29 percent, and the third of us with the highest incomes saw a modest 7 percent increase.[25] **A full 80 percent of the total increase in income went to the top 1 percent.**[26]

At the same time, we have consistently raised our sights in terms of how much more money and possessions we generally think we need. As we explored in chapter two, we have increased the size of our homes, the number of appliances and devices we own, the number and size of our cars and how much we drive them, how much we fly, and so on. The result is that instead of working just a handful of hours each week, we actually worked 115 hours *more* in 2020 than we did in 1975. That's nearly three extra work weeks per year.[27] Clocking in at nearly 1,800 hours a year, **Americans work more than anyone in any other industrialized nation**—137 hours a year more than the Japanese, and 258 hours more than Europeans. We work more on the weekends, we take far fewer vacations, and we retire later.[28] John DeGraaf, the executive director of Take Back Your Time, has calculated that **we work more hours than medieval peasants did.**[29]

What would it be like if we actually prioritized our well-being over ever more production and consumption? Environmentalist Bill McKibben shares this powerful story:

> *In 1930, in the teeth of the Depression, the cereal entrepreneur W. K. Kellogg put his workers on a six-hour day at full pay. Productivity increased dramatically, helping pay for the experiment. Meanwhile, the company's town parks, community centers, churches, and YMCAs all flourished. Researchers who interviewed the townspeople found that their interests had grown and changed; they now asked themselves, "What shall I do?" Not just "What shall I buy?" Indeed, workers looked back on the eight-hour day with a shudder. "I wouldn't go back for anything," said one. "I wouldn't have time to do anything but work and eat."[30]*

What if we had, in fact, chosen to accept our 1975 levels of production and we only worked eighteen hours a week? McKibben notes, "True, we drove smaller cars and lived in smaller houses and ate out less. On the other hand, we ate together more."[31] Or forget 1975. What if we decided we were satisfied with the amount we are producing *right now*? In twenty years we could be working ten fewer hours a week.[32]

Even if at a societal level, we continue in this same vein, pouring gains in productivity into gains in wealth for some of us, while maintaining or increasing work hours, there is often much that we can still do to apply these lessons to our *individual* lives. Robert Biswas-Diener offers the following:

> *People often mistakenly believe they're in an either-or situation. The conflict itself creates distress and gives you very few solutions to choose from. The truth is, you don't need to do something as extreme as quitting your job: You can make small changes—blocking out part of a Saturday afternoon for friends or committing to getting home two hours before the kids' bedtime. Just taking those steps gives you psychological peace of mind, with which you can start planning bigger strategies. You might end up deciding to leave your job, or you and your boss could work out a schedule of 10-hour days but only four days a week. Most likely, there are other solutions that, in the moment, you're not seeing.[33]*

Between 1990 and 1996, 20 percent of all Americans made a voluntary lifestyle change that entailed earning less money. A full 85 percent of them were happy with the change. We're not talking about only people with high incomes who cut back to still-really-high-but-less-than-before earnings: almost half of them earned $35,000 or less to begin with. And we're not talking small changes: 30 percent of them reduced their spending by one quarter, and 30 percent by a half or more.

Perhaps even more telling is that during that same time frame, another 12 percent of Americans saw their incomes drop *involuntarily*, yet fully *half of them* later described it as "a blessing in disguise."[34]

More recently, a study of one hundred thousand working adults published in the *Harvard Business Review* found that those who were willing to give up a little money in favor of having more time—either working less or paying for services that freed up some of their time—experienced "more fulfilling social relationships, more satisfying careers, and more joy."[35]

Joe Robinson, author of *Don't Miss Your Life*, summarizes: "Family, friends, hobbies, passions, travel, exploring … **the realm of nonproductivity is where your life lives.**"[36]

Chapter 8

The Hard Test of
Our Wisdom:
Television and the Internet

*Television is a new, hard test of our wisdom. If we succeed in mastering this
new medium it will enrich us. But it can also put our minds to sleep.*[1]

—Rudolf Arnheim (1935)

HOW MANY MORE hours would we actually need in a day to achieve a sense of time
affluence? Would one extra hour a day do it? Two? More? You've got it.

Yes, there are vast reserves of time available to us by cutting back on work, but that
can be elusive for many of us. It isn't always easy or even possible to reduce our hours,
change jobs, or get by on less income. But there is another reserve of time available to us
that is immediate and free and totally within our control. This land of great promise?
Television.

We spend roughly half of our waking hours that we are not working watching TV. That's
the time we spend watching shows, movies, and live programming, whether it's on an
actual TV or on a computer or phone. Younger and middle-aged people spend between
two to four hours a day watching, and older people watch five to seven.[2]

**This technology that didn't exist for hundreds of thousands of years of human
history, that only came on the scene about one hundred years ago—we spend more time
watching it than doing anything except sleeping and working.**[3] The average home has
more TVs than people, not to mention computers, tablets, and smartphones.[4]

Watching television is relatively cheap, and it's there, ready for us, whenever we want,
no matter how much time or energy we have, whatever we're wearing—no transportation
or advanced planning necessary. That is part of its appeal, and it is also part of the problem.
TV provides significantly less enjoyment than most other leisure activities, and it's the first

37

thing we say we'd cut if something needed to go, yet we still watch. In fact, for every extra hour of leisure time we get, at least half of it goes right into watching more TV.[5]

What is the impact of TV on our happiness? **If we don't watch a lot of it, the impact is minimal, but the more we watch, the less happy we tend to be.**[6] Some of that is because unhappy people tend to watch more TV, but watching TV also contributes to unhappiness.[7]

Part of that is the experience of watching TV itself. We're happiest in activities where we're working our edge—where they're challenging, but doable.[8] Aside from the logistics of navigating multiple remotes and various streaming services, TV mostly doesn't deliver on this one. As Martin Seligman notes, the average "mood state" people are in when watching TV is "mildly depressed."[9]

Roughly half of the effect on our happiness, though, is about the way the content of what we are watching affects our sense of the world and ourselves:[10]

- Those of us who watch a lot of TV are more distrustful of other people and feel less safe, due to the kinds of things that tend to be highlighted in shows and on the news.[11] It has been estimated that by age eighteen, the average person has seen two hundred thousand acts of violence on TV, including forty thousand murders.[12]
- Those of us who watch a lot of TV also place more importance on being rich, we overestimate how affluent other people are and how many millionaires there are, and we have a lower satisfaction with our own financial situation.[13] (We also overestimate how many people suffer from dandruff, bladder control problems, and hemorrhoids.[14])
- The more TV we watch, the less satisfied we are with our body image. This affects women and girls the most, but is true for all of us.[15] Social anthropologist Kate Fox notes that "The current media ideal of thinness for women is achievable by less than 5 percent of the female population," and that young women today are exposed to more of these kinds of images in one day than women saw in their entire adolescence a mere generation or two ago.[16]

(Incidentally, time spent on social media and games has a more mixed or as-yet unclear impact on happiness.[17] Especially for those of us who are isolated or older, social interactions online have been found to boost happiness, while for some of us, preteen girls, for example, time online seems to undermine our happiness.[18] Even gaming, which has sparked concern for decades now, seems to have both positive and negative effects.[19])

Renowned Buddhist monk, teacher, and activist Thich Nhat Hanh, writes eloquently about the power of where we place our attention and the impact TV can have:

> *Our senses are our windows to the world, and sometimes the wind blows through them and disturbs everything within us. Some of us leave our windows open all the time, allowing the sights and sounds of the world to*

invade us, penetrate us. … Do you ever find yourself watching an awful TV program, unable to turn it off? The raucous noises, explosions of gunfire are upsetting. Yet you don't get up and turn it off. Why do you torture yourself this way? Don't you want to close your windows? Are you frightened of solitude—the emptiness and the loneliness you may find when you face yourself alone? Watching a bad TV program we become the TV program. We are what we feel and perceive. If we are angry, we are the anger. If we are in love, we are love. If we look at a snow-covered mountain peak, we are the mountain. We can be anything we want, so why do we open our windows to bad TV programs?20

Part III

Happiness and the Journey Within

The refuge you seek
you will never find
in the outside world.
It is within you ...

Realize that here alone is the force,
here alone is the peace,
and here alone is the refuge
you are seeking.[1]

—Selvarajan Yesudian

Chapter 9:

Free Your Mind

JAMES DOTY IS A neurosurgeon and heads Stanford's Center for Compassion and Altruism Research and Education. He grew up with an alcoholic father and an incapacitated and depressed mother. He often got into fights, went hungry, and had trouble in school. He tells the following story about one of his school experiences:

> I had transferred to this school, this Catholic school, because my father actually had gotten a job there, although it was very brief employment. But, as a result, it allowed me to go to the Catholic school. The problem was the Catholic school was multiple grades ahead of the school I had attended, and I was suddenly thrown into this difficult environment. And I did not know how to behave in that environment. And this particular nun assumed I had done something wrong, when in fact I had done nothing wrong, and just walked up to me and literally slapped me, which I'd never had that done to me before, at least in that context. And my reaction just as a person who has been in fights was to hit them back. And that was my last day in Catholic school.

But then one day Doty had an unexpected experience that set him on a new course. He went into a magic shop (to replace a fake plastic thumb he had lost) and the owner's mother was covering for the owner. They started talking and she wound up telling him about what she called the real and much more powerful magic of the mind. She promised that it would change his life if he wanted to learn about it and would commit himself to practicing every day. She was only in town for six weeks, but Doty went every day and she taught him mental exercises that helped him calm his mind, stay present, and keep his heart open.

Those six weeks did in fact change his life. Before them he felt like "a leaf being blown by an ill wind."

> I had no control over anything, and events would happen, and I couldn't do anything about them. And I felt—and I think it was, in fact, reality—that at that point, when I met her, I had limited to no possibilities. And after that

42

six-week period of time, I suddenly had this vision that anything and everything was possible. … And that vision of possibility was so strong and so deep and so powerful that it was absolutely amazing. And the thing is, though, that my own personal circumstances fundamentally did not change at all.[1]

IN 1992, PSYCHOLOGIST Robert Holden started offering an eight-week course designed to boost people's happiness. By 1995 the course had caught the attention of the BBC and they filmed a documentary about it, following three of the participants from start to finish and regularly measuring their happiness to see how effective the course was.

Very effective, it turns out. The participants showed significant increases in happiness, and one of the three went from mildly unhappy to literally off-the-charts happy. The BBC decided to delay broadcasting the documentary so they could see whether the effect lasted. Sixteen weeks out, the three were still enjoying the boosted levels of happiness, beyond even the 10–15 percent increases that Martin Seligman suggested in chapter five are generally attainable for most of us.[2]

Holden would begin his course with a simple question: "Could you be happier—even if nothing in the world around you changed?"

He would give participants five minutes to think on it, but most needed only a moment. For the vast majority the answer was clear—yes, definitely.[3] And they were right. In fact, the science on this is so clear that they were probably more right than they realized.

You may recall that if we want to predict how happy someone is with their financial situation, it is four times more effective to ask them how happy they are in general than it is to look at any objective measure of their income. This is also true for work. We can predict how happy someone is with their work much better by simply asking them how happy they are in general than by looking at objective criteria like what kind of work they do, how long they've been doing it, or how much they get paid.[4] This also holds true for physical health. Ask someone how happy they are in general and you will be able to much more accurately predict how happy they are with their health than by checking objective measures such as recent illness, physical activity, access to health care, and so on.[5]

The factors that we explored in parts I and II—money, circumstances beyond our control, relationships, health, work, and television—these are largely about the lives we weave in the world around us. They each play an important role in how happy we are, but, of all the things under the sun that affect our happiness, the most significant ones lie within.[6]

Psychologist Sonya Lyubomirsky writes, "A common thread running through the research is that happy and unhappy individuals appear to experience—indeed, to reside in—different subjective worlds."[7] Science journalist Sharon Begley likewise notes, "Outer conditions are important contributive factors to our well-being or suffering. But in the end, our mind can override that. You can retain inner strength and well-being in very difficult situations, and you can be totally a wreck where … everything seems perfect."[8]

And when it comes to the inner landscape of our lives, the research is clear: there is significant room for us to intentionally shift our subjective experiences.[9] In fact, with practice we can change the very form and function of our brains. Sharon Begley summarizes the science of "neuroplasticity":

> *We are not stuck with the brain we were born with but have the capacity to willfully direct which functions will flower and which will wither … which emotions flourish and which are stilled. … Connections among neurons can be physically modified through mental training just as biceps can be modified by physical training.[10]*

If this isn't something you've seen a lot of in your own life or in the people around you when it comes to happiness, that's not surprising. Begley adds:

> *Consider an analogy. You are studying whether measures of cardiovascular health—resting heart rate and blood pressure, for instance—can be improved. You are conducting the experiment in a society that has yet to get the news that there is such a thing as aerobic exercise. You dutifully measure the resting heart rate and blood pressure of your couch potatoes every year for several decades. Except for some change due to aging, their heart rate and blood pressure are, you find, remarkably stable. You win fame and fortune … for discovering the "cardiovascular set point."*

> *There is only one problem. You neglected to see whether resting heart rate and blood pressure can be lowered through a regimen of regular, rigorous, pulse-raising exercise. …*

> *Just as people now see the value of exercising the body consistently and for the rest of their life, it's similar with emotional skills.[11]*

The inner journey to happiness isn't about trying to trick ourselves into being happier, putting blinders on and pretending our way to happiness. It also isn't about just trying to always be happy or forcing happiness. Totally the contrary, it is about living ever more in the truth of who we are, embracing the fullness of ourselves and the world and the human experience, even when that doesn't look or feel much like being happy.

Broadly speaking, it is about walking five different paths:

- Welcoming the breadth of emotions that are naturally part of being human, both the ones we like and the ones we often do not
- Connecting with a greater sense of wonder, gratitude, and presence
- Being more compassionate and loving with ourselves

44

- Releasing attitudes that block us from being happy, including fears of happiness and attachments to not being happy
- Living with greater purpose and more fully manifesting the lives we are drawn to live

Ultimately, it is about living more fully in the inherent sacredness and wonder of ourselves and the world around us

It is to this, and these five paths, that we dedicate the next five chapters of our journey.

Chapter 10:

Nurturing a Positive Relationship with Destruction (and Your Emotions)

Instead of asking ourselves, "How can I find security and happiness?" we could ask ourselves, "Can I touch the center of my pain? Can I sit with suffering, both yours and mine, without trying to make it go away? Can I stay present to the ache of loss or disgrace—disappointment in all its many forms—and let it open me?" [1]

—Pema Chödrön

THE WAY WE conventionally think of age, we are as old as our bodies are. Whether we have circled the sun twenty-three times or forty-seven or eighty-six, that's how old we are. And—so the conventional thinking goes—that is how many years of knowledge and experience we have available to draw on to help guide us in our lives. That includes the knowledge and experience shared with us during those years from other people, through their spoken and written word and their actions.

However, this is profoundly deceptive. In another very important sense, we are as old as life itself. For some four billion years, life has been trying out endless ways of being in the world, and in the process it has honed innumerable brilliant strategies, which it has in turn handed down to us. We are each of us born into the world with a deep storehouse of this wisdom woven into our DNA. Indeed, the knowledge we accumulate in the course of our own lives is only the thinnest of veneers compared to the deep oceans of wisdom about "what works" that we have inherited from our four billion years of ancestors.

This wisdom is most obvious in the countless ways our bodies function in every moment to keep us alive, completely independent of our conscious thought: our hearts beating, our DNA replicating, our lungs transferring oxygen from breath to blood, our nerves passing electrical signals, and on and on.

It is perhaps not as obvious, but one of the ways we most directly experience this deep wisdom on a conscious level is through our feelings. Emotions have been an important part of our ancestors' lives at least since we were mammals, and possibly going back to our reptile ancestors as well.[2] So our emotions have been honed through at least 200 million years of our ancestors learning how to most successfully be mammals, primates, hominids, and humans.[3] It makes sense, then, that our emotions serve very important purposes.

Charles Darwin once noted, "Every single organic being ... lives by a struggle at some period in its life; ... heavy destruction inevitably falls on the young or old." How we relate to the inevitable struggles in our lives plays an important role in our happiness, and our emotions play a key role in this. There are immense benefits in allowing ourselves to experience our natural range of emotions and in letting them do the heavy work they were designed for: helping our bodies and psyches process our experiences of the world in healthy ways.[4]

Not surprisingly, research confirms that we are happier when we are experiencing pleasant emotions rather than unpleasant ones. However, a study by psychologist Maya Tamir found that **even more significant than whether the emotions we are experiencing are pleasant or not is** *the degree to which we are accepting of them.* We are generally happier and more resilient when we embrace our emotions, when we let our heads get out of the way and let our hearts lead for a while.[5] This includes the emotions we tend not to like as much, such as fear, anger, and grief.[6]

THERE ARE DEFINITELY times when it can be helpful to suppress or work around our emotions—if we are in the middle of a job interview, for example, or are landing a plane, or we just need to get dinner on the table for the kids.[7] Some numbing of our emotions can also sometimes be beneficial following extreme events or losses.[8] And there are definitely times when it is helpful to pause and bring a little perspective to a situation, to acknowledge when something actually isn't worth getting worked up over, and to take a deep breath and just let it go.[9]

Beyond these specific circumstances, however, suppressing our emotions is generally *not* a good thing. For one, **we cannot selectively numb our emotions.** Social scientist Brené Brown notes, "You can't say, here's the bad stuff. Here's vulnerability, here's grief, here's shame, here's fear, here's disappointment. I don't want to feel these. ... When we numb those, we numb joy, we numb gratitude, we numb happiness."[10] Renowned Buddhist instructor Pema Chödrön writes, "When we protect ourselves so we won't feel pain, that protection becomes like armor, like armor that imprisons the softness of the heart."[11]

When we open ourselves more fully to our emotions, including grief, anger, and fear, we increase our capacity to feel all of these other emotions as well.

We may fear that our emotions will overwhelm us, that if we give ourselves over to them we will get completely lost in them. Yet the opposite is usually closer to the truth. Psychologist Scott Barry Kaufman notes, "Avoiding feared thoughts, feelings, and sensations … paradoxically makes things worse, reinforcing our belief that the world is not safe and making it more difficult to pursue valued long-term goals." Meanwhile those thoughts and feelings often become stronger for being suppressed.[12] They can build up and then come out in ways that are sometimes dramatic, even explosive, and that are sometimes subtle and complicated.[13]

We live in a culture where emotions tend to be scorned, and we are often encouraged to suppress them. Writing specifically about grief, Sobonfu Somé, a specialist in grief and grief rituals, writes:

> In today's world, most of us carry grief and do not even know it. We have been trained at a very young age how not to feel. In the West we are often taught that to be good girls and boys we have to "suck it up." The consequences are that even with your most intimate and trustworthy friends you might feel like, "I am burdening them." Crying in front of others is too often a forbidden fruit. We learn to compartmentalize our grief. …
> We are born fully knowing how to grieve. … We cry naturally to feel better, to unburden ourselves and take a few pounds off our shoulders and souls. **We need to begin to see grief not as a foreign entity and not as an alien to be held down or caged up, but as a natural process.**[14]

There are physical benefits to embracing our emotions as well. A study done by psychiatrist Benjamin Chapman found that suppressing our emotions was linked with a higher incidence of many diseases, including a 70 percent increased risk of cancer, and a 30 percent higher risk of premature death.[15]

The renowned Buddhist leader, the Dalai Lama, urges us:

> Do not fight against pain; do not fight against irritation or jealousy. Embrace them with great tenderness, as though you were embracing a little baby. Your anger is yourself, and you should not be violent toward it. The same thing goes for all of your emotions.[16]

This being human is a guest house
Every morning a new arrival.
A joy, a depression, a meanness,
some momentary awareness comes

as an unexpected visitor.
Welcome and entertain them all!
Even if they are a crowd of sorrows,
who violently sweep your house
empty of its furniture,
still treat each guest honorably.
He may be clearing you out for some new delight.
The dark thought, the shame, the malice,
meet them at the door laughing,
and invite them in.
Be grateful for whoever comes,
because each has been sent
as a guide from beyond.[17]

—Rumi

ADVERSE EXPERIENCES NEED not simply be things we try to get through as best we can, where the goal is to minimize the negative impact and get back to our "normal" selves and lives. They can actually be catalysts for profound growth, taking us beyond our previous norms—"clearing us out for new delights," as Rumi puts it.[18]

The challenges and pains of the world are, in fact, to some degree *necessary* for our growth, and **to do all that we can to avoid difficult situations, or to suppress our experience of them, is to actually suppress our greatest potential and our deepest callings**. Renowned Buddhist instructor Pema Chödrön writes:

> *A common misunderstanding among all the human beings who have ever been born on the earth is that the best way to live is to try to avoid pain and just try to get comfortable. … A much more interesting, kind, adventurous, and joyful approach to life is to begin to develop our curiosity. … We must realize that we can endure a lot of pain and pleasure for the sake of finding out who we are and what this world is, how we tick and how our world ticks, how the whole thing just is. If we're committed to comfort at any cost, as soon as we come up against the least edge of pain, we're going to run; we'll never know what's beyond that particular barrier or wall or fearful thing.*[19]

Researchers have documented that difficult experiences can benefit us in many ways:

- They often put other things in perspective and help us appreciate the blessings in our daily lives.
- They can help us get a better sense of our priorities and to reorient toward them.
- They can enhance our spiritual and creative growth.

- They can nourish a deeper compassion for the suffering and humanity of others.
- They can strengthen our connections with family and friends.
- When another negative event inevitably comes along, we are often more resilient in dealing with it.[20]

Howard Cutler, coauthor with the Dalai Lama of the book *The Art of Happiness*, once asked participants in a workshop he was leading to name a hard time they had gone through and something positive that had come out of it. A man in the workshop named Joseph had lost a daughter to leukemia and he replied, "I just can't see anything good that came from it in any way, nothing I learned from it except pain. … It was just bad."

The others in the circle took turns and after the last person spoke, Joseph said he'd been thinking about it and that despite all the pain he'd experienced he could think of two good things that had come from his daughter's death. First, it made him stronger. "There's nothing left to fear, because no matter how hard things get, I know that I've survived worse." Second, it made him appreciate his younger daughter even more. "It made me realize what a gift she is, each day, and not to take her for granted, and as a result I think I'm a better father to her."[21]

Hard times can also actually help us to lead happier lives. One study documented that people who had experienced more than twelve major negative experiences in their lives were the least happy. But people who had experienced zero such events were generally no happier than people who had experienced seven to twelve of them. People who had experienced between two and six were the happiest.[22]

Recognizing that every bad situation does invariably lead to *something* good, whether small or large, has actually been documented as one of the most helpful things we can do when we are going through a difficult time.[23]

Life coach and breast cancer survivor Melissa Eppard takes this idea even further:

> *I don't know who first said this simple phrase, but it has been coming to me again and again, like an urgent message asking to be shared. It goes like this, "This isn't happening to you, but FOR YOU."*
>
> *I totally resisted this statement the first time I heard it, while going through cancer treatment 6 years ago. Only much later could I see how that experience cracked me open, how the raw vulnerability deepened my relationships, and the frayed struggle to survive made me grasp all that much harder for the preciousness of each day. I was given this chance to have a more honest relationship with myself, to let go of this idea that my value as a human being had to do with how I looked, or how hard I worked, and finally put down that mantle of people-pleasing and being a "good girl."…*
>
> *When you replace "Why is this happening to me?" with "What is this trying to teach me?" everything shifts.[24]*

IT IS IMPORTANT to distinguish between the positive things that can come out of difficult experiences and the experiences themselves. **The experiences are often simply horrible. They are not to be justified or rationalized away with some growth or positive change or joy that emerges in its aftermath.**

And even if we are better off in some ways for having had those awful experiences, that is not to say we wouldn't be better off in other ways if we had *not* had them. Rabbi Harold Kushner, author of *When Bad Things Happen to Good People*, reflecting on the death of his son, writes:

> *I am a more sensitive person, a more effective pastor, a more sympathetic counselor because of Aaron's life and death than I would ever have been without it. And I would give up all of those gains in a second if I could have my son back. If I could choose, I would forego all the spiritual growth and depth. ... But I cannot choose.*[25]

Psychologist Judith Mangelsdorf, in a systematic review of research on this topic, found that significant *positive* experiences can lead to just as much growth, change, and insight in their own ways as negative ones.[26]

In the end, it is important that we be gentle with ourselves and that we not *expect* that we will always have some kind of wonderful growth or gains after a hard experience. Sometimes we don't.[27] To try to force a positive attitude can be to skip past the time or space these other emotions—and we—need.[28] Or it can add a layer of judgment in an already difficult situation ("Why can't I just be more spiritual and grateful?!?").

Psychologist Barbara Held, in an essay titled, "Combating the Tyranny of the Positive Attitude," writes:

> *I believe that there is no one right way to cope with all the pain of living. ... People have different temperaments, and if we are prevented from coping in our own way, be it "positive" or "negative," we function less well. ... Sometimes a lot of what people need when faced with adversity is permission to feel crummy for a while. Some of my one-session "cures" have come from reminding people that life can be difficult, and it's okay if we're not happy all the time. ... I believe that we would be better off if we let everyone be themselves—positive, negative, or even somewhere in-between.*[29]

This is not about romanticizing pain and loss. It is about recognizing that difficult experiences can, in fact, play a very important role in our lives. It is about *not* living "committed to comfort at any cost," or turning and running "as soon as we come up against the least edge of pain," and missing out on all of the living and growth and joy that awaits us on the other side.

And it is about leaning into the healing power of our emotions.

I draw inspiration from the way little children react to things. If they are frightened or hurt or sad, they don't question their feelings or whether a situation is worth being upset over, or whether they deserve to feel that way, or whether they should try to feel something different. They just cut loose with their feelings. Then the wisdom of millions of years and lifetimes flows forth. And often, pretty quickly, there is relief, and then they are off to whatever the next moment brings.

Chapter 11

Living in Mystery, Magic, and Miracle

Miracles, in fact, are a retelling in small letters of the very same story which is written across the whole world in letters too large for some of us to see.[1]

—C. S. Lewis

T HE BOOK *CHARLOTTE'S Web* tells the story of a pig, Wilbur, who lives on a farm, and a spider, Charlotte, who weaves messages into her web about how wonderful Wilbur is, so that he will not be slaughtered by the owners of the farm. The following conversation takes place among two of the townspeople after the first message appears:

> *"Have you heard about the words that appeared in the spider's web?" asked Mrs. Arable nervously.*
> *"Yes," replied the doctor.*
> *"Well, do you understand it?" asked Mrs. Arable.*
> *"Understand what?"*
> *"Do you understand how there could be any writing in a spider's web?"*
> *"Oh no," said Dr. Dorian. "I don't understand it. But for that matter I don't understand how a spider learned to spin a web in the first place. When the words appeared, everyone said they were a miracle. But nobody pointed out that the web itself is a miracle."* [2]

Just what is a miracle? I looked up a number of definitions and I came up with this: a miracle is "a wondrous or astonishing event that is not explicable by natural or scientific laws."

Which means that *everything* is fundamentally a miracle.

Science has rolled forward at a tremendous pace over the past five hundred years, illuminating with immense detail how so many things work in the world and beyond. Yet, ultimately, there is a limit to what science can explain. Indeed, there isn't anything where our scientific understanding of it doesn't eventually run off the map of not just what is known but of what *can* be known. This is a point that countless scientists have sought to make clear, many of them the greatest names in science, including Einstein, Schrödinger, and Heisenberg.[3] Science simply cannot ever explain the fundamental nature of the universe, or "ultimate reality," as Sir James Jeans, a leading physicist of the twentieth century, put it.[4] Sir Arthur Eddington, another leading physicist, noted that even the most penetrating and powerful scientific ideas are understood to touch on only a "partial aspect of something wider," which science cannot ever discern.[5]

Nobel Prize winner Albert Schweitzer once suggested, "**The highest knowledge is to know that we are surrounded by mystery.**"[6] Put another way, the highest knowledge is to know that we are surrounded by miracles, that we and everything in existence are miracles, fundamentally inexplicable.

I may, like C. S. Lewis and Dr. Dorian, be using the word miracle in a slightly different way than people commonly use it, but I think it's really a better use of the word. It invites us to see the world in a way that is not only more beautiful and alive, but that is also fundamentally more honest and true.

It's like in a story where someone looks at a bookshelf when the light is exactly right, and a door reveals itself, opening onto an entire other magical world. *That* is the universe we live in, one grounded entirely in magic, held together in magic, and recreated every moment in magic. All we have to do is get a slightly different angle on just what a miracle is, and let the just-right light of that concept—miracle—reveal to us the magic all around us. It is an invitation to awe, to presence, to gratitude—and to greater happiness.

THERE IS PERHAPS no better way to highlight the splendidly magical nature of the universe than to consider the fundamental building blocks of which everything is comprised, that is, the foundation of everything that science can describe.

We've rapidly gone from the groundbreaking discovery of the atom, once thought to be the most basic, indivisible unit of matter, to discovering that the atom itself is made up of elementary particles, which we have given appropriately fanciful names such as charmed quarks, gravitons, and tau neutrinos.

It turns out that atoms are mostly empty space. In fact, if all the space in our bodies between atomic particles were removed, we would each be a million times smaller than the smallest grain of sand.[7] Even then, many of these elementary "particles" don't actually have any mass. In fact, most of the mass in the universe comes from particles called gluons, which on their own don't actually have any mass, but then suddenly do when they get together with other elementary particles.[8]

54

And all of these particles can be in more than one place at the same time. In fact, every bit of matter and energy generally "exists in a state of blurry flux, allowing it to occupy not just two locations but an infinite number of them simultaneously." But then the mere act of our observing or measuring them causes the particles to drop into just one location or path. And taking such a measurement affects not just their current and future locations, but their past locations as well.[9]

Oh, and all of these particles occasionally change from one kind of particle into another.[10]

It's with good reason that Neils Bohr, one of the leading figures in quantum mechanics, once wrote that anybody who can contemplate quantum mechanics without getting dizzy doesn't understand it.[11] Einstein himself found some of these ideas so crazy that he cited the wackiest of them as proof that they couldn't be true, in particular the idea that the very act of measuring one particle could instantaneously affect another particle all the way across the universe. However, in the following decades we have confirmed that that is actually true.[12]

This is the magic and miracle that we and everything are made of—particles that are mostly just space, that have little or no actual mass, that generally exist in a blur of infinite possible locations, that occasionally morph into other kinds of particles, and that can interact with each other instantaneously across the universe and affect each other's pasts.

Oh, and many of these particles are constantly flashing in and out of existence, ceasing to exist one moment and then suddenly turning up again.[13] (Is it the same particle? A different one? Does that distinction even have any meaning at that level?)

The universe began nearly fourteen billion years ago, when everything that exists erupted forth from a single point, and has been expanding ever since.[14] In a sense, that same miracle is being recreated in every instant, as many of the particles that we and everything physical are made of spring back and forth from existence into nonexistence. **That fundamental miracle of creation is not just somewhere in the distant past; we are immersed in it, are in every moment being recreated by it, are *of* it.**

Perhaps through this simple practice of paying attention … you begin to realize that you're always standing in the middle of a sacred circle, and that's your whole life.[15]

—Pema Chödrön

THE WORD MIRACLE comes to us from the Late Latin *miraculum*, meaning "a wonder or marvel." Tracing the word back further we come eventually to the root *smei-*, meaning "to smile" or "to laugh." It's a very telling origin. **Appreciating the fundamentally miraculous** nature of ourselves and everything around us is one of the surest ways we can nurture happiness in our lives.[16]

That can mean having amazing transcendent experiences, but much more often it is about simply slowing down and being deliberate about where we place our attention. When we step back from a heightened level of stimulation, of rush and stress and distraction, when we slow our minds and our bodies, we get to experience more of the **breadth and depth of ourselves and the world around us.**

By intentionally nurturing greater presence and gratitude in our lives we can selectively override the default human process of adaptation that we explored in chapter two. Psychologist Sonya Lyubomirsky beautifully refers to these practices as "*intentional reencounters with the world.*"[17]

In his book *Peace Is Every Step: The Path of Mindfulness in Everyday Life*, Thich Nhat Hanh writes:

> *Peace is right here, right now, in ourselves, and in everything we do and see. The question is whether or not we are in touch with it. … We are very good at preparing to live, but not very good at living. We know how to sacrifice ten years for a diploma, we are willing to work very hard to get a job, a car, a house, and so on. But we have difficulty remembering that we are alive in the present moment, the only moment there is for us to be alive. Every breath we take, every step we make, can be filled with peace, joy, serenity. We need only to be awake, alive in the present moment.[18]*

Mindfulness meditation is one very effective method of slowing down and reencountering the world. It involves focusing our attention on our breath and our thoughts and feelings, without judgment, only observation and acceptance. Mindfulness meditation has been found to significantly increase happiness, to reduce anxiety, pain, and depression, to help people with phobias, depression, or post-traumatic stress disorder, and to help with sleep.[19] One study found that people who were trained to be more aware of their thoughts and feelings for a mere two or three months experienced enhanced well-being for several years after.[20]

Buddhist monks with mindfulness meditation practices have recorded a larger increase in gamma wave activity, which is associated with happiness, than has "ever been reported in neuroscience." "Usually the gamma signal lasts for a couple of hundred milliseconds. But in the adepts it lasted five minutes." Much of this gamma activity continued even after they stopped meditating, and a group of non-meditators given a crash course in meditation showed increased gamma wave activity when tested after just a week of practice.[21]

As MUCH MYSTERY and miracle as there is to be found in the tiniest crevices of existence, there is just as much to be found in its greatest expanses. Leaving behind the quantum world, let's consider our own world, this planet.

The earth is incomprehensively vast, and for even the most well traveled of us, our entire lives play out on only the tiniest portion of it. Every living being we know of was born and died somewhere on this single planet. It is the place where every hope, fear, anger, ecstasy, and imagining of every human and all other creatures we know of has ever taken place (excepting the twenty-four people who have been to the moon).

And yet our planet is just a tiny dot on the scale of our solar system. This vast world could fit inside the sun along with a million other earths.[22] And you would have to travel the distance from the earth to the sun more than twenty-five times over to arrive at Neptune, the farthest planet.[23]

Still, a quick glance at the night sky reveals the humble scale of even our enormous solar system. Depending on where we live and how dark the night sky is, we can see anywhere from dozens to thousands of other distinct stars.[24] If it's dark enough we can see a swath of distant stars so vast it creates an effect of milk having been spilled across the sky (or a river of silver, as they say in China). How many stars does it take to create that effect? Tens of thousands? Hundreds of thousands?

The correct answer is between two hundred and four hundred billion. That's how many stars there are in the Milky Way galaxy, and an additional one to seven new stars are born every year. The vast majority of those stars have at least one planet, and an estimated three hundred million of those could potentially host life.[25]

But that's not all we can see. Under ideal conditions we can see as many as nine other galaxies with the naked eye. They are part of a family of no less than fifty-four galaxies that make up what scientists call the "Local Group."[26]

As the name Local Group implies, we're not done yet. As immense as a single galaxy is, and as immense as a group of fifty-four galaxies is, there are many such groups—more than a hundred in fact.[27] That brings our total to somewhere around 5,400 galaxies, each of those with anywhere from billions to hundreds of billions of stars and planets.

As literally incomprehensible as those numbers already are to the human mind, we're not quite done yet. Because that supercluster of 5,400 galaxies, known as the Virgo Supercluster, is not the only one. There are, in fact, quite a few superclusters. How many is "quite a few?" Thirty? Seventy? A hundred?

There are actually an estimated ten million superclusters in the universe, each with thousands upon thousands of galaxies.[28]

In all, there are an estimated septillion stars in the universe.[29] That's a trillion trillion, or a one with twenty-four zeroes.

Let's say you were given the task of counting every one of those stars. But to make the task a little more manageable, let's say we also gave you super counting powers, so that you can count a thousand stars per second. And let's say we also gave you some help. A lot of help. One million other people with the same super counting powers. How many lifetimes of, say, one hundred years, would it take all of you to count all the stars without ever stopping?

Roughly thirty-two million.

But oops, wait. In 2016, after twenty years of compiling photos from the Hubble Space Telescope, scientists estimated that there are about twenty times more galaxies than was previously thought.

So make that 640 million lifetimes.[30]

And most of those stars have at least one planet, and roughly one in every ten of those could potentially sustain life (based on the Milky Way statistics).

Cosmologist Brian Swimme has noted that at the moment of the big bang, if the rate of expansion had been even a millionth of 1 percent faster, the universe would have expanded too quickly for any of these stars and planets to have formed. "There would have been no galaxies, no structure, no life, nothing but dust for all time." Or if it had expanded a millionth of 1 percent slower, after a million years it would have simply collapsed and formed into a massive black hole—again, no galaxies, no structure, no us.[31]

Physicist Brian Greene likewise notes that if any of the four fundamental forces in the universe—gravity, electromagnetism, and the strong and weak nuclear forces—were even slightly stronger or weaker, the universe and life as we know it could not exist. Stars would have burnt out too quickly, or they could have never formed in the first place, or atoms would simply disintegrate, and so on.[32]

This is the profound magic and miracle we are held in: the immensity of this gorgeous planet that gave birth to us, with hundreds of billions of stars spilling across the night sky, all part of the family of a single galaxy, this galaxy one of thousands upon thousands in the Virgo Supercluster, which is one of some ten million superclusters—all of us held in the grace of an exact, essential balance of fundamental forces and expansion.

The greatest of all miracles is to be alive, and when you breathe in, you touch that miracle. Therefore, your breathing can be a celebration of life.[33]

—Thich Nhat Hahn

IT IS A remarkable feature of the ocean that, whatever is going on at the surface, we don't have to drop very far below the surface before we leave it all behind and it has no effect on us. There can be a storm with driving wind and rain and waves, or it can be still and the sun blazing. There can be loud noises, thunder or yelling. But if we just drop a few feet beneath the surface, none of that is present anymore. Even with large waves, we don't have to drop very far before all we experience is the slightest up and down motion. It's like descending into another world, and there can be an immense peacefulness to it.[34]

This is a wonderful metaphor for meditating, or simply slowing down, closing our eyes, and taking some deep breaths. There is often so much going on in the world around us and in our daily lives—so much going places, getting things done, preparing and interacting and reacting. Our minds are often completely caught up in it all. But by simply taking a moment to close our eyes and take a few breaths we can drop into ourselves and largely leave all of that behind. It is its own descent into a different world, and there can similarly be an immense peacefulness to it.

There are hundreds of different types of meditation, most of them with their roots in the Hindu and Buddhist traditions of South Asia, though often today they are entirely decoupled from any "religious" attachment. (The Buddha actually encouraged people to apply his insights to their own religious/nonreligious/cultural contexts.) While each of them has its own approach and purpose, many of them, as well as many yoga practices and stress-relief techniques, center around the breath.[35] Simply taking a few extended breaths causes our heart rate, blood pressure, and cortisol levels to drop and our muscles to relax. Even brief dips into experiences of mindfulness meditation have been found to enhance well-being for years afterward. Regularly sitting and focusing on the breath for longer periods of time offers even more significant and long-lasting benefits.[36]

A common practice in mindfulness meditation is to briefly note what thoughts or feelings are coming up whenever we find our attention has strayed from our breath. This can help us get to know ourselves and our minds better, to become more familiar with the kinds of thoughts and feelings we are inclined to. But it can also help us see more clearly that *we are not these thoughts or feelings*. We can drop beneath them, beneath even the big waves in our own lives and our minds—you are not that fear or that sadness or that joy, you are not that project or that relationship or that problem or that experience. In time we can stay more deeply rooted in this part of ourselves, in some degree of presence and calm, even when we are at the surface, right there amidst the waves and the rain and the sun.[37]

The metaphor of dropping into the ocean is relevant not just for the idea of diving beneath the surface, but also because there is such vastness and richness to our inner worlds. There is indeed an ocean within each of us! It can be stunning and transformative to get even a glimpse of this vastness.

When we dip into this ocean, we also often encounter some of the things that we have ourselves kept beneath the surface. We may find a whole range of emotions that have built up and been held back by the busyness in our lives and our minds. When we create a little space for them they often flow forth. We may start to cry, or feel angry or scared, or any number of emotions. In that case, dropping into ourselves may be an important opportunity to connect with our feelings, to ride these different currents within us.[38]

There may be other things longing for our attention as well. There are often parts of ourselves that we keep at a distance, such as judgments, fears, and hurts, and given the opportunity they will bubble up as well. Slowing down and actually giving them some attention can give us a chance to take down some of the walls and loosen some of the knots we keep them in. While this does generally bring us greater peace and happiness overall,

59

it can be uncomfortable or disorienting also. Jay Michaelson, a meditation instructor and author of *Evolving Dharma*, writes, "If a vacation banishes cares from the mind, a meditation retreat puts them right smack in front of you."[39] Mindfulness instructor Ed Helliwell adds, "Mindfulness has a great many benefits, but they tend to come as a by-product of getting up close to unpleasant experiences like pain, turmoil, and 'negative' thought patterns."[40] It is meditation as a path of liberation, not by boxing up these negative thoughts and feelings or ignoring them, but by creating greater ease and understanding around them.

For some people, the experience of diving deeply into ourselves can also open a door to going beyond ourselves, of connecting with the vast ocean of existence that we are part of. Our sense of self may dissolve, we may find ourselves seeing the world and ourselves in very different and expanded ways, we may have an experience of existing beyond all of this, as pure consciousness. We may have experiences of connecting with something much greater than ourselves—the universe, nature, spirit, love, the divine. The sixteenth-century Hindu mystic Mirabai, recounting her own experience and also inviting us to our own, wrote:

> *In my travels I spent time with a great yogi.*
> *Once he said to me,*
> *"Become so still you hear the blood flowing*
> *through your veins."*
>
> *One night as I sat in quiet,*
> *I seemed on the verge of entering a world inside so vast*
> *I know it is the source of*
> *all of us.*[41]

Author Maya Angelou, in her final written message before she passed away, likewise wrote, "Listen to yourself, and in that quietude you might hear the voice of God."[42]

> *Even the poorest parents give their children vast riches, in the form of*
> *senses, emotions, and mental faculties that have been optimized through*
> *millions of years of product development. They are so reliable, efficient,*
> *intricate, self-growing, and self-repairing that no technology comes*
> *anywhere close to matching them. The human genome is the ancestral vault*
> *of riches, the secret Swiss bank account. … Beyond our true necessities and*
> *luxuries—our biological adaptations—we get only a little added value from*
> *market-traded products.*[43]

—Geoffrey Miller

SOMEWHERE BETWEEN THE delirious magic of the quantum realm and the breathtaking miracle of the universe, entirely immersed in both, is you—a miracle of staggering beauty and wonder, as much as any charmed quark or supercluster of galaxies.

Indeed, for as many stars as exist in the universe, there are far more atoms in your body. Indeed, it would take some *seven thousand* universes for the number of stars to match the number of atoms in your body![44] Or, put another way, **there are more than a billion atoms within you for every second the universe has existed.**[45]

Through these atoms you have an extraordinary and intimate relationship with the universe. Your body is 10 percent hydrogen, and every one of those atoms has been present since almost the very birth of the universe, having formed along with helium and lithium as soon as the universe cooled enough after the big bang for atoms to form.[46]

The heavier elements like oxygen and carbon, which make up 83 percent of you, could only be forged from those initial, light elements within the hearts of stars, where temperatures can reach two hundred million degrees.[47] And the still-heavier elements within you, like zinc, copper, and iodine, could only be formed in the even more extreme heat of supernovas, the miraculous gift of giant stars and white dwarfs that explode at the end of their lives and reach nearly two billion degrees.[48]

In other words, parts of you have existed in their current form since the birth of the universe nearly fourteen billion years ago, while the rest of you is literally made of stardust, having once churned in the hearts of stars and some of it having been blasted into existence in the death explosions of the most massive of stars. Yes, that is you. Take a look at your body; that is what your body has been through, where it comes from.

These elements in turn drifted through vast expanses of time and space until 4.5 billion years ago when they were caught by, and contributed to, the gravitational pull of a huge amount of matter that came together to form our planet.[49] Billions of years later they were woven together into the elegant structures that comprise your body.

Guiding that process of weaving these elements is your DNA, at once inherited and also unique to you. Your DNA first sprang into existence through the union of an egg cell, among the largest of human cells, and a sperm cell, among the smallest, and the combination of twenty-three DNA molecules from each.[50] These molecules and that single cell replicated and divided over and over, eventually giving rise to roughly thirty-seven trillion cells in the adult body.[51]

But they didn't just divide themselves over and over into a blob of all the same cells; from that single cell they differentiated, giving rise to over two hundred kinds of cells able to do extraordinary things:[52]

- Produce bones, which are four times stronger than concrete, and enamel (in your teeth), which is stronger than steel[53]
- Secrete hydrochloric acid for your stomach and mucous to protect your stomach from that acid

- Create a network of nerves that use electricity to send information to your brain at 150 miles an hour[54]
- Grow about ten miles of hair each year, with a full head of hair strong enough to support the weight of two elephants[55]
- Create four hundred types of scent receptors, detecting more than a *trillion* different smells[56]
- Detect even just a few photons entering your eyes from a source as small as a candle over one and a half miles away, or as far away as the Andromeda Galaxy, having traveled 2.5 million years to get to you[57]
- Produce the tiniest bones in your body, mere fractions of an inch in length, delicately attuned to sensing vibrations (sound) in the air.
- And replace some 23 million cells in your body every minute.[58]

That single cell also gave rise to what Nobel laureate James Watson described as "the most complex thing we have yet discovered in our universe"—the human brain.[59] Our brains perform the equivalent of some billion billion operations per second.[60] To be clear, there is a computer in production named El Capitan that around 2023 is expected to perform twice that many operations per second.[61] However, thanks to the complex interconnectedness of the brain and its immense range of functions, it can perform "processes that might take a computer a few million steps" in just "a few hundred neuron transmissions."[62] Meanwhile, El Capitan will take about two tennis courts' worth of space, weigh as much as thirty-five school busses, and need thirty megawatts of electricity, enough to power twelve thousand homes.[63] The human brain requires about *fifteen* watts, the amount of energy consumed by a really dim light bulb. And it all fits right there neatly inside your skull, also craftily forged by your DNA.[64] Our brains can also "rewire" themselves, with neurons disconnecting and reconnecting with each other, even changing their basic features, something no computer can do.[65]

The thirty-seven trillion cells throughout your body are likewise extremely dynamic and adaptable. They are all working with basically identical copies of your DNA, yet each must understand what its own unique purpose is, depending on the kind of cell it is and where it is in the body, so that it can interpret and build from that DNA differently.[66] Science writer R. Phillip Bouchard notes:

> There is no central authority—not in the brain, nor in any other part of the body—that drives the development and maintenance of the human body. Instead, the overall results are an example of an emergent system.[67]

Our cells are also in constant conversation with our environment. Our genes express themselves very differently, turning on and off, depending on our daily experiences of stress, physical activity, joy and touch, disease, pollution, the foods we eat, the foods our biological parents ate, and so on.[68]

We can see this magic, known as epigenetics, at work in identical twins and clones. Living beings that start out with the exact same DNA can nonetheless develop in strikingly different ways, including physically. For example, the "parent" of the first cloned cat was a calico, yet the clone emerged as tabby and white, perhaps reflecting an influence from the host mother who carried the clone in her womb, who was a tabby.[69] Identical twins likewise not only often develop into very different people personality-wise, they can look very different as well.[70] Bouchard notes, "The genetic recipe for a cat will not give you an elephant, but you can't read the DNA of an individual and see a Mini-Me of [their] features."[71] Our cells, our bodies, are that dynamic and adaptable."[72]

You are the vast magic of all of this. You contain within you the big bang and the hearts of stars, along with more than a billion atoms for every second the universe has existed. You are a single cell that became trillions of cells that can detect photons and create acid and bones, that produce and move electricity. You are an emergent system of all of these cells working independently yet also in a constant and intimate conversation with each other and with the world around you.

The moment one gives close attention to anything, even a blade of grass, it becomes a mysterious, awesome, indescribably magnificent world in itself.[73]

—Henry Miller

A LEAF. A wave. Laughter. Stars and quarks. Each of these by itself is vastly beautiful and exquisite. Collectively they paint a picture of breathtaking abundance.

If we take the time to consider this abundance, how can we not be overwhelmed, drunk even (at least momentarily), with the experience of being so fortunate, so wondrously gifted as to exist amid this endless splendor and richness, to be absolutely swimming in miracles in every moment? To be entirely of miracles ourselves?

Psychologist Robert Emmons notes, "The ability to notice, appreciate, and savor life is a crucial determinant of well-being."[74] Taking time at the end of the day to write about three things that went well, counting our blessings once a week, writing a letter of gratitude to someone—each of these has been found in various studies to boost people's happiness for anywhere from three to nine months.[75]

Robert Holden, the psychologist whose course on happiness was documented by the BBC, writes that "**Gratitude is one of the highest forms of prayer.**" His students rate an activity called One Hundred Gratitudes as among the most powerful in his course. It involves creating a list of one hundred things you are grateful for and briefly explaining why you are grateful for each. One student wrote that this activity "showed me that my past was better than I realized; my life right now is better than I realized; and I think my future just got a whole lot better, too." Another wrote, "All my life I have felt a huge sense of lack. … Now I realize that the lack I experienced was due to my lack of gratitude."

For many people the activity helps them "discern between what is important … and what is not." Holden writes that people "no longer complain about the little things with quite the same conviction." Research confirms that as we experience more gratitude in our lives, we are less likely to be materialistic, acquisitive, and to feel like we are lacking something compared to others.[76]

Robotically doing gratitude exercises will not increase our happiness. Indeed, it can actually make us less happy, particularly those of us who are clinically depressed, perhaps because it adds stress and judgment ("I should be more grateful," "Why am I not more grateful?").[77] Yet practices related to wonder and gratitude, these "intentional reencounters with the world," do help most of us. Indeed, **they generally have a greater effect on our happiness than our circumstances.**[78]

Holden emphasizes that the One Hundred Gratitudes exercise doesn't just focus on what is "positive," but what is "significant." Creating this list helps us see "the truth, the learning, the healing, and the love" in our "past mistakes, heartbreaks, illnesses, divorces, bereavement, failures, and dark nights of the ego."[79]

Holden actually describes three different kinds of gratitude. The first is appreciation for the good things in our lives. The second is what he calls "unconditional gratitude." This is about being open to "the possibility of blessings in every moment," even the ones that "come wrapped in fears, pain, and tears." And then there is the third form of gratitude, which "stems from the holy realization that you are what you seek."

> *It gives you a glimpse of the light that is your original, divine nature. This type of gratitude is based on a deep spiritual realization that you were created perfectly, and that everything you have chased after and longed for—love, happiness, peace—is already yours. It is, therefore, a type of salvation, a homecoming, and a chance to finally rest.*[80]

IN THE 2009 science fiction movie *Avatar*, a corporate executive trying to mine rare metals on a distant planet named Pandora is frustrated by how intimate the relationship is between the local "humanoids" and nature. He complains, "You throw a stick in the air around here and it's gonna land on some sacred fern!" He meant it as an insult, but of course he was actually naming a beautiful truth about the way those beings understood their world, that, yes, everything truly is sacred.

And it's not just true on Pandora; it's true on Earth, too. It's a matter of opening our eyes and our hearts to the miraculous nature of ourselves and everything around us, small and large. It's a matter of nourishing awareness and breath and space in our lives despite so much that presses us the other way. Thich Nhat Hanh writes:

> *The real miracle is not to fly or walk on fire. The real miracle is to walk on the Earth, and you can perform that miracle at any time. Just bring your mind home to your body, become alive, and perform the miracle of walking on Earth.*[81]

Chapter 12

The Heart of Happiness:
Loving Ourselves

Once a young woman asked Hafiz
[the epic Persian mystic and poet],
"What is the sign of someone knowing God?"
Hafiz remained silent for a few moments and
looked deep into the young person's eyes, then said,
"Dear, they have dropped the knife.
They have dropped the cruel knife most so often use
upon their tender self and others." [1]

IN THE COURSE I teach on Reclaiming the Sacred, I do an activity where I stand in front of everyone and one by one place sticky notes on myself. Written on each note is one of my deepest insecurities or shames—the things I least want people to know or think about me. These are ideas I have internalized about myself over the years based on messages I have gotten from other people and society about the things that are not good about me.[ii]

Regardless of what I may *think* of these things—whether I believe I *should* feel bad about them, or whether I believe they are actually even true—they have become woven into my felt sense of myself. On a deep level I *feel* they are true and I feel shame about them. Even now, five decades into my life, and having done a lot of deep and intentional work to heal around them, they each still have some power, some of them a *lot* of power,

---

ii I was inspired in this by a powerful video I saw of two people at Be Present, Inc., doing something similar. (Allen, E. [loveisgood]. (2011, March 21). *The Be Present Empowerment Model Part 1* [Video]. YouTube. youtu.be/KDQks_UqNYE?t=169)

and they can each still flare up in the right circumstances and hurt deeply. They go something like this:

1. *I'm too skinny, I'm very bony, I'm not attractive or sexy.*
2. *I was bullied as a kid and didn't fight back. (I think most people experience some stigma when they are bullied or abused, but I also know that for me part of the shame here was related to gender. As a boy, being bullied or being weak was like waving a flag that said I was "less than," a "loser." And not fighting back was the worst; it was confirmation that I was indeed weak and pathetic.)*
3. *I'm too sensitive and gentle; I should be more assertive, more aggressive. I shouldn't cry. I need to grow up, toughen up.*
4. *Those last two are deeply related to this next one: I'm not a real man. So I deserve to be excluded, made fun of. (Again, I didn't actually think this, but the feeling existed within me.)*
5. *Feeding further into that one: I didn't have a girlfriend until after high school and didn't have sex until I was twenty. (I had the idea—pressed on me by popular culture—that these were both late and that they were proof that I was unattractive, not manly, less than.)*
6. *I didn't drink or do drugs or party growing up. (Again, I internalized the idea that was just part of growing up, at least for people who were "cool," or attractive, desirable.)*
7. *The next three are all very similar, but they are each completely true on their own: Humans are selfish, thoughtless, destructive, bad—you can't get around it; that's just what humans are, and so on some level that is unavoidably what I am.*
8. *Men are selfish, thoughtless, destructive, bad—that's just what men are, and so on some level that is unavoidably what I am.*
9. *White people are selfish, thoughtless, destructive, bad—and so on some level that is unavoidably what I am.*
10. *And then on top of all of that, I feel shame about my shame. I feel ashamed that I get stimulated or triggered, that these can't just be facts about me and my past and that I am OK with that. It feels like yet further evidence of how unworthy, unattractive, less-than I am, otherwise I wouldn't be so affected.*

The journey to happiness is a profound one. It gets to the core of our sense of ourselves, and of our relationships with this world and this life. **And we have arrived at the heart of that journey—our own hearts**.

Psychologist Tal Ben-Shahar writes:

> *Enlightened nations set up political structures … to protect our right to freely pursue happiness. Yet nothing external can protect us from what I have come to believe is one of the greatest impediments in our pursuit of this ultimate currency—our feeling that we are somehow unworthy of happiness. … We must appreciate our core self, who we really are, independent of our tangible accomplishments.*[2]

Psychologist Roy Baumeister notes that "one's level of self-esteem is among the strongest predictors of happiness,"[3] and that "people with high self-esteem are significantly, substantially happier than other people."[4] Sonya Lyubomirsky, author of *The How of Happiness*, writes that happiness and self-esteem are "so intimately related" that "happiness may not be possible or realizable without a healthy dose of self-confidence and self-acceptance."[5]

Oprah Winfrey, on the grand finale of her show, spoke poignantly about one of the most important lessons she learned in her twenty-five years on the air:

> *The show has taught me there is a common thread that runs through all of our pain and all of our suffering, and that is unworthiness. Not feeling worthy enough to own the life you were created for. … What I got was we often block our own blessings because we don't feel inherently good enough or smart enough or pretty enough or worthy enough.*[6]

Whether any of the shames on my own list resonated with you or not, in a general sense this is us, this is you and me and everyone around us—the vast majority of us have things we feel ashamed of.[7] Shame is not a passing sense of regret or embarrassment; shame is the feeling that there is something fundamentally wrong with us, that we are in some way unworthy or bad or unlovable.[8]

Perhaps those words ring very true for you. Or perhaps "shame" isn't a word you usually associate with yourself, and feeling "fundamentally unworthy" or "unlovable" sounds foreign to you or extreme. Yet social scientist Brené Brown, along with many others, suggests that shame "is universal, we all have it … and the less you talk about it, the more you have it."[9] Experts indicate that shame is "far more pervasive and powerful than most of us imagine," it is "the concealed driving force behind many of our daily interactions," and it is even "a primary force in social and political evolution."[10]

The word *shame* **actually derives from a word meaning "to cover" or "to hide."**[11] These are the things we want to hide about ourselves, and we go to great lengths to do so, often unconsciously. Brown writes:

> *Shame often produces overwhelming and painful feelings of confusion, fear, anger, judgment, and/or the need to hide. It is difficult to identify shame as the core issue when trying to manage these intense feelings. Even when the [research] participants were able to identify shame, the silencing and secret nature of shame made it very difficult to identify and act on the choices that would facilitate change.*[12]

In this way, shame is like a black hole—it is often not directly visible to others or even ourselves, yet it is apparent for the powerful effect it has on everything around it. The following are some of the ways shame often shows up. Shame isn't always involved, but it often is, and so these behaviors and feelings can be clues to underlying shame.

Assuming the negative. We may assume people have negative opinions of us or see us as flawed in the same ways we feel we are flawed. We can simply walk into a room or share some work we've done, for example, and we imagine all kinds of negative things people are thinking about us without anyone saying a word. We may misinterpret what people are saying, tending to hear criticism when none is intended, and minimizing or simply not believing praise that is sincere.

Numbing or distraction. Some of us seek to escape our feelings of insecurity through drugs and alcohol. Some experts have argued that shame and drug/alcohol addiction are "inseparable." Others of us overeat, compulsively shop, or watch TV. Some of us throw ourselves into work, which can be a way of distracting ourselves or trying to prove our worthiness to ourselves and others. The pursuit of fame or other kinds of "success" can play a similar role.[13]

Pleasing others. Some of us feel driven to constantly please others, to be very agreeable and not assert ourselves. We don't want to call attention to ourselves, or we feel we need to make up for what we think are our flaws. We may feel that our own needs or opinions aren't as important as others'.[14]

Conforming. Our sense of being flawed is often part of the way we conform to how other people dress, speak, and act, and with the "masks" we often wear, hiding our full selves and showing only what we believe people want to see.[15]

Competition and perfectionism. Shame can likewise fuel extreme competitiveness and perfectionism, again as a way of proving something to ourselves and others, or avoiding feelings of being flawed.[16] Numerous experts argue that shame is actually at the root of narcissism, that the total elimination of an awareness of our flaws at a conscious level is the ultimate attempt to avoid shame.[17]

Jealousy. Shame is often at its most raw and obvious when it rears its head as jealousy. Our feelings of inadequacy are right in our face and can cause us to feel completely overwhelmed, far out of proportion with anything that is actually going on. It can cause us to act in ways that hurt and distance us from the very people we care about and whose attention and respect we desire, even while we torment ourselves, sometimes long after the original incident has passed.[18]

Cutting others down. Sometimes we critique others or try to bring them down a notch because we feel inadequate compared to them. These can be people around us or people we don't even know, like celebrities. We may cut them down to make ourselves feel better when we perceive them as more successful, more attractive, more accomplished, and so forth.[19]

Being sensitive to conflict or provoking conflict. When conflict arises, or if we are criticized or belittled in some way, shame can cause us to feel an overwhelming discomfort out of scale with what happened. We may want to just disappear or we may keep beating ourselves up over it long after the experience. Alternately, shame sometimes provokes intense anger and we strike out at others to channel or crush the awful feelings that come up, or to transfer our attention and the attention of others away from the source of shame.[20] James Gilligan, an expert on violence, believes that shame is "the primary or ultimate cause of all violence, whether it is toward others or toward the self."[21]

Holding back. Some of us tend to keep some distance in relationships or keep secrets about ourselves so we don't risk other people discovering how "flawed" we are.[22] We may try to play small or hide ourselves, in a literal or figurative sense.[23] We may wear our hair or clothes or makeup in ways that hide us, or we may make ourselves smaller through our posture and voice. We may minimize our ambitions based on assumptions about how we will be perceived and received.

Dependency. A sense of inadequacy can also feed an over-dependency in relationships.[24] Shame is one of many complex reasons people often stay in abusive relationships.[25]

The things that cause us to feel shame and behave in these ways are different for each of us.[26] They can be related to our abilities or intelligence or behaviors, our likes or dislikes, our physical features, our families, things we have or haven't done, things that were done to us or things that were withheld, being or having "too much" or "not enough" of something.

Brown has found that while "men and women are equally affected by shame," "the primary trigger for women" (in the US, anyway, where this research has generally been conducted) seems to be appearance and body image. Motherhood is a close second. "Society views womanhood and motherhood as inextricably bound," and whether a woman has kids or not, there are endless judgments placed on her related to being a mother.

Meanwhile, Brown suggests that "men live under the pressure of one unrelenting message: Do not be perceived as weak." Another powerful source of shame for men is a concern about sex appeal and sexual performance. In her research, one man reported, "When it comes to sex, it feels like our life is on the line."[27]

Transgender and gender nonbinary people can fall into these same male and female traps, but also often suffer shame specifically related to being trans or nonbinary.[28] People of marginalized identities in general—related to race, class, physical ability, sexual orientation, and so on—often internalize shame related to those identities based on the negative or limited ways they are often perceived and treated in our society.[29]

ROBERT HOLDEN HAS articulated what he calls the "Self Principle," which he says is the central principle of his work. It states that **"The quality of your relationship with yourself determines the quality of your relationships with everyone and everything."** That includes "family, friends, colleagues, God, the taxman, and everyone else in the world." It also includes your relationships with money, time, love, success, and, yes, happiness.[30]

In this way, shame affects—and undermines—almost every aspect of our lives. That list of the ways shame often shows up in our lives conveys some of this: the ways we try to numb or distract ourselves, the ways we hide parts of ourselves or keep people at a distance, the ways we try to be who we think people want us to be, or are compulsive in trying to be successful or please others, and the ways we play small or are combative.[31]

Then there are all the feelings that go along with these behaviors: the self-judgment and fear, the insecurity and disconnect, the sense of just not being good enough—even if we aren't conscious of them, even if they are knotted deep within us, protected by walls or barbed wire.[32]

The quality of our relationship with ourselves also directly affects our ability to experience love. Other people may love us dearly, may communicate that love in many ways, and yet if we do not love ourselves we can only receive so much of it. When we do not believe we are intelligent, kind, attractive, good, worthy, it is hard to believe that others see these things in us. We are always braced, unbelieving. At best we may imagine that others *think* we are these wonderful things, but they are not seeing our real selves. Either way, there is a degree of love and belonging available to us that we are closed to, which is a profound loss for both us and the people who love us.[33]

"i love myself."
the
quietest.
simplest.
most
powerful.
revolution.
ever.[34]

—Nayyirah Waheed
(from her beautiful book of poetry, *Salt*)

SOME PEOPLE DO not speak of self-love; they use terms like self-acceptance or self-compassion. For some people "self-love" just sounds too selfish or touchy-feely. For me, when I think about the highest, best feeling or relationship I can have with another person, it isn't compassion, and it sure isn't acceptance. Acceptance is extremely important, and is sometimes the most we can really extend to someone. But at the same time it feels like the most basic thing I could offer someone, almost insulting even—that somebody isn't worthy of anything better than me just "accepting" them.

Likewise, compassion is amazing, but is still only a part of what I would hope to be able to extend to other people, other creatures, and the world.

Love. That is the highest, best way I feel I can hold another person, to truly celebrate and honor them—love. So how can I justify aspiring to anything less for myself?

It can be very difficult, if not impossible, to be compassionate and accepting of ourselves, or parts of ourselves. Every step any of us takes in the direction of any of these—self-acceptance, self-compassion, or self-love—is a blessing and deserves our respect.

And still, my wish for you and for all of us is that self-acceptance and self-compassion are waystations, not destinations, that all of us, to some degree, get to experience nothing less than a rich self-love.

If you feel some discomfort around the idea of "self-love" and "loving yourself" you are not alone. Sharon Salzberg, a meditation teacher and the author of *Real Happiness*, writes:

> *I've found that many, if not most, of the people with whom I've spoken, feel the greatest sense of struggle around the question of cultivating love for oneself. We are conditioned to associate self-love with selfishness, and self-deprecation with virtue.*[35]

Scholar and activist bell hooks likewise notes in her book *All About Love* that the chapter on self-love was the hardest to write and the hardest to discuss with people:

> *Self-acceptance is hard for many of us. There is a voice inside that is constantly judging, first ourselves and then others. That voice enjoys the indulgence of an endless negative critique. Because we have learned to believe negativity is more realistic, it appears more real than any positive voice.*[36]

And psychologist Kristin Neff, author of *Self-Compassion*, writes, "The number-one reason people give for why they aren't more self-compassionate is the fear that they will be too easy on themselves."[37]

I would encourage all of us to strive to get more comfortable with the language and practice of loving ourselves. Indeed, I would suggest that anything less is a reflection of a violent and deceptive edge to our culture. Because **loving ourselves is not only a critical factor in our being happy, it is also the surest path to being our highest, best selves.**

Brené Brown writes, "Shame is highly correlated with addiction, violence, aggression, depression, eating disorders, and bullying. **Researchers don't find shame correlated with**

71

positive outcomes at all."[38] On the contrary, research shows that the more loving and accepting we are of ourselves:

- The more motivated we are to change ourselves and our lives for the better
- The more we are able to honestly assess our strengths and weaknesses, and to receive both positive and negative feedback from others, without exaggerating it, minimizing it, or denying it
- The less fearful we are of failure
- The better we are at sticking to meaningful goals
- The faster and more effectively we rebound when we run into obstacles[39]

Self-love isn't about letting ourselves off the hook or condoning negative behaviors or habits. *It's the opposite.* When we are locked down in a position of defensiveness, protecting and trying to avoid our wounds, we are less able to move forward and evolve. Each critique or shortcoming threatens to trigger our shame and to upend our precarious sense of worthiness. The more we embrace the truth that we are fundamentally wondrous, beautiful, brilliant, the freer we are to acknowledge things we want to change about ourselves. Psychologist Carl Rogers once wrote, "The curious paradox is that when I accept myself just as I am, then I can change."[40]

And when we do something that is wrong or hurtful, instead of retreating into a paralyzing place of shame, we can move through the productive experience of *guilt*. Brown writes, "The difference between shame and guilt is best understood as the difference between 'I am bad,' and 'I did something bad.' Guilt is just as powerful as shame, but its effects are often positive." When we feel shame, our reaction is usually to shut down, strike out, redirect, and so forth, because we feel too threatened. Guilt can motivate us to apologize for something we've done, make amends to others, or change a behavior we don't feel good about.[41]

bell hooks writes, "Once we begin to replace negative thinking with positive thinking, it becomes utterly clear that, far from being realistic, negative thinking is absolutely disenabling. The more we accept ourselves, the better prepared we are to take responsibility in all areas of our lives."[42]

Another very important consideration is that, like our ability to receive love, **the quality of our relationship with ourselves also affects our ability to** *give* **love.** Whether it's exactly true, as is sometimes said, that we can only love others as much as we love ourselves, our ability to love others is definitely undermined by limitations in loving ourselves. Brown writes:

> *I think it's because in order for us to tolerate imperfection and vulnerability in other people, we have to be able to accept what is imperfect in ourselves. If there are prerequisites for worthiness that we carry either knowingly or unknowingly within us, then we apply them to ourselves as well as other people.*[43]

As Brown noted in chapter ten, we also cannot selectively numb our emotions.[44] **When we numb our ability to love ourselves, we numb our ability to love, period.** It's as if, in blocking some of the flow of love into ourselves, we limit how much is available to flow out. Or, in excluding ourselves from our own circle of love, we are to a degree excluding ourselves from the realm of love in general.

We also have a tendency to resent and even resist (consciously or not) other people getting things we want for ourselves but have been denied, whether that's praise or a promotion or a home, or joy or belonging—or love. On a subtle level we may be blocked from giving the absolute fullest of our love when part of us feels left out of that experience ourselves.[45]

The flip side of this is also true and immensely important: The more deeply we love ourselves, the greater our capacity to love other people. The more we experience our own fundamental wondrousness, it automatically, necessarily, deepens our appreciation for the wondrousness of other people as well. It is a way of seeing and loving others that we cannot know if we lack the direct experience of it in ourselves.

Brown writes, "I hated this part of the research. I wanted to believe that I could love my kids more than I loved myself. But all the research finds that we really can't offer people more compassion than we have for ourselves."[46] She goes on to note that, for this reason, "Where we are on our journey of living and loving with our whole hearts is a much stronger indicator of parenting success than anything we can learn from how-to books."[47]

Your task is not to seek for love,
but merely to seek and find
all the barriers within yourself
that you have built against it.[48]

—Rumi

IN THE ACTIVITY where I share my deepest shames, it is important that they are written on sticky notes and placed on me so that I can visually represent the fact that they come from outside of us. Our shames are not inherent to us, and no matter how deeply they may lodge inside us, no matter how much we may experience them as a core part of ourselves, **they are not, in fact, us.**

Our shames are based on ideas we have internalized about the way we "should" be—what is admirable, respectable, attractive, good, and what is not. Sometimes they are based on things people have said to us or others, or ways they behaved with us. Sometimes the messages came to us through shows, movies, books, songs, advertising, and so on.

David Simon, cofounder of the Chopra Center for Wellness, has written, "Ninety-nine percent of people looking at a baby would say that of course this baby deserves to be happy, healthy and deserves to be loved."[49] Yet, as psychologist Leon Seltzer notes, "As a result of what most mental health professionals would agree reflects a subtle form of emotional abuse, almost all of us come to regard ourselves as only conditionally acceptable."[50]

The truth is that we do not need to become anything other than what we are in order to be entirely worthy of love. We are each of us wondrous, beautiful, sacred. This is true for all of us. *This is true for you.*

This is a radical and liberating truth, because it means that the path to releasing these painful, limiting beliefs is, as Rumi notes, not to seek love, but to instead seek and take down the barriers within us that have been thrown up against love. **Self-love does not need to be so much cultivated or pursued as simply allowed, freed**.

After placing all the sticky notes on myself and explaining the ways they came to me and how they've affected me, I then get to remove them, one by one, and share what the journey of softening and releasing my shames has been like for me. There are three paths we can walk to heal our places of shame, and central to each of them is this truth that our shames are not inherent to us and that clearing them allows our self-love to flow naturally. Those three paths are *transcending shame, transforming shame,* and *positively adapting to shame.*

And the day came
when the risk to remain tight in a bud
was more painful
than the risk it took to blossom.[51]

—Elizabeth Appell

WHEN I WAS in my mid-twenties I went to a party with my girlfriend and found myself over the course of the evening overwhelmed by feelings of insecurity. I didn't have an awareness of this then, but it had everything to do with some of the shames from my list. My girlfriend didn't do anything, it was just the situation, but in response to those feelings I completely disconnected from her and closed myself off. This wasn't a well-thought-out response—it wasn't thought out at all; it was just where my feelings took me.

This probably won't come as a surprise, but me pushing my girlfriend away didn't help the situation. I didn't feel any better and she became frustrated and upset. (It didn't help any of the other times I reacted that way either … Sigh.)

I couldn't stand the feelings or my behavior, so I went to see a therapist a friend recommended. I remember telling the therapist there was this wounded kid inside me who kept causing me to act in these ways I didn't like, but I wasn't that wounded kid anymore. I was happy with myself and my life, I was proud of who I was, and I had every reason to feel confident. But this little kid kept taking over and causing me to act out, and it was incredibly frustrating. I wanted to strangle that little kid. Those were my words, I wanted to strangle him.

My therapist suggested we might find some other way of approaching the issue, or at least find some different language I could use. But I was insistent. Strangle that little kid was what I really wanted to do. I needed to be released from him and his painful, jeering world and these embarrassing and not-helpful responses.

Brené Brown tells a story that is similar in certain ways (though less violent). She says the first time she went to a therapist she told the therapist she was struggling with vulnerability. She had done a lot of research and recognized that she was not living the way she wanted to, and that vulnerability seemed pretty key to it all. "But," she added, "here's the thing: no family stuff, no childhood shit. I just need some strategies."[52]

Of the three paths we can follow to heal our shames, **the first path, transcending our shame**, is, in fact, sort of like the magic switch that Brené Brown was perhaps hoping for. It is definitely the kind of switch I was looking for. That's because the first path doesn't involve going anywhere near our places of shame; it only requires that we dive right past them and live directly in the truth of our wonderfulness. Which really does feel as good as it sounds.

We might have these kinds of transcendent experiences in many different contexts: from meditating or praying, to dancing or singing, from walking in the woods or looking at the stars, to making love or experiencing birth or death close-hand. We might have them when we are in a religious ceremony or taking psychedelic medicines like mushrooms and peyote, or when we are listening to music or reading a poem or seeing an inspiring movie—anything that dives right past our shames and opens us to a direct experience of the sacredness and wonder of ourselves and of everything.[53]

Anything that nourishes a positive sense of ourselves and promotes greater confidence, groundedness, purpose, and so on, all of these can boost our resilience in the face of shame.[54] But these kinds of singular experiences and their clarity and rightness—the often glistening, joyful obviousness of them—are a powerful contrast to the messages our shames tell us about ourselves and the ways they train us to feel, and so these experiences weaken the hold of our shames. Even as the experiences fade, having known those feelings we can return to them more easily in the future, and the memory of them can be a powerful anchor for us.

In each of us, there is a young, suffering child. We have all had times of difficulty as children and many of us have experienced trauma. To protect and defend ourselves against future suffering, we often try to forget those painful times. Every time we're in touch with the experience of suffering, we believe we can't bear it, and we stuff our feelings and memories deep down in our unconscious mind. It may be that we haven't dared to face this child for many decades.

But just because we may have ignored the child doesn't mean she or he isn't there. The wounded child is always there, trying to get our attention. The child says, "I'm here. I'm here. You can't avoid me. You can't run away from me." We want to end our suffering by sending the child to a deep place inside, and staying as far away as possible. But running away doesn't end our suffering; it only prolongs it.[55]

—Thich Nhat Hanh

OUR SHAME LIVES in very deep places within us and is very complex. This means it is very resistant to change.[56] We tend to keep it locked up and at a distance because we do not want to go anywhere near it or the feelings it stirs up. This does, in certain ways, protect us from it, but it also protects it from us. In fact, avoiding our shame tends to feed it.[57] Because we hold our shames so deeply, transcendent experiences alone are rarely, if ever, adequate to completely heal and transcend our shame.

At some point, if we want to live more joyfully and freely, we must explore some of the "family stuff and childhood shit"; we must invite the wounded child within us closer, not exile (or strangle) it. We must also walk the **second path, that of transforming our shame**.[58] This path is about feeling the deep desire to avoid our shame, and nonetheless courageously and lovingly drawing toward it and exploring it—gently, patiently. In so doing, we find that our shame slowly dissipates, and the lies give way before the truth.

I say "lies" because **if we are not living in the truth of our wondrousness and worthiness, then we are living a lie**. And, indeed, every shame contains within it a lie. Sometimes we discover that a belief we've internalized about ourselves just isn't true. Maybe it was never true—we were never dumb or ugly or selfish—it was just the meanness of other people and their own pain and distorted ways of thinking that planted those ideas within us. Or maybe there is some truth to the belief we're holding, but it's not true the way we have internalized it. Maybe we feel deep down that we aren't smart, and perhaps we did have trouble in school, but that has nothing to do with how smart we are in general. Or—using myself and my first sticky note as an example—while I am definitely still very skinny, I was even skinnier as a teenager, and it is that sense of myself that has stuck with me, even though it's just not true the same way anymore.

76

More often, though, we find that these things we judge in ourselves are true in certain respects, and *the lie is instead in how we relate to these aspects of ourselves*. In other words, these are things that are, in fact, to some degree true, but they aren't "bad" or "wrong" in the way we have come to feel about them. Part of my journey around being skinny has been (and still is) to recognize that I am indeed really skinny, but that I can be skinny and beautiful, that my skinniness is, in fact, part of my beauty. Similarly, I am most definitely very sensitive and gentle (sticky note number three). *The lie is that there is something wrong with that*, that I should feel bad about it. It has taken a lot of work to release the messages that were beaten into me, but I now love my sensitivity and gentleness and think they are perhaps the most wonderful things about me.

Indeed, the things we feel ashamed of are often actually positive, are things that other people actually appreciate about us. *Or if they are not positive, they are, at the very least, understandable, forgivable, complex—part of our being human, part of the vast, desperate fabric of life and humanity and this society, and we deserve to be compassionate with ourselves. We are still beautiful and sacred and worthy of love.*

As Rumi suggests and research confirms, a critical step in the path of transformation is to **identify our shame**.[59] Are there things you think are wrong with you, or things about yourself that you don't want other people to know or notice, that perhaps you can't even bear to think about or notice yourself? Maybe some of those behaviors listed earlier rang true for you and can give you some clues as to beliefs you have internalized about yourself, even if you aren't totally clear on a conscious level what they may be about.

Once we have identified some shames, it can be helpful to explore how these judgments got into us, where they came from. This can help create a little space between us and our shame, highlighting the fact that these judgments are not, in fact, set in stone or fundamental to who we are. What do you generally think of the people or aspect of our culture that planted these ideas in you? What do you think of their values, the way they treat others? Do you respect their opinions when it comes to these kinds of things? Would you want people you love—your friends or children—to turn to these people or the norms of this culture to understand how to feel about themselves? Are these people perhaps passing on some of their own pain and shame? Perhaps this is a wound that has been passed on for many generations.

Personally, I recognize that many of the messages I internalized about sexuality and masculinity (sticky notes four, five, and six) came from the media, especially the movies and videos I saw growing up. Reflecting on it now, I recognize that so many of the ideas they were communicating—no matter how much they may sit in me as "the way things really are"—are entirely shallow and wrong and hurtful, and the people behind them are not people I would ever turn to for guidance. This point was made quite vividly for me when it was recently revealed that one of the most significant film producers of our time had been sexually assaulting and harassing women for thirty years. That is the person inside my head telling me what it means to be a real man and what it means to be sexually desirable and fulfilled?!?

We can also consider if we have any evidence that seems to contradict this judgment we have of ourselves, any points of information that seem to point to the lie. Perhaps people have told you things or you have had experiences of yourself that counter this idea you have about yourself. When a feeling of shame emerges, perhaps you can have a few of these counter-facts in mind to help anchor you a bit. "I know that people consistently tell me they value my friendship even though I feel like a crummy friend," "I know this person repeatedly appreciates my comments and work even though I think it's never good enough," "I know this person admires me and cares so much about me, even though I keep doubting it."

It can also be helpful to consider what we would think of someone else in our own shoes. We may find that we are judging ourselves in a way that we would never judge other people. For example, when I imagine a child being bullied and not fighting back (sticky note number two), I am clear that I do not have any negative judgment of them. I feel only a deep well of compassion and appreciation. My deepest concerns are that somebody step in to end the violence and that the child not take on the layers of shame that I have. In fact, I see their gentleness, their lack of comfort with fighting, perhaps their fear, all of them as part of their beauty, their humanity. This doesn't just automatically transfer wholesale to me, but it definitely creates a little ease and softness in me where before things were just hard and set.

It can also be tremendously helpful to involve other people in our journey of transformation.[60] We can do a lot on our own—reflecting, reading, journaling, watching helpful videos. However, Brené Brown argues that, ultimately, "the only way to resolve shame is to talk about it," that shame "can't survive being shared."[61] Shame is about the way we feel others see us or judge us, so there is tremendous power in experiencing compassion and affirmation from others around our shames.

In opening up to others, we create the opportunity for them to potentially open up as well.[62] Brown notes that "one of the most important benefits of developing empathy and connection with others is recognizing how the experiences that make us feel the most alone, and even isolated, are often the most universal."[63]

We can talk about our shames with a trusted friend or family member, a counselor or therapist, a spiritual counselor, an AA group—anyone we trust and feel can support us. Therapy isn't necessarily for everyone, but a review of thousands of studies found that 75–80 percent of people benefited from therapy, and a study by psychologist Christopher Boyce found that four months of therapy, costing $1,300, had the same impact on a person's happiness as suddenly receiving *at the very least* $41,000. In other words, therapy was at least thirty-one times more cost-effective than receiving money.[64]

Lastly, it can be helpful to consider why we tend to take on these kinds of judgments in the first place, and why there is part of us that is always at the ready to put us down in these ways. The shame that I feel about having shame (my final sticky note)—that's a common one. We can be feel totally embarrassed and self-conscious, and can be very harsh on ourselves for the ways we feel and react when it comes to these deep seated feelings.

As humans, we have always been dependent on our families and communities. Their opinions of us matter tremendously and can, in fact, be a matter of life and death (and reproduction).[65] So we are very sensitive about aspects of ourselves that we think others might be critical of. In other words, as unhelpful and wrong and painful as we may find the part of ourselves that is so sensitive and critical of us about these things, this part of us is actually doing its best to keep us safe and help us meet our needs.

This is almost entirely subconscious. Gerald Fishkin, author of *The Science of Shame*, notes, "Shame isn't associated with cognition at all. **At the precise moment shame is triggered, we are emotionally hijacked, and there's no prefrontal activity.**" And, "because it's a type of stress response," it can also lead to blushing cheeks, increased body temperature, sweating, or queasiness, and a quickened pulse.[66]

This helps to explain why our responses can be so powerful, can actually feel like a matter of survival, and why our shame can be so tenacious. It also helps to explain why what we *think* of ourselves can be so different from what we *feel*. As I noted with regard to my own shame, I may not actually believe in my mind that these are things I should be ashamed of, or that they are necessarily even true in the first place, yet still they reside deeply within me.

Recognizing all of this, we can perhaps be a bit more patient and understanding of ourselves. Instead of judging ourselves harshly for having these reactions—instead of wanting to strangle this wounded child within us, this wounded child that is just as bewildered and hurt as we are, that is trying to somehow make the best of things, to help us even—we can strive to hold these parts of ourselves, and our entire selves, with greater care and compassion.

It is no easy task to be self-loving.
Simple axioms that make self-love sound easy only make matters worse.
It leaves many people wondering why, if it is so easy,
they continue to be trapped by feelings of low self-esteem or self-hatred.[67]

—bell hooks

WHEN SOMEONE ELSE talks about their shame, it can seem obvious that this idea they have about themselves is just not true, that they are absolutely beautiful, smart, capable, and so on. Or this thing that they feel badly about may be something we don't think matters in the least, or is something we actually like about them. We can often see clearly that it is really just about the hurt people and culture that pushed these ideas onto them.

It is completely different when the shame is our own. Then it feels so awful and true and unapproachable. Drawing near to it and exploring it can leave us feeling exposed and disoriented, ill even. This is natural. We are letting ourselves more directly encounter these things that feel so significant and threatening, that we have gone to great lengths to keep at a distance. **It is normal that we feel ungrounded for a time, because we are, in fact, abandoning the old ground we have stood on and are creating new ground beneath us.** (We should definitely be gentle with ourselves and take our time around these things, especially around traumatic experiences, so that we do not retraumatize ourselves. A therapist or other professional can be very helpful with this.)

The journey of healing is at times uncomfortable, and it also takes time. An image that has come to mind in relation to my own journey is that of me as a huge ship moving through an ocean at night. At the front of the boat are these beautiful luminous fishes that keep kissing one side of the bow. Those kisses are each of my incremental experiences of transcendence and transformation—my insights, my steps forward. Each of them is wonderful and important. The shift of even a single degree in the direction the boat is heading sends me on an entirely different life trajectory, and the additional joy and love of every degree adds up to a lot over time. Still, it takes a long time for fishes kissing a ship to change its course, and we have to be patient.

Indeed, for most of us, our shame will be with us to some degree for the rest of our lives. As I mentioned earlier, every one of my shames still holds some power within me, and can at times cause me to feel anywhere from uncomfortable to totally overwhelmed, despite having dedicated a lot of time to healing them.

This is where the third path comes in: **positively adapting to our shame.** We all adapt to our shame; each of the behaviors we listed earlier was an adaptation—numbing or distracting ourselves, pleasing others, keeping our distance. We can consider them successful to the degree they have helped us get by in the world; they are a testimony to how smart and complex we are.

But those kinds of adaptations tend to hurt us and limit us, as well as the people around us. There are ways we can adapt to our shame that are more intentional and that can make our lives better, that can even contribute to our healing. **Once we acknowledge that we have these sensitivities and that we tend to feel hurt and reactive when they come up, we can be intentional about limiting our exposure to them, and we can create support for ourselves around them.**

Maybe we're triggered by our romantic partner's ex. Rather than trying to pretend we don't have that sensitivity and repeatedly diving into situations where we know they are going to be there—because we think that is how we should be or want to be—we can gently, lovingly, acknowledge to ourselves that there is indeed a wound there for us, and that it is fine for us to just decline getting into those situations.

Or maybe our shame really comes up when we go to a certain kind of social event, like a party or conference, but we need to go to one for some reason. Perhaps we can ask a friend to please generally stay close to us, or to check in with us every fifteen minutes, or occasionally take our hand and give it a squeeze. Or we can make sure we get some time with a trusted colleague before or after the event, instead of just pretending everything is fine, or rushing into it, or heading straight into something else afterward.

These intentional strategies may sometimes resemble the unconscious adaptations we have taken on (like avoidance or distraction), but in this case we are not denying our shames, we are acknowledging them. We are acknowledging that they are part of who we are, for now anyway, and that it is OK that we have them, that everyone has them, and that we deserve care and thoughtfulness around them. And unlike the unconscious adaptations, which tend to lock our shames in place and feed them, these positive adaptations generally contribute to our healing.[68] They are acts of self-love.

To offer another personal example from later in my life, very early on in a romantic relationship I intentionally sat down with my girlfriend and shared with her about the shame that comes up for me related to sex and masculinity, and to my being very sensitive and having been bullied. It was the first time I had enough awareness and presence to be able to talk with someone that way. It was incredibly hard for me to open up so vulnerably and share about these painful places. It felt like putting out a big flashing neon sign pointing out all the things that are most deeply inadequate about me. But I know that when I get triggered around these shames it is really painful for me, and that I completely shut the person out and create distance, and I wanted as little of that as possible for both of us. And I wanted her to understand the kinds of things that hurt me so she could be sensitive and supportive around them as well.

I also wanted to invite this person into a deep level of intimacy and connection with me. Psychologist John Welwood writes eloquently about this.

> *When we reveal ourselves to our partner and find that this brings healing rather than harm, we make an important discovery—that intimate relationship can provide a sanctuary from the world of facades, a sacred space where we can be ourselves, as we are. … This kind of unmasking—speaking our truth, sharing our inner struggles, and revealing our raw edges—is sacred activity, which allows two souls to meet and touch more deeply.[69]*

While Welwood and I are both referring to a romantic partner, this can be true in any type of relationship.

For three days after I shared all of this about myself, I felt completely unmoored, not myself. I felt insecure, exposed, flawed. I literally felt ill. Three days. But I was deeply rewarded. My girlfriend did hear me, with tremendous empathy and love. She was able to affirm me and hold me, and it did bring us closer. Sharing that way and being affirmed by her made me less reactive, less defensive, less self-conscious around those wounds. And my vulnerability opened space for her to share about her own wounds and needs as well. It was very powerful and healing and helpful.

The change of which I speak is the change from living life as a painful test to prove that you deserve to be loved, to living it as an unceasing "Yes" to the truth of [your] Belovedness.[70]

—Henri Nouwen

IN 1990, MEDITATION teacher Sharon Salzberg went to a conference in India with the Buddhist leader and Nobel Peace Prize laureate, the Dalai Lama,. When it came her turn to suggest a topic for discussion, she asked, "What do you think about self-hatred?"

> *I was eager to get directly to the suffering I'd seen so often in my students, a suffering I was familiar with myself. The room went quiet as all of us awaited the answer of the Dalai Lama. Looking startled, he turned to his translator and asked pointedly in Tibetan again and again for an explanation. Finally, looking back at me, the Dalai Lama tilted his head, his eyes narrowed in confusion. "Self-hatred?" he repeated in English, as though trying out the words. "What is that?" ...*
>
> *I explained to him what I meant by the term—talking about the cycle of self-judgment, guilt, unproductive thought patterns. ...*
>
> *During the remainder of the session, the Dalai Lama repeatedly attempted to explore the contours of self-hatred with us. "Is that some kind of nervous disorder?" "Are people like that very violent?" "But you have Buddha nature. How could you think of yourself that way?" At the end he said, "I thought I had a very good acquaintance with the mind, but now I feel quite ignorant. I find this very, very strange."*[71]

It's a beautiful thing when I arrive at the point in the demonstration where I have removed the sticky notes and they are all lying on the floor around me. Of course, in real life it isn't that clean or complete. What is definitely true though is that all of my shames have significantly softened. The fishes' kisses have indeed shifted the course of my life and my feelings!

I think it's important to recognize how deeply entrenched our shames are, and to understand that for most of us they will be with us to some degree for the rest of our lives. In time, hopefully they will fall upon us more like gentle rains than thunderstorms, but they will probably never dissipate entirely.

But I also think it's important to consider a very different possibility suggested by the Dalai Lama (a possibility Rumi would wholeheartedly endorse): that we can live entirely free of shame, or very close to it, that our natural state is one of self-love, self-reverence, self-wonder.

It's the difference between "Shame is universal, now make the best of it" and "Your natural state is love, now see how free you can get."

Perhaps the more we walk these three path of healing our shames, and the more that people around us are also walking these paths—all of us transforming ourselves and, at the same time, transforming our culture—the more thoroughly we can decondition ourselves, take down these barriers, and realize Henri Nouwen's beautiful vision, each of us living our lives as unceasing "yeses" to the truth of our belovedness.

The time will come
when, with elation,
you will greet yourself arriving
at your own door, in your own mirror,
and each will smile at the other's welcome,
and say, sit here. Eat.
You will love again the stranger who was your self.
Give wine. Give bread. Give back your heart
to itself, to the stranger who has loved you
all your life, whom you ignored
for another, who knows you by heart.
Take down the love letters from the bookshelf,
the photographs, the desperate notes,
peel your own image from the mirror.
Sit. Feast on your life.[72]

—Derek Walcott

Chapter 13

The Scourge of
Rampant Happiness:
You Have to Actually Want It

The psychological world is now abuzz with a new field, positive psychology, devoted to finding ways to enhance happiness through pleasure, engagement, and meaning ... Mainstream publishers are now ... printing thousands of books on how to be happy and why we are happy. The self-help press fills the shelves with step-by-step plans for worldly satisfaction. Everywhere I see advertisements offering even more happiness, happiness on land and by sea, in a car or under the stars. ...

*Aren't some of us so smitten with the American dream that we have become brainwashed into believing that our sole purpose on this earth is to be happy? Don't we fear that **this rabid focus on exuberance leads to half-lives, to bland existences, to wastelands of mechanistic behavior?**[1]*

—Eric G. Wilson

IN HIS BRILLIANT book, *Be Happy!*, psychologist Robert Holden notes that in his experience as a therapist and coach he has worked with many people who have done a lot of work to get themselves out of a hole, and seem to be on the brink of being truly happy, when they "suddenly turn around and run back into the hole. Sometimes it's a different hole, perhaps a slightly more interesting one, but it is still a hole."[2]

Just as we can only be so happy if we do not fully love ourselves, we can only be so happy if we do not actually want to be happy. Our happiness will be limited if consciously or subconsciously we are fearful of moving toward happiness, if we have concerns or doubts about happiness, or if we are in some way attached to unhappiness.

There are legitimate reasons to be skeptical of happiness. Eric G. Wilson offers one example in that opening quote. But often these barriers to happiness do not actually serve us; they are utterly unnecessary and block some degree of happiness that could otherwise be ours.

We have only two chapters left in our exploration of the factors that most significantly contribute to human well-being. In the next chapter we will look at the meeting place of our inner and outer worlds. For this final chapter dedicated to our inner worlds, we will explore four important forms these barriers to happiness often take. And we will begin where Wilson left off.

Happiness and Purpose

> *Terrors, deprivations, impoverishments, midnights, adventures, risks and blunders are as necessary for me and you as their opposites; … the path to one's own heaven always leads through the voluptuousness of one's own hell.*
>
> *Should you … refuse to let your own suffering lie on you even for an hour and instead constantly try to prevent all misfortune ahead of time; should you experience suffering and displeasure as evil, hateful, deserving of annihilation, as a defect of existence, then you have … in your hearts … the religion of comfortableness.*
>
> *How little you know of human happiness, you comfortable and benevolent people, for happiness and unhappiness are sisters and even twins that either grow up together or, as in your case, remain small together!*[3]
>
> —Friedrich Nietzsche

WE HAVE TWELVE chapters under our belts and through it all we've taken it for granted that happiness is generally a good thing. We've acknowledged that allowing ourselves to experience a whole range of emotions is best overall, but even then, we were considering that in the context of how to generally support our happiness. But what if happiness is not something we should necessarily prioritize? What if being happy means living "half lives" and "bland existences," as Wilson writes? What if other things are more important?

Nietzsche believed that the greatest human goal is not happiness, but creation, and specifically creation in service of the advancement of human culture.[4] And he felt that that kind of advancement requires that we live full, vigorous lives, rich with challenges and emotional strain.[iii]

A century removed from Nietzsche, Buddhist teacher Pema Chödrön offered the warning we considered in chapter ten: "If we're committed to comfort at any cost, as soon as we come up against the least edge of pain, we're going to run; we'll never know what's beyond that particular barrier or wall or fearful thing."[5]

Chödrön considers the greatest human goal to be spiritual awakening:

> *Our life's work is to use what we have been given to wake up. If there were two people with exactly the same—same body, same speech, same mind, same mother, same father, same house, same food, everything the same—one of them could use what he has to wake up and the other could use it to become more resentful, bitter, and sour.*[6]

Albert Schweitzer won the Nobel Peace Prize in 1952 largely for developing a philosophy he called "Reverence for Life." He argued that "The purpose of human life is to serve and to show compassion and the will to help others."[7] Ralph Waldo Emerson is often quoted as having said something similar: "The purpose of life is not to be happy. It is to be useful, to be honorable, to be compassionate, to have it make some difference that you have lived and lived well." (Though this quote is actually from author Leo Rosten.)[8]

These are just a few of the many different ideas people have expressed about the purpose of life: creativity, the advancement of human culture, spiritual awakening, service. Yet every one of these people just quoted has also expressed a deep respect for happiness. Not happiness as something to be pursued rabidly, and definitely not to the exclusion of life's other rich and important experiences, but **happiness as a rich and important experience also.**

Robert Holden writes that some people resist happiness because they perceive it as superficial or disconnected from the deeper, more important things in life. He jokes that these people think "happiness is best suited to intellectual lightweights, dizzy blondes, and little dogs, but that the more serious-minded human should study Russian poetry and lament the hopeless sufferings of the world."[9]

Even Wilson, (in his book which is actually titled *Against Happiness*) acknowledges that he isn't really so much against happiness in general as he is against "shallow happiness," "immediate gratification," and "static contentment." He reserves a special scorn for people who would pursue "happiness at any cost" and completely abandon the

---

[iii] Nietzsche, by the way, has something of a reputation as having been pro-Nazi. He was actually vehemently anti-Nazi, but his works were manipulated by his pro-Nazi sister who survived him by nearly 40 years. (Prideaux, S. (October 5, 2018). "Far right, misogynist, humourless? Why Nietzsche is misunderstood." *The Guardian.* theguardian.com/books/2018/oct/06/exploding-nietzsche-myths-need-dynamiting)

experiences of sadness and melancholy, people who are willing to "have an essential part of their hearts sliced away and discarded like so much waste."[10]

Nietzsche, Chödrön, Schweitzer, and Rosten would all certainly agree, but so would some of happiness's greatest advocates. The Dalai Lama, for example, believes that happiness is the result of an "enlightened mind" and involves a life fundamentally grounded in love, compassion, and a commitment to being in service to all sentient beings.[11] And still he writes:

> I believe that the purpose of life is to be happy. From the moment of birth, every human being wants happiness and does not want suffering. Neither social conditioning nor education nor ideology affects this. From the very core of our being, we simply desire contentment. I don't know whether the universe, with its countless galaxies, stars and planets, has a deeper meaning or not, but at the very least, it is clear that we humans who live on this earth face the task of making a happy life for ourselves.[12]

Being happy also actually contributes in significant ways to these other important purposes. For example, happy people tend to be more altruistic and charitable, and to "perceive members of other groups with less prejudice."[13] Researchers at UC Berkeley found that the number one predictor of how children fare emotionally, socially and academically is the emotional and social well-being of their parents.[14]

There is a fair amount of evidence that unhappiness and adversity *do* contribute to creativity.[15] Wilson, in his book, profiles a large number of people who transformed their "despair into unforgettable gold," among them Ludwig van Beethoven, Vincent van Gogh, Virginia Woolf, Franz Kafka, Ernest Hemingway, and Florence Nightingale. However, *there is also a fair amount of evidence that happy people are just as creative, or more so*, and that they are also more persistent and motivated.[16]

Research suggests that happy people tend to produce higher quality work in general, to be more productive, and to suffer less burnout.[17] (Not surprisingly, *really* happy people tend to be a little *less* productive because they tend *not* to care as much about things like grades, attendance, promotions, and salaries, but they still fare better when it comes to relationships and volunteer work.[18])

It is also important to acknowledge that people often pass on their pain and unhappiness not in the form of contributions to culture and society, but in the form of further pain and unhappiness. People who are unhappy are more likely to be hostile and abusive to others and to abuse alcohol and drugs.[19] Wilson himself notes that it is "perhaps easy to admire these creators from afar," while it was not so easy for their friends and family "who suffered all those long and tearful nights mourning or medicating those broken-down souls." And many of the creative people Wilson profiles communicated very clearly how they felt about being so unhappy, through their addictions and, all too often, through taking their own lives.[20]

As we noted in chapter ten, all of our emotions have gifts for us. That includes sadness and melancholy. It also includes happiness. We should be careful not to romanticize the benefits of pain and despair, nor to fall into crude and superficial dismissals of happiness. The people around us, society at large, and we ourselves will be better off if we do not unnecessarily wall ourselves off from happiness that could otherwise be ours.

Placing Conditions on Happiness

OFTEN OUR RESISTANCE to happiness operates at a much more subconscious level than these considerations about purpose. Robert Holden has found that each of us has what he calls a happiness contract, which is essentially a list of conditions we place on ourselves about when and how much we are allowed to enjoy happiness. The conditions come in nine varieties. Perhaps some will sound familiar to you:

- We must make sure that everyone else is happy first.
- We must have 100 percent approval from others that our happiness is OK.
- We must work to earn happiness.
- We must suffer to deserve happiness.
- We must be perfect to deserve happiness.
- We must be good (and kind and thoughtful and patient) to deserve happiness.
- We must be in complete control to deserve happiness. (You can be happy "once you have sorted your life out and figured out what your purpose is and know with 100 percent certainty what you want to be when you grow up.")
- We must be enlightened to know real happiness. (In this case "happiness is no longer a natural experience, it is a problem to be solved.")
- We must do everything on our own to deserve happiness.

Holden points out that not only are these clauses all very hard to fulfill, but the reverse of them is usually closer to the truth. He offers the following examples:

- Work? Happiness will help you identify the "real work" of your life, and it will grace your work "with a new level of inspiration and creativity."
- Perfectionism? Forget about trying to be perfect. A perfectionism that would be worthy of us, one that could be part of a "true spiritual path," would be one that shows you how to see your perfection and to celebrate the perfection of creation.
- Independence? Sometimes taking responsibility for your life means "being willing to ask for help. You will experience greater happiness each time you let go of any separation and connect wholeheartedly with life, nature, people, God, and the universe."

- Martyrdom? Rather than deferring your happiness until everyone else is happy first, think of your happiness as your "gift to the world," which will actually help others be happier as well.

The most important thing these clauses have in common is that they are self-imposed. None of them "exist anywhere in the universe except inside your own mind." You don't deserve to be saddled with any of them. You are wondrous, miraculous, inherently deserving of happiness, and nothing you or anyone else says to the contrary is true. You can rewrite your contracts or tear them up altogether.[21]

Fear of Happiness

There comes a point in a person's life journey when the only block left that stands between you and a whole new level of happiness is your fear of happiness.[22]

—Robert Holden

HOLDEN DEDICATES AN entire session of his eight-week course to what he describes as happiness fears. What is there to fear about happiness? Often, it's basically a fear of change, or a fear of the unknown. Sometimes it's a fear of disappointment, a fear that we will make a bunch of changes and still not be happy.

A common fear is that of being selfish, that to be happy is to be insensitive to the suffering of others, or to be less caring. (This is, again, empirically not true—happy people are consistently more caring and active in supporting others.)

Another is that people will be envious or attack us if we are "too happy." As we noted in our exploration of shame, people do often feel jealous of other people's happiness and do try to cut them down. The alternative, though, is rather unappealing: to beat them to the punch by limiting ourselves and our happiness in the first place. Author Alice Walker writes, "No person is your friend (or kin) who demands your silence, or denies your right to grow and be perceived as fully blossomed as you were intended."[23]

Another common fear is what Holden refers to as the "fall from grace." So many of us have been in love, been successful, been happy—and then not been. It's natural that "you may fear being happy again (just in case you lose it again); you may fear being open again (just in case you get hurt again); you may fear trusting in life again (just in case you get betrayed again). Your fears whisper to you, 'Keep your broken heart,' 'Keep your defenses,' 'Keep away from happiness.'"[24]

Of course, love, happiness, and trust are not lifeless, unchanging things, and being alive to them means also being alive to their "twin sisters," as Nietzsche put it, such as hurt, sadness, and disconnect. It's entirely fair for us to decide at times that we have had enough of these, and to wall ourselves off a bit—just so long as we don't go too far and wall ourselves off permanently, locking ourselves in place, and locking out the things we want along with the things we don't.

Holden also points out that there isn't always a fall from grace. There is no "great big cosmic invoice for your pleasure," no "pantheon of gods" preparing to exact a price for your happiness. "The universe has no need to repeat itself, and that is especially true if you are willing to tell a new story."[25]

Attachments to Being Unhappy

> *you ask*
> *your heart*
> *why it is always hurting.*
> *it says*
> *"this is the only thing you will allow me to say to you.*
> *the only feeling you are willing to feel."* [26]

—Nayyirah Waheed

ARGUABLY THE MOST powerful of our happiness fears is what Holden calls *the fear of the loss of suffering.* This is about being attached to not being happy and to the benefits that come with it. Because there can indeed be benefits to not being happy. It can draw sympathy and solidarity from others. It can help drive us and inform our work and our creativity. It can be a source of learning and personal growth. It can shield us from attacks from other people, or help us hide parts of ourselves we may be insecure about, or give us a place to retreat into if we are scared about fully putting ourselves out there or taking risks.

Holden writes that at some point we often "become identified with and attached to" our suffering. "No one has suffered exactly the way you have. Your suffering is special. It is your own story. You may also believe that your suffering is what makes you interesting, complex, unique, quirky, erudite, and mysterious." It can become interwoven with our sense of purpose and fuel our politics, our sense of morality, or our sense of righteousness.[27]

I personally remember in my teens trying to present myself as a quiet, withdrawn person in the (only slightly conscious) hope that people would not see my "real," more vulnerable and perhaps inadequate self. I hoped they would find me mysterious or sympathetic or perhaps even cool in a misunderstood outsider kind of way.

There are many identities people adopt that often seem to preclude being happy: the important person, the martyr, the intellectual, the activist, the strong person, the spiritually enlightened person, the cool or sexy person, the disillusioned person. There is nothing inherently wrong with any of these identities. The question is the degree to which they cause us to stifle our happiness to match our expectations of what we think it looks like to be these kinds of people. **Perhaps our calling is to shed an identity we have adopted that is not true to our nature**. Or perhaps we need to expand our identities so they can coexist with us being happy: the open-hearted strong person, the joyful intellectual, the "I don't want to be part of your revolution if I can't dance" activist, the giddy spiritual person, the cool *and* joyful person, the sexy *and* silly person.

It's no small task to release our attachment to suffering. Holden writes, "You will have to encounter the full force of your identification with and attachment to feelings like anger, grief, and guilt. You will encounter your resistance to letting go. You will even have to meet the pleasure of holding on to resentment or grievances."[28]

But ultimately, of course, you deserve nothing less.

OCCASIONALLY THERE ARE headlines in the news about how *trying* to be happy actually makes people *less* happy. And it is true that being overly attached to happiness can in fact be a barrier to happiness.[29]

One reason is that we sometimes have unrealistic expectations about what being happy feels like, which leaves us disappointed or self-critical when we don't feel as happy as we think we should. The reality is that most of us report being moderately happy overall, and even the happiest of us aren't happy all the time, and rarely experience "euphoria" or "ecstasy."[30]

Also, research shows that frequently thinking about happiness—thinking about whether we're happy, whether we could be happier, what could make us happier—often inclines us to look at our experiences through a critical lens. It also takes us out of the present moment.[31] (**Zen master Po-Chang was once asked about seeking the joy of the Buddha. He replied, "It is much like riding an ox in search of an ox."[32]**)

Psychologist Jonathan Schooler has concluded that "intermittent attention may be critical to maximizing happiness."[33] In other words, it can be very helpful to periodically reflect on how we are living our lives and to make changes we think will be helpful, but to not make it something we obsess about, and to not think we can just will ourselves to be happier.[34]

Another key point researchers make on this subject is that we absolutely can "achieve lasting increases in happiness" *if we build our efforts around an understanding of what factors really do boost happiness*, rather than chasing so many of the things that do not. Which is, of course, the purpose of our current exploration.

And so, with that as our inspiration, let us turn to the final critical happiness factor.

Chapter 14

Dancing with the Universe: Uniting Our Inner and Outer Worlds

There is a vitality, a life force, an energy, a quickening that is translated through you into action, and because there is only one of you in all of time, this expression is unique. And if you block it, it will never exist through any other medium and it will be lost.[1]

—Martha Graham

I MAKE THE point repeatedly in my course on Reclaiming the Sacred that my goal is not necessarily that participants emerge happier. My goal is that they emerge living more fully in the truth of themselves, in their unique and profound wondrousness and worthiness, and with a greater capacity to manifest the lives they are drawn to live.

This doesn't always translate into greater happiness. To live fully and authentically means opening ourselves to a range of emotions and experiences that can include change, risk, complexity, grief, confrontation, and more. These are, in many ways, their own reward, each of them a teacher.

The truth is also, though, that to live in this way is indeed to generally, overall, live happier.

Part of this comes back to what we explored earlier, that people who are more accepting of a range of emotions in themselves are generally happier. But even more significant is *the role of purpose and meaning*. Parts I and II of this book were dedicated to happiness in our lives and the world around us, while part III has been dedicated to happiness and the journey within. Purpose and meaning represent the place where these come together, where our inner worlds intersect with our outer worlds, where our values and perceptions shake hands with our circumstances and our actions. They are where the miracle of each of us gets to dance with the miracle of the universe.

THE SCIENTIFIC TERM for happiness is *subjective well-being*. It's subjective because it's all about our personal experience. There are certain objective measures scientists can use to try to assess a person's happiness—the brain scans they did on those monks in chapter eleven, for example. But it is still ultimately something we get to decide for ourselves—nobody gets to override our sense of how happy we are based on some external measures.

The importance of meaning to our happiness is enshrined in the very definition of subjective well-being, which is generally regarded as having three components.[2] The first two are positive and negative affect, basically good and bad feelings. At any time we can experience a mix of these—generally, the more good feelings, the happier we are, the more bad feelings, the less happy.

But the third component can actually be more important than either of these, overruling and making far more complex any simple equation of good and bad feelings. That third component is *life satisfaction*.[3] We filter our experiences through an assessment of whether our lives are matching up with our sense of what it means to live a good life. Psychologist Tal Ben-Shahar refers to positive and negative affect as the emotional components of happiness, and life satisfaction as the "evaluative component." It is "the meaning we attribute to an experience."[4]

When we have meaning, it boosts our satisfaction in almost every aspect of our lives, and when we don't, it undermines it.[5] The effect is so powerful that someone who feels they are succeeding in life is twice as likely to be "very happy" as someone who doesn't feel successful.[6] **One study found that people's sense of accomplishment counts for more than six times their income in regard to happiness.**[7]

On the one hand, meaning makes the hard times seem worth it. We're willing to make sacrifices in the name of an important goal and we can accept the challenges as something we just need to get through, or that are even worthwhile.[8] Author Laurens van der Post writes, "Once what you are doing has for you meaning, it is irrelevant whether you're happy, or unhappy—you are content—you are not alone in your spirit—you belong."[9]

The effect is so powerful that when we feel our lives have meaning we actually experience less pain.[10]

Holocaust survivor and author Viktor Frankl tells the story of two men he knew in the concentration camps who were suicidal. "In both cases, it was a question of getting them to realize that life was still expecting something from them." One of them found that in the form of a young child he had living in another country. The other found it in a series of books he hadn't yet finished.

Frankl argued that the fundamental uniqueness of each of us imbues our lives with inherent meaning:

> When the impossibility of replacing a person is realized, it allows the responsibility that person has for their existence to appear in all its magnitude. A man who becomes conscious of the responsibility he bears toward a human being who affectionately waits for him, or to an unfinished work, will never be able to throw away his life. He knows the "why" for his existence, and will be able to bear almost any "how."[11]

Meaning doesn't just help us through the hard times, though; it further elevates the good ones too. It allows us to release some of our uncertainty about the big picture and focus on the here and now. We feel better about ourselves. And we experience less stress and anxiety. Meaning basically boosts our "well-being each step of the way."[12]

The distinction between meaning, on the one hand, and positive and negative affect, on the other, helps explain why things that may not be particularly enjoyable in the moment can still have a positive impact on our happiness, or at least a neutral one. Kids are a good example. Parents experience "less frequent positive emotions and more frequent negative emotions than their childless peers," but they "report feeling a greater sense of purpose and meaning in their lives," which helps make up for at least some of the emotional toll.[13]

Meaning also helps to explain the fact that people in the US who describe themselves as "spiritual" or "religious" generally report being slightly happier than those who don't. Part of this is due to the social aspect of many religious activities, but psychologists Ed Diener and Robert Biswas-Diener have found that meaning plays a critical role. Religion and spirituality can help connect us with something larger than ourselves.[14]

A WEALTH OF research has found that we are happier when we feel our jobs are worthwhile.[15] To cite just one example, a survey of lawyers found that those who do high-pay, high-status work are less happy than public service lawyers. (The high-status lawyers also drink a lot more.)[16]

The research on volunteering is even more striking.[17] In one remarkable study, social scientist Carolyn Schwartz followed over a hundred people with multiple sclerosis for two years. The purpose of the study was to compare the impact of teaching these people coping skills versus providing them with a monthly support phone call. The surprise finding was that while both strategies were effective, the five women who were randomly selected to make the support calls (and who themselves had MS) actually enjoyed the greatest benefits—*seven times the impact of the people receiving the trainings and calls.*[18]

A study of people over age fifty-five by Doug Oman, a professor of spirituality and health, found that people who volunteered extensively were half as likely to die during the five-year study period. **That's twice the impact of exercising four times a week, and nearly as great as the impact of quitting smoking.**[19]

A body of work by psychologists Michael Norton, Elizabeth Dunn, and Lara Aknin, including the book *Happy Money: The Science of Smarter Spending*, by Dunn and Norton, shows that we are also happier when we use our money to help others. Indeed, while spending money doesn't do anything for our happiness in general, it is related to "significantly greater happiness" when we spend it on charity or gifts for others. In fact, **the impact of spending money on others is stronger than increasing our income.** This is true regardless of how much money we make.[20]

One study found that for people who received a bonus at work, the question of whether they donated part of their bonus to charity was a stronger predictor of happiness than how much of a bonus they received in the first place.[21] And an international poll found that **people who had donated to charity in the previous month were as happy as people with *twice* their income,** all other factors being equal.[22] It is true that the happier we are, the more inclined we are to help others, but researchers have documented that the giving also makes us happier.[23] Purpose plays a key role in the boost in happiness we get in each of these situations, when we are doing meaningful work, volunteering, or using our money to help others.[24]

There is one more fascinating finding that speaks to both the power of giving and the power of our subjective experience: **when we give money away, *we feel richer*** (presumably within certain limits).[25] Indeed, giving money away is a much stronger predictor of how wealthy we feel, dollar for dollar, than income. A pair of studies found that donating $500 had the same effect on how wealthy we feel (our "subjective wealth") as unexpectedly receiving anywhere from $1,600 to $10,000. That means that dollar for dollar, giving has anywhere from three to twenty times the impact as receiving.[26]

Interestingly, the same is true for our time. When we give away some of our time, say, volunteering or pausing a moment to help someone at work, we actually feel like we have *more* time (again, presumably within limits).[27] It seems that with both time and money, in giving some of it away, **we snap ourselves out of what might otherwise be a sense of scarcity and remind ourselves that we really *do* have all we need, more even, and that we can give some away and are still going to be fine.** It takes what can be an abstract awareness of abundance in our lives and gives us a concrete affirmation of it and orients us toward it, leaving us actually feeling wealthier and more time affluent.[28]

> *I remember in the novitiate, there was a young novice who would get up in the morning at 6:30 and pray all the time. And I thought, "Well, gee, to be holy, I guess I have to do that." So I'd get up, and I'd pray, and I was falling asleep all the time. And then there was another novice who was super quiet, so I thought, "Oh, I have to be really quiet, and diffident, and sort of soft*

spoken." And my spiritual director said to me, "What's wrong with you?
You're so quiet." I said, "Well, so-and-so's quiet. And he's really holy." And
he said, "In order to become holy, you don't become someone else. You just
become yourself." [29]

—Father James Martin

IT IS IMPORTANT to note that the findings about purpose and happiness are not about being successful on society's terms but about being successful on our own terms. The more we live in alignment with our own inherent needs and wants, the happier we tend to be, versus when we try to please others or meet other people's standards.[30]

Each of us has the opportunity to be in the world in a way that reflects the truth of our deepest selves. That's true whether we're talking about our capital "P" Purpose—like staying alive so we can be reunited with a child, or doing work we feel is our calling—or whether we're talking about our lower case "p" purpose, which can be what we choose to do with this day or this week, or with a dollar or a hundred dollars, or what we do with our next interaction with another being, or how we experience our next breath.

What we do know
is that when we have traveled the world over
in search of the Holy Grail, ransacked the texts
of the saints and sages for clues to its whereabouts,
pleaded with the night to yield it to us,
abased ourselves in rituals, followed the light paths
of romantic love and relationships, and the dark ones of drugs and disorder,
and done
the million and one things it appears we have to do before finally we grind
to a halt in exhaustion,
our happiness still tantalizingly out of reach;
then we see with a snort of absurdity
that there is nowhere left to reach out to;
that all of the words have gone.
The running, brother and sister, is done.
The answer is contained here in this frail being
in this dark night on this lasted breath;
here or nowhere.
The Holy Grail is us.[31]

—John Pepper

Happiness Coda

F OR EIGHT YEARS Bronnie Ware, a nurse in Australia, recorded the reflections of people who knew they were near the end of their lives. When asked if they had any regrets, five key themes emerged.

I wish I had let myself be happier.

I wish I had stayed in touch with my friends.

I wish I'd had the courage to express my feelings.

I wish I hadn't worked so hard.

I wish I'd had the courage to live a life true to myself,
not the life others expected of me.[1]

ROBERT HOLDEN TELLS the story of someone who once did an activity in his course called "The Happiness Walk-Through." Holden asked her to rate her happiness on a scale of one to ten. She answered six. He then had her close her eyes, take a deep breath, and describe what it would take for her to go from six to seven. She thought, then replied, "A raise, or paying off the mortgage."

He then asked what it would take to go from seven to eight. She took a little longer that time, then answered, "Both my daughters would be at university. And be happy."

And to go from eight to nine? She took a minute to think on that one, then answered, "I would own a vacation home somewhere by the sea. And I would paint with my oils all day long. And I would read great literature. And I would enjoy every sunset with my husband."

Holden pointed out that all of her answers were in the distant future. They then did the exercise again, but this time focusing on what would have to happen now for those shifts to occur.

The new frame not only caused her answers to move to the present, but also to things that were much more in her immediate control. "I will enjoy watching my daughters grow up." "I promise to take all my vacation days this year." "I am going to start painting again." "Tomorrow evening I will take my husband out on a date, down by the river, to watch a beautiful sunset together."

She found herself powerfully impacted by this experience, and she wrote Holden about it:

> *Robert, you have saved me 20 years of my life. There I was, busy saving up for some imaginary happiness pension, and dreaming my life away in the meantime. What "The Happiness Walk-Through" has shown me is that to have a happy future I have to start now.*[2]

There is a prayer
that lives at the center of your heart.
If you pray it, it will change your life.
How does it begin?[3]

—Matthew Anderson

Part IV

Dueling with Death, Destiny,
and Other Minor Demons:
The Psychology of Money

Chapter 15

Money Strikes Back (Sort Of): Long Life, Prosperity, and Power

Considering that we so earnestly believe
money to be the root of all earthly ills,
and that on no account
can a monied man enter heaven …
Ah! how cheerfully we consign ourselves
to perdition! [1]

—Ishmael
(from Herman Melville's *Moby Dick*)

FOR ME, IT'S a surreal experience to have dived so deeply into the profound waters of parts II and III and to return now to the topic of money. Yes, we need money to meet our essential, basic needs. Beyond that, though, it seems so obviously inadequate to the task of nourishing our well-being. How can money ever compare with love and companionship? Or with having time to enjoy life and do things that feed a deeper sense of purpose? How much can money lift us if we are in some way attached to unhappiness? Or if on some level we just feel inadequate or unworthy, must hide parts of ourselves, or feel we must move in the world with an energy of apology?

I don't have to step away from my writing and cast myself into mainstream culture for very long, though, before I start to lose that clarity. Money and material things hold spectacular value in our society. This is reflected in how most of us spend our days, the things we are willing to do to get money, and how many possessions we tend to buy. We are constantly bombarded with messages about the importance of money and possessions in songs and ads and movies. Our personal income and the gross national product are often considered synonymous with how we are doing as individuals and as a society, and the more the better. It's just a given, a central part of our politics, the news, our personal lives. Before long, it can seem almost ludicrous to consider that money *doesn't* play an essential role in our happiness, even our very purpose. Political psychologist Robert E. Lane notes, "If you question the money-happiness nexus, people think you are not so much dangerous as mad."[2]

We're not delusional to think that money plays a key role in being happy. **Wrong, yes, but not delusional.**

Money is undeniably powerful, and it can have a significant impact on our lives. Psychologists Ed Diener and Robert Biswas-Diener have documented that those of us with higher incomes generally

- Live longer
- Have lower rates of infant mortality
- Have better physical and mental health, in some respects
- Experience fewer stressful life events
- Are given lighter prison sentences for the same crimes[3]

Findings by other researchers add to the picture:

- Those of us with higher incomes have higher rates of marriage, lower rates of divorce, and a much lower incidence of domestic violence, all of which are strongly affected by financial stress.[4]
- The children of people with higher incomes are less likely to drop out of school or become pregnant as teens, more likely to perform better in school, and more likely to attend and complete college.[5] (A child's parents' wealth is such a strong predictor of how well the child will do on the SATs that a writer for the *Wall Street Journal* suggested that SAT could stand for Student Affluence Test instead of Scholastic Aptitude Test.[6])
- Financially wealthy people also have vastly greater influence over the political system.[7] A study by Martin Gliens, Professor of Public Policy at Princeton University, found that the preferences of the average American have a "near-zero, statistically non-significant impact upon public policy," which is driven instead by "economic elites and organized groups representing business interests."[8]
- No surprise, then, that the wealthiest among us also receive the most government benefits. The poorest of us receive more entitlement benefits, but the richest of us more than make up for that in expenditure benefits—that is, deductions and exemptions. People in the top 1 percent receive nine times more benefits than someone in the bottom 20 percent, and nearly five times more than someone in the middle 60 percent.[9]

And what a blessing to have some extra money to be able to buy dinner out if we're feeling overwhelmed, or get a babysitter for an evening, or take a vacation or visit relatives. Or be able to fix the roof or the car when something goes wrong, or take care of ourselves or our family if someone gets sick or has an accident.

In all of these ways, money can be incredibly useful. **But money can not only be useful, it can also be sacred.** It can be as profound an expression of life and miracle and beauty as our bodies and the sun and water. How could it not be, when it is central to so much of what we need to survive and thrive, from the food that sustains us and becomes us to the clothing and shelter that warm us and hold us in some of the most significant moments of our lives?

Money and the things we buy with it can also be vibrant expressions of our identity and creativity, they can be part of how we connect with other people through shared values and interests.[10] A frequent writer on materialism and consumption, James B. Twitchell notes, "Tell me what you buy and I will tell you who you are and who you want to be."[11]

DIENER AND BISWAS-DIENER note that, given all the benefits of having more money, "we should not be surprised if wealthier people are substantially happier than others."[12]

Yet, we are stuck with the stubborn fact that they decisively are not. Depending on which studies you read, those of us who are rich tend to be only slightly happier than the average American, or actually have a "better-than-even chance of being *un*happy."[13] **Rates of depression seem to be highest among the financially wealthy, and are twice the national average among corporate executives.**[14]

Meanwhile, the children of wealthy people actually tend to be less happy than those of poorer ones, at least while adolescents.[15] They also have "higher rates of depression and anxiety," are more likely to engage in "certain delinquent behaviors," and have "elevated levels of substance abuse (of *every* kind of substance)."[16]

Again, for those of us living below the abundance point, money is critical to meeting our basic needs, and it does clearly impact our happiness. Yet, as we noted before, even for the poorest of us, the role of money is still often limited. In a study spanning nineteen countries, Diener and Biswas-Diener found that a full **69 percent of people living in serious poverty were still satisfied with their lives. For wealthy and upper-middle-class people the statistic was only 17 percent higher, an incredibly slight difference for going from serious poverty to tremendous financial riches.**[17]

Despite the many advantages of having more money, we are left with the unwavering fact that started us on our journey: income accounts for a mere 2–4 percent of our happiness.[18]

And it comes with some rather unfortunate side effects as well.

Mother Teresa once noted what she called "the deep poverty of the soul" that afflicts the wealthy, and said that the poverty of the soul in America was deeper than any poverty she had seen anywhere on earth.[19]

—Lynne Twist

THERE IS A violence in having been given the gift of everything we need to sustain and delight ourselves and to find it wanting. And we pay a price when we give ourselves over to this violence.

- Those of us who place great importance on money and material things "typically report less loving, more conflicted relationships with friends and romantic partners" and are "more competitive and Machiavellian."[20]
- We tend to have a more fragile sense of self-worth, frequently comparing ourselves with others and feeling we are worthy only if we can make the next sale or reach the next milestone.[21]
- We are less likely to actualize ourselves and reach our full potential.[22]
- And we have a higher risk of depression and anxiety.[23]

Ed Diener summarized the research on this topic by noting, "**Materialism is toxic for happiness.**"[24]

Ironically, psychologist Carol Nickerson quantified the impact of all of this in terms of money. She found that it would take an income of about $116,000 per person to make up for the happiness lost by being materialistic.[25] And $116,000 is actually four times beyond the saturation point, the point at which additional money no longer makes any difference to our happiness, which means **there actually isn't enough money in the world to ever make up for the costs of being materialistic.**

Simply having a lot of money, not even particularly giving it a lot of importance, has significant consequences as well. In a study titled "Money Giveth, Money Taketh Away," psychologist Jordi Quoidbach documented that **the more money we have, the more that joy, awe, contentment, and gratitude are fleeting.** Wealthier people are simply less able to "savor positive emotions and experiences."[26] The classic study of lottery winners provides evidence of this as well: People who won big actually ended up getting less enjoyment from life's simple pleasures.[27] It's a matter of adaptation, and taking simpler, less novel things for granted. It's like having the richness and wonder of everything covered in a layer of gauze.

A number of researchers, including psychologist Dacher Keltner, who hosts *The Science of Happiness* podcast, have also documented that **having more money strongly undermines our ability to empathize with others.**[28] Those of us who are wealthy are more likely to shoplift, exaggerate how hard we work and the hardships we have endured, cut off other drivers, ignore pedestrians in the crosswalk, engage in unethical behavior at work, and be less sensitive to people we are interacting with.[29] We're even more likely to steal candy from a baby. (Well, to take candy from a bowl that has a sign saying it's for kids, anyway.)[30]

In one study where people were competing to actually win $50, people with incomes of over $150,000 were *four times* more likely to cheat than people with a fraction of their income, for whom $50 would actually mean more.[31]

Psychologist Paul Piff summarizes, "While having money doesn't necessarily make anybody anything, the rich are far more likely to prioritize their own self-interests above the interests of other people."[32]

EARLY ON IN writing this book I did an Internet search on "the psychology of money" to start to get an idea of how money so powerfully hooks us, despite generally doing so little for our happiness. It was very telling that what I got back was wave after wave of articles and books and websites about the psychology of *making* money—how to invest, how to negotiate, how to be smart about spending, where to place items in a store to sell more, and so on. It was a stark reflection of the power of money in our society that there was so little about how to *not* get caught up in money and so much about how to dive right in.

In order to actively craft the lives and society that we want and deserve, it's vital that we recognize the very weak connection between money and happiness above the abundance point. And we need to know what actually *does* nourish our happiness, regardless of our material circumstances. And it is also very helpful to know that having lots of money and being materialistic both tend to come with some not-so-desirable side effects, like fleeting joy and muted empathy.

Still though, it's not enough. Trying to live in a healthy relationship with money and possessions and ourselves and the world is like swimming upstream—up a torrential waterfall more like it. In order to free ourselves from the hooks of money—and the constant messages of friends and family and songs and movies and ads and news anchors and on and on—we need something more. We need to understand why money so "obviously" seems like it *does* matter so much, why so many people and institutions dedicate so much of their lives and very purpose to accumulating money and possessions, seemingly ever more of it. **We have to understand *this* psychology of money—not the ways to make more of it, but the ways it hooks us.**

Despite my initial Internet results, significant research has, in fact, been done on this topic, and it turns out that money works on us from five different angles. Sure enough, they don't have much to do with our actual happiness or well-being. In fact, they have more to do with a *lack* of happiness and well-being, as we shall soon see.

It is to *this* psychology and these hooks that we dedicate part IV of our journey.

Chapter 16

The Dopamine Reward System, Hijacked

That moment. That instant when your fingers curl round the handles of a shiny, uncreased bag—and all the gorgeous new things inside it become yours. What's it like? It's like going hungry for days, then cramming your mouth full of warm buttered toast. It's like waking up and realizing it's the weekend. It's like the better moments of sex. Everything else is blocked out of your mind.
It's pure, selfish pleasure.[1]

—Becky Bloomwood
(from Sophie Kinsella's *Confessions of a Shopaholic*)

I DON'T SKI a lot, but I did as a kid, and a few years back I got a chance to work for a season at a big resort. I own some used skis and boots that totally do the trick, but I got a lot of comments about them while I was there. Nothing snarky, just honest: my skis were absolutely the best skis a person could own—twenty years earlier. Everyone kept telling me I wouldn't believe the difference newer skis would make. I'm sure it's true, but I was happy with my skis, and I just nodded and smiled.

Then one day I was in a sporting goods store and I was stopped in my tracks by a display of new skis. I was in awe. Their shapes and colors were so beautiful and seductive. I actually started talking with a salesperson about them and found myself gravitating toward a few specific pairs. I didn't need someone to tie me to a mast and sail me out of there, but for days they kept popping into my mind and calling to me.

Perhaps what I'm describing is familiar to you. Perhaps you've had a similar experience when looking at clothes or cars or smartphones or anything. My desire to buy those skis was not an intellectual exercise. I wasn't sitting there calculating whether I would actually be happier if I bought them. It was a basic and powerful emotional response. There was just a deep draw to them. Their beauty would be part of my life, part of me. Owning them would be an accomplishment, something to feel good about. My life would be more complete. There was a real feeling of it as "meant to be," those skis and me together.

Again, Becky Bloomwood puts it well:

> As I stare at it, I can feel little invisible strings, silently tugging me toward it. I have to touch it. I have to wear it. It's the most beautiful thing I've ever seen. … I have to have this scarf. I have to have it.[2]

There are many ways that money and possessions hook us, but at the foundation of them all is a basic physiological response. Whenever we perceive something we like, including something we might want to buy, the "pleasure center" of our brain fires, releasing dopamine and other chemicals that make us feel good.[3] We get a similar burst of pleasure when we receive money. In fact, it can cause as much activity in the pleasure center of our brain as sex, food, and drugs.[4]

The high diminishes quickly once the purchase is completed, which is part of why we often feel deflated once we get home with our new possessions, and why we sometimes even regret the purchase ("buyer's remorse").[5] Still, it's a powerful draw. It helps explain why we generally buy more and are willing to pay higher prices when we're sad, depressed, stressed, or bored: for all the things shopping may or may not do for us, it does often give us a brief dopamine rush that momentarily distracts us from whatever else is going on in our lives or the world.[6]

We have built-in mechanisms to help keep these inclinations in check. For example, when we see the price tag or become aware of other potential costs of buying or pursuing something, a part of our brain called the insula gets to have a say in what we do next. The insula is related to unpleasant emotions and the anticipation of loss or pain. If the insula fires, it is unlikely that we will pursue things any further.[7] Another built-in mechanism is the more conscious stage of "deliberation" before we pursue something, but the initial subconscious reactions largely drive what we ultimately consciously decide.[8]

This interplay of attraction and reaction has been honed over millions of years and generally works well. It's just not very well adapted to today's world. These parts of the brain developed long before the invention of money, not to mention before the advent of many of the vast range of things we can buy today.[9]

Consider the trouble our bodies are having adapting to the modern world. In 1990 not one state in the US had an adult obesity rate above 15 percent. By 2020 all fifty states had rates of 20 percent or higher, and sixteen states have rates of 35 percent or higher.[10] One in seven of us suffers from diabetes, and one in three of us will die of heart disease.[11] Our deepest instincts about eating are simply not adapted to an environment where we have unlimited access to burgers, donuts, and carbonated sugar water.[12]

To get the same amount of sugar in a twelve-ounce can of soda (which is less than the smallest cup you get at most fast-food restaurants and gas stations today), our ancestors would have had to gnaw through more than three feet of sugarcane. Today we can down that in mere minutes.[13]

Our brains have similarly not adapted to the modern whatever-you-want, whenever-you-want-it marketplace, where the "dopamine reward systems of the brain are easily hijacked," as psychiatrist Peter Whybrow notes in his book *American Mania: When More Is Not Enough*.[14] So, we shouldn't be surprised that we suffer the equivalents of obesity, diabetes, and heart disease in the realm of money and possessions—always drawn to more, even when it's not in our best interests, even when the initial rush quickly gives way to whatever we were feeling before, or worse.

In the end, spending money contributes to our happiness in only a few limited situations where it ties in with one of the aspects of our lives that do impact our happiness. Spending money on other people, for example, tends to boost our happiness (far more than receiving money) because it nourishes our sense of connection and purpose.[15] Spending money on experiences, as opposed to possessions, can likewise boost our happiness because those experiences often have a social aspect to them.[16] And spending money on things like learning new skills can boost our happiness by challenging us and also contributing to our sense of purpose.[17]

With the exception of these kinds of spending, though, which represent only a tiny fraction of most of our spending, research confirms that once we live beyond the abundance point, **the amount of money we spend has *zero* relationship with how happy we are**, despite any initial dopamine rush.[18]

ROBERT E. LANE suggests that the most important reason we tend to believe money is important to our happiness is that we have so many experiences where it *does* make us happy, even though they are usually brief and sometimes lead to unhappiness afterward.[19] We may even be aware that the pleasure of money and possessions is fleeting, may actually remember previous experiences where the happiness didn't last, yet the draw of the dopamine kick can be very compelling.

Yet, psychologist Steve Taylor notes that "there is no evidence that—apart from some food-hoarding for the winter months—other animals share our materialistic impulses," despite having inherited the same basic dopamine reward systems as us. Taylor argues that our drive to accumulate money and possessions is best understood in psychological terms related more distinctly to our humanness than to this basic dopamine high.[20]

Granted, no other animals have the ease of access to such an array of temptations that we humans have. Still, drawing inspiration from Taylor's idea, let's move on to the four other ways money hooks us, building from the more basic and physiological toward the more psychological and intellectual.

Chapter 17

Sex, Status, and Stuff:
Being Conspicuous
with Our Consumption

WOULD YOU RATHER live in a world where your income is $50,000 a year and everyone else's income is half that, or a world where your income is $100,000 and everyone else's is twice that?

A study by economist Sara Solnick found that more than half of people surveyed prefer a lower income so long as it is twice everyone else's. In other words, a majority of people cared more about their "relative income" than their "absolute income." The results were even more striking when it came to appearances: 75 percent of people would prefer to be less attractive overall but more attractive than everyone else. People felt the same when it came to intelligence and praise at work.

Studies reveal that we pretty consistently care about our status in the eyes of other people, so much so that we sometimes actually prefer to be worse off in absolute terms so long as we are better off relative to other people (as in the case of preferring less income so long as it is twice everyone else's).[1] This is rooted in how much we as a species have always depended on our families and communities for safety, protection, companionship, sex, and so on, and it translates directly to our feelings about money and possessions.[2]

If this doesn't sound relevant to you, don't be surprised. At least one study found that the vast majority of us don't perceive ourselves as trying to make money or buy things to impress other people, even though our behavior clearly reveals otherwise.[3] For example, one study in the US found that we tend to increase our consumption, even beyond our means, when the incomes of our friends, family members, and neighbors go up.[4] We also generally care far more about buying brand names when the products are highly visible, such as clothing, cars, and watches, than when they are not, like shampoo, blankets, and home telephone service. Which is why it is often referred to as **conspicuous consumption**.

108

It's about buying things that will be visible to others as a signal to them that we are successful, have good taste, are cool or intelligent—that we are worthy of their respect, friendship, love, and so on.[5]

We're not only willing to pay more for highly visible things, but we're also more likely to keep buying them even if the price goes up.[6] In fact, what makes many of these things so attractive is the fact that they *do* cost more—it makes them less accessible to others and makes them a stronger signal to others of our financial wealth, whether it's an expensive home or car or vacation or clothes.[7]

Researchers have found that some people likewise seek status by playing up how busy they are and how little leisure time they have, as a way of signaling how important and valued they are—what we might call conspicuous *production* rather than *consumption*.[8]

Extensive research by Ed Diener and others strongly suggests that, despite our frequent desire to be relatively and conspicuously wealthy, it has zero impact on our happiness when we are.[9] Psychologist Geoffrey Miller offers an intriguing insight that helps explain why this is so: despite what we might imagine (and what advertisers dearly want us to believe), **money and possessions actually have very little effect on our status with other people.** Research indicates that we are very good at assessing people within minutes or even seconds of meeting them in terms of the things we really care about: age, gender, personality, intelligence, physical and emotional states, and attractiveness. We are not deceived by what they own, by what they're wearing, or by how much money they have. And we get better at these assessments with age.[10] (This perhaps helps to explain why income has zero relation to how often we have sex or how many people we have sex with.[11])

Not everyone agrees with Diener. Economist Richard Layard argues that relative income actually has more of an impact on happiness than absolute income, at least for people above the abundance point. His research suggests that we actually tend to be less happy, despite increases in our income, if other people's incomes rise faster than ours.[12] There are also studies suggesting that our sense of whether we are successful or not is affected by how we seem to be doing compared to the people around us. For example, there are two studies that found we tend to be more satisfied with our lives when we are financially better off than our neighbors or people in nearby neighborhoods.[13]

What is absolutely clear is that, while relative income may or may not have any impact on our happiness, *caring* **about our status relative to other people is consistently associated with being** *less* **happy**. Making "frequent social comparisons" is related to having low self-esteem and a lack of a strong sense of self, which we already know from chapter twelve powerfully undermines our happiness.[14] People who make frequent social comparisons also tend to brag and ingratiate themselves to others, gossip about others, bully them, or even sabotage them.[15]

Anne Cushman, a trainer in mindfulness yoga and meditation, writes, "It's as if we're hardwired to believe that there's only so much happiness to go around and that if someone else gets too big a chunk of it, there won't be any left for us." Or, we might add, it's as if we aren't adequate and worthy, in and of ourselves, and we need other people to come up short so we can look better.

Cushman writes, "**As we become more appreciative of our own blessings,**" and more secure "**in our own abundant happiness, the joys of other people, instead of being a threat, naturally start to feed our hearts as well.**"[16] The Dalai Lama notes, "If I am only happy for myself, many fewer chances for happiness. If I am happy when good things happen to other people, billions more chances to be happy!"[17]

Chapter 18

Retail Therapy:
Taking Our Insecurities
and Fears to the Mall

*Modern man is drinking and drugging himself out of awareness,
or he spends his time shopping, which is the same thing.*[1]

—Ernest Becker

IN 2010, NIRO Sivanathan of the London Business School conducted an experiment where participants were given the task of crossing out all the vowels on a page. They were told that this was to test their "information processing and concentration ability." One-third of them were *randomly* told that they had performed in the bottom 10 percent, another third the top 10 percent, and another third received no feedback at all.

In what they were told was a separate study, the same participants were then asked how much they would hypothetically pay for different photographs, some of which they were told were rare and available for a limited time. The people who had been given negative feedback on the vowel exercise were willing to pay *four times* as much for the more prestigious pictures than people who received positive or no feedback.

Over the course of the 1900s, there was a significant shift in how marketers and academics thought about people's spending decisions. Previously, the idea had been that people behaved "rationally," and the focus was on the "utility" of products and services. Since then, it has become universally recognized that people are often driven by "emotional, experiential and non-conscious mechanisms," as is demonstrated in Sivanathan's study.[2]

A speaker at a convention of salespeople in 1923 told them:

Sell them their dreams. Sell them what they long for and hoped for and despaired of having. … Sell them dreams—dreams of country clubs and proms and visions of what might happen if only. After all, people don't buy things to have things. … They buy hope—hope of what your merchandise will do for them. Sell them this hope and you won't have to worry about selling your goods.[3]

As Sivanathan's study highlights, we often buy things to make up for a lack of something else in our lives that is the thing we actually want or need. **This is referred to in the scientific literature as compensatory consumption, more popularly known as "retail therapy."** We tend to buy more and place more importance on money and work when we feel unhappy, stressed, or lonely, when we feel inadequate, scared, or rejected, and when we feel a lack of purpose or meaning in our lives. It's our way of distracting ourselves from these negative feelings, or numbing them, or trying to make up for them.[4]

(James B. Twitchell noted several chapters back, "Tell me what you buy and I will tell you who you are and who you want to be." We could equally say, "Tell me what you buy, and I will tell you *how you feel about yourself and the world.*")

Psychologist April Lane Benson, in her book *To Buy Or Not to Buy: Why We Overshop and How to Stop*, notes that roughly 6–9 percent of us are "shopaholics," where we are so attached to buying things that it actually interferes with our ability to function, causing us to spend inordinate amounts of time shopping, thinking about shopping, and racking up debt.[5] Therapist Bryan Robinson, in his book *Chained to the Desk*, likewise notes that roughly 25 percent of us are workaholics, meaning we often "can't stop thinking about, talking about, or engaging in work," and our work habits often interfere with our everyday lives, families, and health.[6]

Benson and Robinson note that compulsive shopping and compulsive working come down to basically the same thing:

Benson: *Overshopping is a coping mechanism, a way you temporarily distract yourself from authentic personal needs that aren't being met.*[7]

Robinson: *Workaholics overextend themselves to fill an inner void, to medicate emotional pain, and to repress a range of emotions.*[8]

Our desire to buy more or work more can be stimulated by almost anything that affects us negatively, from someone giving us a rude look to living in a war zone, from reading about something disturbing in the news to seeing people waving signs that are counter to our own core values, from the loss of someone we care about to … participating in a study where we are randomly told that our vowel-crossing-out skills are in the bottom 10 percent.[9]

Another important one is being reminded of our mortality. Arguably one of the most pervasive and profound of human insecurities is the fear of death. Ernest Becker wrote extensively about this in his Pulitzer Prize-winning book, *The Denial of Death*:

The idea of death, the fear of it, haunts the human animal like nothing else; it is a mainspring of human activity—activity designed largely to avoid the fatality of death, to overcome it by denying in some way that it is the final destiny for all of us.[10]

How deeply we fear death varies by individual and by culture, but it is particularly pronounced in the US, where we go to great lengths to not face our mortality. The impact of this fear is significant.[11] It has been found to influence everything from birthrates to politics to the environment to art: when we are reminded of death in some way, we have a stronger desire to have children, we are more drawn to politicians who convey an image of strength, we prefer art that is clear and not abstract, and we like nature that is well-kept, not chaotic or wild.[12]

It also affects our relationships with work and money. Reminders of our mortality stimulate us to work more and buy more things, as a way of trying to create order and control in our lives and distract ourselves from our fear.[13] In his book *Lead Us Into Temptation*: *The Triumph of American Materialism*, James B. Twitchell writes, "Rather than cast our lot with the moody existentialists and admit we are but mites on dust bunnies skittering around the universe, we prefer a touch of order" which is, thankfully, available for purchase in "the Marketplace of Objects."[14]

SEEN THROUGH THE lens of compensatory consumption, the importance we give to work and material wealth can be seen as an elaborate dance that most of us give ourselves over to with great abandon. We agree to talk about and pursue certain things (namely work and money and possessions), and we agree to not remind ourselves or each other too much about other things, like the fragility of life, or feeling flawed or disconnected or lacking in purpose.

This completely flips on its head the typical narrative we tell ourselves in the US about our massive levels of production and consumption, which are normally heralded as symbols of success and progress, signs that we are exceptional, blessed, doing things right. They are instead revealed as signs that something **is very wrong, as indicators of tremendous sadness, disconnect, fear, lack of meaning. They tell the story of a people so lost to ourselves and the world that we are driven to sacrifice huge portions of our lives, along with the well-being of our families, communities, and the planet, in an attempt to fill an emptiness inside ourselves and our culture.**

There is something profoundly sad, cruel, and dystopian about a society that so often denies us meaning and connection and dignity, that denies us the inherent wonder and worthiness of ourselves and the world, but then sells back to us the possibility of some degree of relief—just enough to keep us going—in the form of trillions of dollars worth of products and shows, food and pills and alcohol, *while keeping everything else the same*, while

113

urging us to continue to channel our lives into simply producing and consuming ever more, to accept that this is just the way life is.[15]

PLATO ONCE NOTED, "We cannot lie to the deepest part of ourselves."[16] Or, as theologian Roger Corliss put it, "Trying to be happy by accumulating possessions is like trying to satisfy hunger by taping sandwiches all over your body."[17] While buying things and working more can indeed provide some temporary comfort and sense of accomplishment and control (as well as a brief dopamine high), research confirms that **compensatory consumption doesn't make us any happier.**[18]

Indeed, researchers have documented that it can often leave us less happy. Leaning into work and shopping represents the kind of avoidance that, as we saw in chapter twelve, tends to actually lock in and feed our insecurities and fears. Moreover, we sometimes judge ourselves negatively for having bought something we know we didn't really want or need. Sometimes the things we buy actually remind us of the negative feelings and experiences that led us to buy them in the first place.[19] Author George Monbiot summarizes the research succinctly: "**The material pursuit of self-esteem reduces your self-esteem.**"[20]

(The term *retail therapy* was actually first used in a 1986 newspaper article as a term of mockery, as if shopping could somehow address our real needs and ills, but it found its way into common usage as a tongue-in-cheek but generally positive expression.[21])

Ultimately, experts agree that what will actually make us happier, and relieve these deep concerns, is more directly addressing our true needs, tending to the things that really do nourish human happiness, rather than trying to cover them up or compensate for them.

Are you a workaholic? Robinson says we should try to please ourselves instead of others, accept human limitations without feeling flawed, appreciate the people and things around us, allow ourselves to feel the pain of past hurts, and nurture a positive outlook on life.[22]

Are you a shopaholic? Benson advises that we seek self-acceptance and compassion, find hobbies and physical activities that we enjoy, celebrate everyday life, and nurture connections with other people and animals.[23]

These same prescriptions are actually helpful even with regard to a fear of death. Because while death may be inevitable, a fear of it is not, and these same pursuits have been documented to relieve our fear. **Indeed, for people who have a strong sense of self and self-acceptance, reminders of mortality have little to no negative effect.**[24] Meanwhile, people who have actually had near-death experiences often emerge with not only very little fear of death, but also with a diminished concern for impressing others, and a decreased interest in wealth and possessions, which they often come to see as "empty and meaningless."[25]

There was actually a round two to Sivanathan's study. It was done with a new set of participants, but this time *everyone* was told they had performed in the bottom 10 percent.

Then half of them were asked to briefly write about a value that is important to them. This was designed to give those people a chance to release that negative judgment and connect with their sense of self. The people who got to do the brief writing exercise were 30 percent less motivated to buy a prestigious watch afterward than those who didn't.[26]

Benson notes that the immediate rewards of shopping are so "persuasive, even captivating," that they can "discourage a peek over the wall to see what else is there."[27] Shopping and working can be quick, easy, and familiar compared to loving ourselves more, tending to our relationships, cultivating greater presence and gratitude, feeling all of our emotions, and so on.

Yet as Sivanathan's study and many others make clear, it's not all-or-nothing. Even small shifts in these areas—even a brief moment writing about a value that is important to us—can not only boost our happiness, but can have an immediate and important impact on reducing our vulnerability to compensatory consumption.[28]

Chapter 19

The Great Breath Retention Contest: Security at All Costs

The brain is a beautifully engineered get-out-of-the-way machine that constantly scans the environment for things out of whose way it should right now get. That's what brains did for several hundred million years—and then, just a few million years ago, the mammalian brain learned a new trick: to predict the timing and location of dangers before they actually happened.

Our ability to duck that which is not yet coming is one of the brain's most stunning innovations, and we wouldn't have dental floss or 401(k) plans without it. But this innovation is in the early stages of development. The application that allows us to respond to visible baseballs is ancient and reliable, but the add-on utility that allows us to respond to threats that loom in an unseen future is still in beta testing.[1]

—Daniel Gilbert

BRUCE SCHNEIER, A security expert at Harvard University, notes that "Security is both a feeling and a reality. And they're not the same." Just as we can feel secure and not realize that a threat is looming, we can also feel insecure when we're actually quite safe. And in both cases, we can do things that make us feel more secure when they actually do very little for us, or even leave us less secure.[2]

For example, Schneier asks:

Why is it that, when food poisoning kills 5,000 people every year and 9/11 terrorists killed 2,973 people in one non-repeated incident, we are spending tens of billions of dollars per year (not even counting the wars in Iraq and Afghanistan) on terrorism defense while the entire budget for the Food and Drug Administration in 2007 is only $1.9 billion?[3]

Or consider that China, Russia, Iran, and North Korea's combined military expenditure is only 40 percent of that of the US, while, by one estimate, a mere two and a half weeks of annual US military spending could eliminate hunger—not only in the US but worldwide.[4]

The desire for security is another way that money hooks us. We are inclined to believe that money can buy it for us, and when we feel insecure (in the sense of threatened or unsafe), we tend to buy more things and value them more. *Not that we buy things or value possessions that actually make us more secure, just that buying and owning things becomes more important to us.*[5]

The add-on part of our brain that tries to "duck that which is not yet coming" has a number of built-in flaws that help to explain the often-misguided ways we approach security. These include:

- We exaggerate spectacular-but-rare risks and downplay ones that are common or evolve slowly over time.
- We overestimate risks that are highly publicized.
- We care a lot more—roughly twice as much—about the risk of *losing* something as we do about the possibility of *gaining* something.
- We exaggerate risks to our children and downplay risks to ourselves.[6]

As we already know, we also tend to greatly exaggerate the impact that any single event will have on our happiness. Psychologist Daniel Gilbert summarizes it this way:

> *From field studies to laboratory studies, we see that winning or losing an election, gaining or losing a romantic partner, getting or not getting a promotion, passing or not passing a college test, on and on, have far less impact, less intensity, and much less duration than people expect them to have. This almost floors me—a recent study showing how major life traumas affect people suggests that if it happened over three months ago, with only a few exceptions, it has no impact whatsoever on your happiness.*[7]

Another significant mistake is that we tend to exaggerate how much security our money can buy. In my conversations with people about money and happiness, security often emerges as an important concern. It goes something like this:

> *I believe the research about money having very little relationship with happiness. And, no, I am not interested in pursuing money and possessions at the expense of important things like friends and family, time to enjoy life, and so on. But I am willing to work hard and sacrifice my time and happiness to make sure that my family and I are safe and secure.*

This sentiment reflects a love and commitment that are entirely admirable, but it also reflects a very mistaken sense of what security is really about. *And it sets up a false dichotomy between security and happiness.* **It suggests that security and happiness are contradictory, when they are actually intimately related and deeply complementary.**

The abundance point, you will recall, is the point where we have enough money to meet our basic needs. *Which is another way of saying the point where we have enough money to be basically secure.* Up until that point, more money does generally buy us greater security, in the form of food, shelter, clothing, and so on. But beyond that it generally only buys what Schneier refers to as "security theater," things that may create a sense of greater security, but really do little or nothing for us. **If having more money really made us more secure, the abundance point would simply be higher. Security is built right in; it's not a separate topic.**

In this sense, the abundance point could also be called the security point. And, as is so often true with happiness, we tend to exaggerate the importance of money to our security, and *we often prioritize money to a degree that actually compromises our security.*

One of the most obvious examples of this is when we neglect being physically active. Very small investments of time in physical activity not only boost our happiness, they provide significant protection from disease, physical decline, and death. Nurturing our social lives also contributes to health and length of life, and it can also translate into critical support during challenging times.[8] *Yet these are both things that we often neglect or compromise in our pursuit of money.* Tending to our relationship with ourselves is also an important way to affect our relationship with security: people who have a solid sense of self tend to make better decisions about their security, while people with a low sense of self-worth tend to exaggerate threats and be very reactive to them.[9]

There *are* definitely some ways that money could provide most of us with some additional security. For example, financial advisors generally suggest that we have enough savings to cover 3–6 months of expenses, in case we lose our job or face unexpected costs.[10] Yet a quarter of us have no savings, and another quarter have less than three months of savings.[11] Meanwhile, nearly half of Americans have no money set aside for retirement.[12]

Still, at the end of the day, as with happiness, once we are living above the abundance point, money tends to buy only marginal increases in security, and other factors, often much more accessible factors, tend to have a much greater impact.

SECURITY IS NOT just about freedom from danger and threats; it is also about resilience in the face of danger and threats. As we documented at the opening of part IV, money can definitely help us in the face of some of life's challenges. It can buy us better health care, give our marital relationships more stability, give us an advantage in dealing with the legal system, and so on.

There are also some critical ways, though, in which we are actually much more resilient when we live simply. We can much more easily meet our needs when they don't cost us much, and there is a greater variety and number of jobs that can pay us what we need to meet those simpler needs. We likewise do not have to work as many hours, and we can go longer without working.

Someone who has higher overhead expenses could theoretically downsize in the face of a crisis, and that is, in fact, what many of us end up having to do. But downsizing is not a quick or easy process, particularly in the midst of a crisis. It takes a logistical and a psychological toll. For example, it takes time to find, move, and adjust to a new home, not to mention sell our old one if we own it. It has a significant emotional impact as well, especially if we are also experiencing stress in our family relationships, in our job situation, and so on.

And the person who lives simply also already knows how to get by with less, and has already adapted to that way of life.

You may recall the study from chapter three by economist Juliet Schor that 25 percent of people with incomes of more than $170,000 reported that they did not have enough money to meet their basic needs. What she also found, though, was that 30 percent of people with incomes of $30,000 a year or less, *did* have enough to meet their basic needs.[13] In all of the ways we just explored, **the 30 percent of us with the lower incomes who are able to buy all we really need are far more secure than the 25 percent of us with the vastly larger incomes who are not able to meet our basic needs**.

> *Security is mostly a superstition.*
> *It does not exist in nature, nor do*
> *the children of men as a whole experience it. …*
> *Avoiding danger is no safer in the long run*
> *than outright exposure. …*
> *Life is either a daring adventure, or nothing.*[14]
>
> —Helen Keller

ROBERT E. LANE notes that for those of us who are financially poor, "the sense of not having enough money to provide food, clothing, shelter, and medical care to one's family, is profoundly distressing." However, it turns out that once we are above the poverty line, or even well above it, "*money does not reduce worrying; it simply changes the subject.*" "There are virtually no differences associated with socioeconomic status"—**we tend to worry just as much about our security when we have lots of money as when we have little of it**.[15]

Consider, for example, that large numbers of adults with investments of over $1 million report that they do not feel secure. Not only that, but in order to feel secure they indicate that they would need twice as much money as they now have. To be clear, people with over $1 million in investments are among the wealthiest on the planet, indeed the wealthiest to have ever lived, and yet **for a third of them having enough money is "a constant worry."**[16]

Part of what makes the question of security so challenging is that we can never be totally secure. Insecurity is inherent to life, no matter what we do. It can be very seductive to want to do everything we can to insulate ourselves from possible threats. To do anything less can seem downright irresponsible. But as the research makes clear, this is a delusion; it is self-defeating. It gives money far too much credit in terms of how much security it can generally buy, and it ignores how vitally important so many other things are to our security, like our physical health, relationships, and emotional well-being.

In his essay "Beyond Security Theater," Bruce Schneier writes, "The way we win over terrorists is to not give in to fear. … Terrorism's goal isn't murder; terrorism attacks the mind. … By refusing to be terrorized, we deny the terrorists their primary weapon: our own fear." The more we overreact to unlikely or overblown risks, "the more we're doing the terrorists' job for them."[17] Likewise, once we have passed the abundance (security) point, we are basically secure. Beyond that, the way to win is not to try to utterly eliminate our insecurity, but to not give in to feelings of insecurity.

(In the weeks after 9/11, Schneier was asked how we can prevent something like that from ever happening again. He replied, "That's easy, simply ground all the aircraft." As in, forever. His point was that prioritizing security over everything else, or trying to eliminate all risks, can actually be self-defeating.[18])

Going one better, rather than simply not giving in to our feelings of insecurity, we can embrace our insecurity. In his book *The Wisdom of Insecurity*, philosopher Alan Watts writes:

> *There is a contradiction in wanting to be perfectly secure in a universe whose very nature is momentariness and fluidity. But the contradiction lies a little deeper than the mere conflict between the desire for security and the fact of change. If I want to be secure, that is, protected from the flux of life, I am wanting to be separate from life. Yet it is this very sense of separateness which makes me feel insecure. To be secure means to isolate and fortify the "I," but it is just the feeling of being an isolated "I" which makes me feel lonely and afraid. In other words, the more security I can get, the more I shall want.*

> *To put it still more plainly: the desire for security and the feeling of insecurity are the same thing. To hold your breath is to lose your breath. A society based on the quest for security is nothing but a breath-retention contest in which everyone is as taut as a drum and as purple as a beet.*[19]

Instead, Watts urges us to embrace our insecurity.

> *When a cat falls out of a tree, it lets go of itself. The cat becomes completely relaxed, and lands lightly on the ground. But if a cat were about to fall out of a tree and suddenly make up its mind that it didn't want to fall, it would become tense and rigid, and would be just a bag of broken bones upon landing.*

In the same way, it is the philosophy of the Tao that we are all falling off a tree, at every moment of our lives ... and there is nothing that can stop it.

So instead of living in a state of chronic tension, and clinging to all sorts of things that are actually falling with us because the whole world is impermanent, be like a cat. Don't resist it.[20]

Chapter 20

The Soulcraft of
Modern Materialism:
Consumption, Culture, and Ideology

The chief business of the American people is business. They are profoundly concerned with producing, buying, selling, investing and prospering in the world. I am strongly of the opinion that the great majority of people will always find these are moving impulses of our life.[1]

—President Calvin Coolidge, 1925

BUFFALO FIRST ARRIVED in North America about three hundred thousand years ago, crossing the Bering Strait from Siberia.[2] More accurately referred to as bison, these creatures were resilient and, despite drought, cold, fire, disease, and predators, for most of those three hundred thousand years they thrived.[3]

Fifteen thousand years ago, following the same route from Siberia, a new predator arrived on the scene—humans.[4] Native Americans would occasionally kill more bison than they actually needed, during feast times, for example, but they were generally very thoughtful in their killing and made use of nearly the entire body of the bison.[5] And so, even as the population of these new arrivals grew into the millions, the bison continued to thrive for most of the next fifteen thousand years. As recently as 1800 there were 30 million of them.[6]

But then another predator arrived on the scene, also human, but a very different breed of human, this one emerging from the east, from Europe. Between 1600 and 1800 their numbers grew to about five million. In the blink of an eye everything changed for the bison.[7]

In fourteen short years, between 1870 and 1883, the bison were hunted to such an extreme that only 320 remained. Yes, 320. From 30 million just seventy years earlier. Many were killed in the earlier 1800s, but more than a million a year were slaughtered during those peak years.[8]

They were killed for their hides, so the leather could be used as machine belts in the burgeoning factories back east. They were killed for their tongues, which were considered a delicacy. And they were killed just for the sport of it. There were shooting sprees from trains and forts, as well as bison-killing contests. The winner of one contest took 120 bison's lives in just forty minutes.[9]

After two infantry companies "wantonly slaughtered buffalo," Chief Satanta of the Kiowa rebuked them, asking, "Has the white man become a child, that he should recklessly kill and not eat? When the red men slay game, they do so that they may live and not starve."[10]

The Europeans treated the Native Americans with the same callous disregard and violence as the bison. That disregard and violence, along with the diseases the Europeans brought with them that were previously unknown in the Americas, took the lives of between five and twelve million Native Americans, leaving only 250,000 of them by the time the bison were being exterminated.[11]

The two exterminations were, in fact, related. The Plains Indians depended on the bison for food and clothing. Colonel Richard Dodge argued that "Every buffalo dead is an Indian gone." General Philip Sheridan noted that the bison hunters "have done in the last two years, and will do more in the next year, to settle the vexed Indian question, than the entire regular army has done in the last thirty years."[12] Chief Plenty Coup of the Crow mourned, "[When] the buffalo went away, the hearts of my people fell to the ground. ... After this, nothing happened. There was little singing anywhere."[13]

It is a complex and sensitive matter to try to size up different cultures, to try to understand their worldviews and weigh their motivations. I don't mean to romanticize the Native Americans or to demonize the Europeans, and I don't mean to ignore the breadth of perspectives among both of them. Some Native Americans participated in the slaughter of the bison, as well as the slaughter of other Indigenous people, and many European Americans actively lobbied to defend the Native Americans and the bison. They even got legislation protecting the bison through Congress once, only to have it vetoed by President Grant.[14]

Yet, for all these complexities, we have the stark fact that the new Americans did to the bison in the span of fourteen years something absolutely inconceivable to the Native Americans prior, and for all the factors that were involved, a primary one is that the new Americans were vastly more materialistic than the Native Americans.

(Incidentally, Indigenous people today represent 5% of the global population, yet 80% of the world's remaining biodiversity is on lands where they reside.[15])

MATERIALISM IS ABOUT how important material things are to us. One measure of this is the variety of things we want and how much of them we want. **We Americans have proven that we want a lot of things**. The average American's "ecological footprint"—that is, how much land we need to provide the resources we use and to absorb our waste—is 70 percent more than the average European, and 700 percent more than the average African.[16] We would need four earths if everyone consumed as much as us.[17]

- The average new home built in the US is nearly twice the size of a French or Japanese home, almost three times a British home, and four times an urban Chinese home.[18]
- The average new car in the US is 43 percent larger than in Europe and 76 percent larger than in India.[19] We have fifty million more registered vehicles than licensed drivers, and we discard seven million cars a year.[20]
- We create half of the world's solid waste and hazardous waste, even though we're only 5 percent of the population.[21] **Just counting our personal trash, we create enough to cover the state of Texas two and a half times each year.**[22] (Sorry, Texas.)
- We consume twice as much energy per person as someone in the UK, and more than someone in Italy, China, Brazil and Indonesia combined.[23]
- We use about 50 percent more raw materials per person per year than the average European.[24]

Our materialism is also clear from our expressed values and opinions:

- A third of us believe money is "the best sign of a person's success." **Note, that's one in three of us who believe not simply that money says *something* about our success, but that money is *the number one indicator* of someone's success.**[25]
- A vast majority of us, 80 percent, admire "people who make a lot of money by working hard," compared with 69 percent of us who admire "people who take a lower paying job in order to help others."[26]
- A whopping 70 percent of us believe we will not be satisfied being anything less than upper class, while only 15 percent of us say we'd be satisfied "living a comfortable life."[27]
- Roughly 80 percent of us report that it is "absolutely essential" or "fairly important" for us to have a high paying job, a beautiful home, a new car, and nice clothes.[28]
- In 2014, the number one ranked life goal for first-year college students was to become "very well-off financially," with 82 percent reporting that this was "very important" or "essential" to them. That's up from 48 percent in 1970 when the number one goal was to "develop a meaningful philosophy of life" (which had dropped to 45 percent by 2014).[29]

Another gauge of how important material things are to us is what we are willing to do or give up in order to get them. **We Americans have proven that we are willing to go to incredible lengths in this sense as well.** We are willing to cause tremendous harm to other people, creatures, and the planet to get those things, not to mention harm to ourselves. What we did to the bison and what we did (and continue to do on a different scale) to the Native Americans has been very much the rule, not the exception:

- We subjected ten million people to slavery, their lives and humanity stolen for the purposes of profit.[30]
- We've created 150 million pounds of nuclear waste, which will be lethal to humans and other creatures for 250,000 years.[31]
- We've overthrown at least fifteen governments worldwide, in part or entirely because they threatened American financial interests.[32]
- We force ten billion animals a year to live out their lives in the pain and confinement of factory farms.[33]
- We've cut some 98 percent of American old-growth forests.[34]
- We've contaminated more than half of US waterways to the point where they aren't healthy for drinking, fishing, or recreation.[35]
- We've brought as many as 35,000 plants and animals to the brink of extinction in the US alone.[36]

IN HIS ARTICLE "There Must Be An Alternative," Psychologist Barry Schwartz poses a question very familiar to us by now: "Why do people insist—despite massive evidence that above subsistence, additional material wealth makes a trivial contribution to well-being—that the only thing standing between them and perfect happiness is a few more dollars in their pay envelopes?" He believes the answer lies in our cultural norms:

> *[The answer is] twenty pages every day in the newspaper devoted to the health of financial institutions, and none devoted to the health of civic institutions. The answer is cabinet-level financial ministers but no cabinet-level well-being ministers. The answer is a measure of per capita GDP, but no measure of per capita GDW (gross domestic well-being).*

In short, he says, "There is no alternative."[37]

Psychologist Ed Deiner likewise speculates that beliefs linking money to happiness "may be firmly entrenched … regardless of our actual experiences" simply because "money is such a central topic of concern" in our society, it's just so indoctrinated in us.[38]

Culture is not only a reflection of who we are; it helps make us who we are. **All of these facts about our history and behaviors and values tell us a lot about who we are, but they also reveal a lot about who we are continually being influenced to be.** We internalize materialistic values at multiple levels: we tend to be more materialistic if our parents are materialistic, if our friends are materialistic, if our workplaces and colleagues are materialistic, and, yes, if our society is materialistic.[39]

It's the nature of socialization that we often aren't even aware that we've internalized certain values or perspectives. Economist Juliet Schor compares consumerism in the US to the air—it's "so ingrained it's hard to recognize within us and around us."[40]

We in the US, as a result of being exposed to materialistic values and levels of consumption vastly beyond what most people in the world have ever experienced, tend to perceive this degree of materialism as simply normal, correct, and anything short of these levels of consumption can be seen as unusual or unacceptable, perhaps even a bad reflection on us or our families, or on our politicians or the system at large. Anything less can actually leave us feeling shortchanged, demeaned, deprived of what we deserve.

Materialism is a spiritual catastrophe, promoted by a corporate media multiplex and a culture industry that have hardened the hearts of hard-core consumers and coarsened the consciences of would-be citizens. Clever gimmicks of mass distraction yield a cheap soulcraft of addicted and self-medicated narcissists.[41]

—Cornel West

IN HIS CLASSIC book *Democracy in America*, Alexis de Tocqueville commented on how materialistic Americans were in 1835, noting, "**A passion stronger than love of life goads [them] on.**"[42] But our materialism has been spurred to ever greater heights over the past hundred years by a powerful new force: the juggernaut of advertising.

Advertisers spent an estimated $560 billion worldwide in 2019, and roughly $240 billion of that was spent right here in the US.[43] That's $730 per person and translates into each of us being exposed to a hundred ads per day.[44] **By age two the average American starts to develop brand loyalties,** and by age three we can recognize more than a hundred corporate logos, a number that grows to a thousand by the time we're adults.[45]

Psychologist Geoffrey Miller writes:

> *Marketing is not just one of the most important ideas in business. It has become the most dominant force in human culture. ... In principle, marketing responds to pre-existing consumer preferences. In fact, marketers sometimes refer to their work as "cultural engineering"—the intentional creation and dissemination of new [ideas and stories] through advertising, branding, and public relations.*[46]

126

And while each ad may have its particular angle on which product or service will make our lives so much better, collectively they all promote the idea that the secret to happiness (and the cures to anything that makes us smell, worry, or age) is the things that we can buy. Most advertising is specifically designed to tap into and even increase our anxieties, and in each case a very simple and clear solution is presented: a product or service that is available for sale.[47]

April Lane Benson, who we met earlier for her work related to compulsive spending, notes, **"We're immersed, cradle to grave, in 'buy messages' that, with greater and greater psychological sophistication, misleadingly associate products we don't need with feelings we deeply desire."**[48]

Research confirms that the more media we take in, the more we become concerned about financial success and acquiring possessions.[49] One study found that for each additional hour of TV we watch per week, we spend an average of $218 more per year. "Five extra hours a week … raises your yearly spending by about $1,000," a result of both the advertising and the materialism in the shows themselves.[50]

Money is sacred, as everyone knows.
So then must be the hunger for it
and the means we use to obtain it[51]

—Delblanc
(from Barry Unsworth's *Sacred Hunger*)

IT WAS THE first day in my college economics class, and the professor, who was from India, started off by introducing the following concept, a cornerstone of Western economics: *"Human wants are unlimited but resources are finite."* He quickly added, "Of course, in my country and throughout the world there are entire cultures and religions based on the very opposite idea. But that's not what I'm being paid to teach you." And that was the last we heard of that.

For all of the values and behaviors that are held subconsciously within a culture, there are also those that are consciously acknowledged and promoted—by the education system, the media, politicians, and so on. Some even rise to the level of being elaborately articulated as overarching philosophies.

Such is the case in the US when it comes to money and possessions. **It's perhaps the single most fundamental tenet of classical economics that economic activity, money, and possessions are good, and that more is better.** Possessions are actually referred to in economics as "goods." Our well-being and progress as individuals and as a nation, it is believed, can actually be measured by our economic activity and growth.[52]

127

James Gustav Speth, cofounder of the Natural Resources Defense Council, notes, "Economic growth has been called 'the secular religion of the advancing industrial societies.' Leading macroeconomists declare it the *summum bonum* [the highest good] of their craft."[53]

By this point in our journey, it should be obvious (if it wasn't already) that these ideas are not just deeply misguided, they are also profoundly harmful, both to those of us who buy into them, and to all of those, human and other, who have to live with the consequences of a society living and dying by this creed.

The former chairman of the Federal Reserve, Ben Bernanke, echoing much of what we already know, has noted, "the material things of life" and "economic growth" are only a means to an end. They are only purposeful to the degree they actually promote human well-being, namely "happiness" and "life satisfaction"—and they don't do much of that. "As countries get richer, beyond the level where basic needs such as food and shelter are met, people don't report being any happier."[54]

Even back in the 1930s, when gross domestic product (the primary measure of economic activity in the US) first came into use, the chief architect of the US accounting system, Simon Kuznets, warned, "The welfare of a nation can … scarcely be inferred from a measure of national income," not to mention a measure of national production.[55]

The former US Attorney General and presidential candidate Bobby Kennedy put an even finer point on the matter:

> *Our Gross National Product … counts air pollution and cigarette advertising, and ambulances to clear our highways of carnage. It counts special locks for our doors and the jails for the people who break them. It counts the destruction of the redwood and the loss of our natural wonder in chaotic sprawl. It counts napalm and counts nuclear warheads and armored cars for the police to fight the riots in our cities … and the television programs which glorify violence in order to sell toys to our children. Yet the Gross National Product does not allow for the health of our children, the quality of their education or the joy of their play. It does not include the beauty of our poetry or the strength of our marriages, the intelligence of our public debate or the integrity of our public officials. It measures neither our wit nor our courage, neither our wisdom nor our learning, neither our compassion nor our devotion to our country. It measures everything, in short, except that which makes life worthwhile.*[56]

Speaking to the environmental piece Kennedy referred to, a collection of economists and ecologists have attempted to assign a dollar value to the benefits we get from nature, to help give them a little more credibility in this system that is so focused on numbers and bottom lines. Looking at only the most immediate and tangible of these benefits, such as filtering water, controlling floods, producing food, and sequestering carbon dioxide, they came up with $33 trillion per year as an absolute minimum. That's about double the entire global GDP. *Yet none of that is reflected in our economic measures. Nor are the loss of these benefits when we destroy them.*[57] Herman Daly, the former senior economist at the World Bank, noted, "The current national accounting system treats the earth as a business in liquidation."[58]

(To be clear, that $33 trillion figure doesn't even attempt to measure the value of the life of a butterfly or an otter or a redwood or a river for their own sake, independent of whatever value they happen to have for humans.)

A person could look at the United States' GNP and other common measures of economic success like the stock market and earnings reports, and think that things are totally fine, great even, because they keep going up.

- A person could look at our GNP and have no idea that 85 percent of the earth's forests have been destroyed or degraded, that deserts and droughts are expanding rapidly, that half the wild animals in the world have been killed off in the last forty years.[59]
- A person could follow the Dow Jones Industrial Average and the S&P 500, and have no idea that the global temperature increase that is now projected will likely lead to "failed states, conflict, and mass migrations," that it will be "devastating to the majority of ecosystems," and that it will quite possibly be "beyond humans' ability to adapt" to.[60]
- A person could follow reports on earnings and inflation and jobs and have no idea that depression rates in the US have increased tenfold, that we work more hours than any other "industrialized nation," that the number of us with no close friends has quadrupled in thirty years, that 15 percent of us live in poverty, that every year one in twenty-five of us seriously considers suicide.

But here comes the part that is absolutely mind-blowing: **many, many people actually do just this.** Many people actually believe that our well-being and progress as individuals and as a nation can be measured by our economic activity and growth, that economic production and consumption are, in fact, the highest good, that if production and consumption continue growing, then most other things will naturally work themselves out, or even if they don't it's for the better.

Environmental scholar Edward Goldsmith writes:

> *Economics has developed in isolation from other sciences, in particular those that concern the living world. As a result, what is necessary to preserve our planet's life processes is all too likely to be "irrational" from an economic standpoint. **The choice is simple: to rewrite economics, or to destroy the natural world.**[61]*

Money Coda

MORE THAN 250 years ago, Adam Smith, often referred to as the "father of capitalism" or the "father of economics," advised that economic advancement often does not actually matter very much to people's well-being.

> *In the most glittering and exalted situation that our idle fancy can hold out to us, the pleasures from which we propose to derive our real happiness are almost always the same as those which, in our actual though humble station, we have at all times at hand and in our power.*[1]

It was a point he made frequently in his writings, which had a moral and philosophical bent as much as an economic one.

The extensive research we explored in the first parts of this book, led by economists, psychologists, political scientists, and many more, is much more forceful and nuanced in making clear this disconnect between money and happiness. Still, for those of us who live in this system and are constantly bombarded with messages telling us that material things and producing and consuming are vitally important, are even the essence and joy of life, I think it is powerful to hear the "father of capitalism" speak to us across the span of centuries, to hear him state so simply, so nonchalantly, that economic advancement actually often doesn't matter that much, that it is often more of a distraction, a deception.

As we have seen over the course of the last five chapters, when we look at what is really behind the obsession with money and possessions, when we go beyond the facade and the bright lights and bluster, what we are left with is a craving for dopamine and a desire for status, a longing to numb and distract ourselves from the things missing in our lives, the inertia of culture and advertising, and a science that "measures everything except that which makes life worthwhile."

We find ourselves face-to-face with a lot of fear and shame, a lot of loneliness and pain and lack of purpose. We find ourselves in many ways looking at the mirror opposite of what we explored in parts II and III, all the things that really matter to human well-being. And we find ourselves confronting the lengths to which people will go to avoid feeling these absences, to try to fill the holes in ourselves and our culture.

In this respect, Adam Smith had another warning for us:

[None of our economic pursuits] deserve to be pursued with that passionate ardour which drives us to violate the rules either of prudence or of justice; or to corrupt the future tranquility of our minds, either by shame from the remembrance of our own folly, or by remorse from the horror of our own injustice.[2]

When we give ourselves over to the pursuit of money and possessions, we distance ourselves from much of what is sacred and often right at hand, immediately available to us—our lives, our hearts, our joy, breath, light, sky. The consequences go far beyond our own lives, though.

With a general understanding of the psychology of money now under our belts, and these words of caution from Adam Smith to help guide us, we turn now to the next part of our journey.

Part V

Living the
Great Lie:
Scarcity

One of the greatest obstacles
to the achievement of liberation
is that the oppressive reality
absorbs those within it
and thereby acts to submerge
human beings' consciousness ...

Once a situation of violence
and oppression has been established,
it engenders an entire way of life
and behavior for those caught up in it.[1]

—Paulo Freire

Chapter 21

Abundance Denied: An Illustrative Tale

D ECEMBER 3, 1984

It's midnight, and nearly eight hundred thousand people are asleep in "The City of Lakes" in the heart of India. It's cold out, and some of those who are still awake are burning coals and trash to keep warm. The stars are shining brightly through a thin veil of smoke that lingers from cooking fires. It's quiet.

Then, in the northern part of the city, a thin white plume of vapor starts to shoot into the air from a high structure. The vapor returns to earth and begins to gather in a fog. It soon spreads through the surrounding neighborhoods. Then the screams begin.

> *"At about 12.30 am I woke to the sound of my baby coughing badly. In the half-light I saw that the room was filled with a white cloud. I heard a lot of people shouting. They were shouting 'run, run.'"*

> *"It felt like somebody had filled our bodies up with red chilies, our eyes tears coming out, noses were watering, we had froth in our mouths. The coughing was so bad that people were writhing in pain."*

> *"Those who fell were not picked up by anybody, they just kept falling, and were trampled on by other people."*

> *"No one knew what was happening. People simply started dying in the most hideous ways. Some vomited uncontrollably, went into convulsions and fell dead. Others choked to death, drowning in their own body fluids. ... The force of the human torrent wrenched children's hands from their parents' grasp. Families were whirled apart."*[2]

The fog continued to spread until over half the people in the city were exposed. More than eight thousand people died that night or soon after. Twenty thousand more died in the months and years that followed. One hundred thousand suffered chronic illnesses and disabilities.[3] As of 2021, "almost four decades on, survivors suffer breathlessness, body pains, weeping sores, failing kidneys, liver disease, ruined eyes, menstrual chaos, and cancers. ... Children are born in high numbers with physical and mental impairments."[4]

The city was Bhopal. The white plume was methyl isocyanate, used for making pesticides. The company responsible for the leak was Union Carbide, today Dow Chemical. It remains the worst industrial disaster of all time.[5]

FOR TWENTY YEARS Union Carbide and Dow denied responsibility for the disaster and did almost nothing to help the victims or remediate the land and water. Then in 2004, on the twentieth anniversary of the disaster, something absolutely remarkable happened. In an interview with BBC World News, Jude Finisterra, a spokesperson for Dow, reported the following:

> *Today is a great day for all of us at Dow, and I think for millions of people around the world as well. It's been twenty years since the disaster and today I'm very, very happy to announce that for the first time, Dow is accepting full responsibility for the Bhopal catastrophe. We have a $12 billion plan to finally, at long last, fully compensate the victims—including the 120,000 who may need medical care for their entire lives—and to fully and swiftly remediate the Bhopal plant site.*
>
> *When we acquired Union Carbide three years ago we knew what we were getting. Union Carbide is worth $12 billion. We have resolved to liquidate Union Carbide, this nightmare for the world and this headache for Dow, and use the $12 billion to provide more than the $500 per victim which is all that they've seen, a maximum of just about $500 per victim. That is not "plenty good for an Indian," as one of our spokespersons unfortunately said a couple of years ago. In fact, it pays for just one year of medical care. We will adequately compensate the victims.*
>
> *Furthermore, we will perform a full and complete remediation of the Bhopal site, which ... has still not been cleaned up. When Union Carbide abandoned the site 16 years ago, they left tons of toxic waste on a site that continues to be used as a playground for children. Water continues to be drunk from the groundwater underneath.*
>
> The news anchor for BBC World asked, *"Does this mean you will also cooperate in any future legal actions in India or the USA?"*
>
> Finisterra replied: *Absolutely ... We are also going to engage in unprecedented transparency. We are going to release, finally, the full composition of the chemicals, and the studies that were performed by Union*

Carbide shortly after the catastrophe. This information has never been released, Steve, and it's time for it to be released in case any of that information can be of use to medical professionals.

And, finally, we are going to fund research—any interested researcher can contact Dow's Ethics and Compliance office. We are going to fund, with no strings attached, research into the safety of any Dow product. There are many Dow products about whose safety many competent scientists have raised significant doubts. We don't want to be a company that sells products that may have long-term negative effects on the world.

This is a momentous occasion, and our new CEO, Andrew Liveris, who has been our CEO for less than a month, has decided to take Dow in this unprecedented direction.

This announcement was truly extraordinary.

Indeed, it was a bit too extraordinary. It was a hoax.

Jude Finisterra was not a spokesperson for Dow. Nor was that his real name (a nod to the patron saint of lost causes, Saint Jude). He was one of the Yes Men, a group of media activists who set up the fake interview in order to draw attention to Dow's continued neglect of the people it had harmed and killed in India.[6]

Dow quickly responded by circulating a press release repudiating everything the Yes Men had claimed in the interview. Among other things, this is what it stated:

Dow will NOT commit ANY funds to compensate and treat 120,000 Bhopal residents who require lifelong care. The Bhopal victims have ALREADY been compensated; many received about US$500 several years ago, which in India can cover a full year of medical care.

Dow will NOT remediate (clean up) the Bhopal plant site …

Dow will NOT release proprietary information on the leaked gases, nor the results of studies commissioned by [Union Carbide] and never released …

Dow will NOT fund research on the safety of Dow endocrine disruptors (ECDs) considered to have long-term negative effects.

Dow's sole and unique responsibility is to its shareholders, and Dow CANNOT do anything that goes against its bottom line unless forced to by law.[7]

Dow's statement was also rather extraordinary. All of this was consistent with Dow's previous statements, but they had never been so forthright about it.

That was because the press release was yet another Yes Men hoax.[8] Everything in the release was absolutely faithful to what Dow actually did; it just wasn't worded as evasively as they might have done themselves. Indeed, Dow chose not to formally respond at all.

LET'S BE CLEAR; this is not like a random person who, say, scrapes another car in the parking lot but doesn't have any insurance and is out of work and just drives off. It's not even like a person who has been given urgent warnings ahead of time that their car has "61 hazards, 30 of them major" (as Union Carbide had been in regard to their Bhopal plant), and who then drives recklessly through the parking lot anyway until twenty thousand people are dead—but who has no insurance or job and so just drives off.[9]

Union Carbide just drove off even though it had significant insurance and an insanely lucrative job. Union Carbide was the twenty-fourth largest company in the world at the time in terms of assets, and the year following the disaster it did $9 billion in sales.[10] Its CEO drew a salary of $840,000.[11]

What would it have taken for a company like Union Carbide to do right by the people and place it devastated? If Union Carbide had paid the victims in India the equivalent of what they paid asbestos victims in the US around the same time, it would have come to roughly $10 billion.[12]

What might it have looked like for Union Carbide to come up with that amount? Even for the twenty-fourth largest corporation in the world, would that have been possible? And—without pretending that it would in any way be comparable to the devastation suffered by the victims in Bhopal—what might the impact have been on Union Carbide's employees and shareholders? Perhaps we can find a company with a similar profile that needed to urgently raise a similar amount of money for an unexpected purpose, to help us imagine what it would have taken?

Indeed, we can, and we don't have to look very far. We have Union Carbide itself.

Less than a year after the Bhopal disaster, another company sought to buy Union Carbide, against the wishes of Union Carbide's management. In response, Union Carbide sold off its Consumer Products Division, "a profitable group that included Glad trash bags, Eveready batteries, Prestone, and STP automotive products, for $840 million. The corporation borrowed $2.8 billion, netted $3.6 billion in asset sales, and repurchased $4.4 billion in stock."[13]

In other words, they very quickly pulled together $11.6 billion, $1.6 billion more than our hypothetical amount it would have taken to address the devastation in Bhopal. They did that simply to prevent another company from buying them out, and they did so with zero related layoffs or salary reductions.[14]

Contrast that with what Union Carbide did in Bhopal, which was to pay the Indian government $470 million, about $2,200 for each family that lost someone in the disaster, and about $500 each for surviving victims.[15] (That was when the head of Union Carbide's PR department infamously noted, "$500 is plenty good for an Indian."[16])

Fully half of the $470 million that Union Carbide paid didn't even come from Union Carbide; it was covered by Union Carbide's insurance.[17]

137

In 2001, Dow Chemical bought Union Carbide. As part of that purchase, Dow sought to settle some of Union Carbide's outstanding issues. They set aside over $2 billion to resolve asbestos claims in the US, and they paid $10 million to a family in the US whose child had suffered brain damage after exposure to a Dow pesticide.[18] **They paid zero for anything related to Bhopal.**

Dow's CEO wrote that what Dow "cannot do and will not do … is take responsibility for the accident."[19]

As recently as 2012, industry analysts with the Motley Fool, a financial services company, circulated a proposal to Dow's top institutional investors, urging Dow to face up to its responsibilities in Bhopal. They suggested that it could cover over half of the expense just by creating "a 1.5 percent dilutive stock offering."[20]

Alas, the Yes Men did not circulate a mock press statement that time, so we do not have any clear statements summarizing Dow's position, because Dow didn't ever respond. We have only Dow's silence and inaction to testify to its utter disdain for the idea.

DOW CLEARLY HAS the capacity to very quickly mobilize the resources necessary to provide medical care to the Bhopal victims and to clean up the Bhopal site. Why wouldn't they, starting first thing tomorrow morning, dedicate themselves to making that happen with at least the same dedication Union Carbide gave to repelling a hostile takeover? Why wouldn't they immediately set things in motion to create a 1.5 percent dilutive stock offering? Or sell off Union Carbide? Or divert a portion of their annual profits? Or any of a hundred other things that would allow them to care for these people and this place they so profoundly harmed?

What could have possibly kept them from doing that, first thing every morning, for the last 13,542 days (as of 2022) since the accident?

The answer—so obvious that neither Union Carbide nor Dow had to ever actually say it, yet it was right there beneath the silence and the tortured responses they did offer—is that *they want as much money as they can get for themselves.*

Not because they would die if they didn't get more money. Not because they or their children would suffer chronic health problems. Not because it would have any real impact on them at all. Dow's average net profit from 2010 to 2019 was $2.7 billion a year.[21] That's after everyone's salaries were paid, including the CEO's more than $18 million.[22] Dow's directors and major shareholders are not only living beyond the abundance point, they are living light-years beyond the abundance and saturation points.

In other words, they already have not only all the money they need to maximize their happiness, but they have many, *many*, times more money than can make any impact on their happiness at all. They have so much money that their materialism and pursuit of more money is most certainly undermining their well-being in some of the significant ways we explored earlier.

Still, at the end of the day, what stands between Dow and a fraction of their wealth being directed to provide desperately needed relief to the people they harmed—often a matter of life and death—is that *they just want as much money as they can get for themselves.*

THIS IS THE world we live in: a world where businesses consistently cause great destruction and violence in their pursuit of money—despite the fact that they, their employees, and their investors generally already have all the money they need to be as happy as they can be, and pursuing more money actually tends to undermine their well-being.

This is also the world most of us in the US inhabit in an emotional and psychological sense. It is a world characterized by a sense of scarcity—a world where we feel there is always something more we need to do or acquire in order to be happy and secure, to prove our worth or be purposeful, to distract us from our fears and insecurities, and disconnection— all the things we explored in part IV.

And this is how most of us act—on a smaller scale, anyway. We prioritize the pursuit of money and possessions despite the fact that it often distances us from the very things we really want and need. And, like the people at Dow, most of us do this while generally turning a blind eye to the pain and destruction on the other end of our consumption.

WE ARE BUILDING up to the final part of our journey, which is dedicated to drinking deeply of the love and abundance that are inherent all around us and within us. It is dedicated to liberation and sweetness and belonging.

There is one more stop we must make before that, though, a stop that is essential to fully realizing that sweetness. Part V of this book is dedicated to the butterfly, the otter, and the redwood. It is dedicated to the bison, the Native Americans, and the Africans that were enslaved. It is dedicated to the children of Bhopal. And it is dedicated to you.

It is dedicated to everyone whose lives and well-being have been, and continue to be, daily sacrificed in the name of economic growth. And it is dedicated to the love and transformation that can emerge when we do not turn away from the world, but sit with it reverently, when we let ourselves see and feel the anguish of the world.

It is dedicated to all of this:

That we may see more clearly the lies we are told about materialism and consumption.

That we may more consciously and actively be part of transcending this system and its violence and destruction.

That we may live in ever-more right relationship with this miraculous world and the other beings here with us.

And that we may live in ever-more right relationship with ourselves and our own hearts.

Interlude

When you touch a flower you can touch it with your fingers, but better yet, you can touch it mindfully, with your full awareness. "Breathing in—I know that the flower is there; breathing out—I smile at the flower." While practicing in this way, you are really there, and at the same time, the flower is really there.

If you are not really there, nothing is there. The sunset is something marvelous, and so is the full moon, but since you are not really there, the sunset is not for you. From time to time I let myself look at the full moon. I take a deep breath in and a deep breath out, and I practice: "I know you are there, and I am very glad about it." I practice that with the full moon, the cherry blossoms. ...

We are surrounded by miracles,
but we have to recognize them;
otherwise there is no life.[23]

—Thich Nhat Hanh

Chapter 22

The Fourth Sister—
and Bottlenecks,
Toxic Waste, and Sorcery

MANY PEOPLE HAVE come to think of humans and human activity in relation to nature as inherently bad, with nature thriving best—or only—when devoid of humans.[1] The truth is much more complicated. We can very much live in reciprocal and mutually beneficial relationship with the natural world. Indeed, we have done so for most of our history.

We humans are of nature, and we have co-evolved with the plants and creatures around us.[2] There have been costs as well as benefits for these plants and creatures, as is true in the relationships between most species. We have protected and spread the seeds of plants that provide us with food and helpful materials, we have helped keep populations of other species in balance, we have provided nitrogen and other nutrients to the soil where we leave our urine and feces, we have provided carbon dioxide to the plants, and so on.

Robin Wall Kimmerer writes eloquently of the relationship between humans and corn, beans, and squash. These plants are often referred to as the Three Sisters because of the ways they support each other's growth. But Kimmerer notes that there is a fourth sister who is vital to their flourishing:

> She's the one who noticed the ways of each species and imagined how they might live together. … We [humans] are the planters, the ones who clear the land, pull the weeds and pick the bugs, we save the seeds over winter and plant them again next spring. We are midwives to their gifts. We cannot live without them, but it is also true that they cannot live without us. Corn, beans, and squash are fully domesticated.[3]

Nonetheless, it is also true that the impact of humans on the environment has been for some time, overall, extremely negative, devastating even. And we have taken this to a whole other level in the past few decades. The impact we are having on the planet is so massive that many argue we have entered a new geologic time period, which they have dubbed the Anthropocene epoch (*anthropo* meaning "human"). To give some context, the previous epoch, the Holocene, began nearly twelve thousand years ago when much of what is today the United States was under ice and was still roamed by wooly mammoths.[4]

We humans consume half of all living material (biomass) the earth creates every year, and half of its readily available fresh water.[5] As was noted earlier, we have killed half the population of wild animals in the last forty years, and we have destroyed or degraded 85 percent of the world's forests.[6] We have transformed more than 70 percent of the earth's land surface, and there will soon be more plastic in the oceans than fish.[7]

Most significant of all, our consumption is driving temperatures to the highest they've been in millions of years.[8] (*At only 4 percent of the world population, we in the US have caused as much as one-third of this.*[9]) **Between 40 and 80 percent of all species are at risk of going extinct before this century is out, including our own.**[10]

The powerful strain of materialism in our culture (and the globalization of that materialism) plays a central role in all of this, but there are three other factors that have also been critical in our shift from a relationship defined primarily by mutuality to one defined primarily by destruction.

The first is the sheer number of people on the planet. For most of the three-hundred-thousand-year history of *Homo sapiens*, there were roughly one hundred thousand of us.[11] Seventy thousand years ago our population fell to fifteen thousand or less. The causes are uncertain, but whatever they were, we pretty effectively put them behind us, because our population not only rebounded, it soared, until around twelve thousand years ago there were two million of us.

> Then seven thousand years ago there were ten million of us.
> Two thousand years ago, three hundred million.
> Around 1820 we passed one billion.
> Then around 1925, two billion.
> By 1974, four billion.
> Around 2000, six billion.
> Around 2025 we will be at eight billion, and our numbers may well continue to ten billion by 2060.[12]

The impact is the difference between all fifteen thousand humans eating a meal seventy thousand years ago and eight billion of us doing that today—plus clothing ourselves and housing ourselves and everything else we do. Even if we tread as lightly on the planet as our ancestors did back then, just given our numbers we would still have a tremendous impact on the world around us.

*And we most decidedly do **not** tread as lightly as our ancestors.* We have used the second factor, **advances in technology,** to fuel our profound materialism and to consume vastly more per person.[13]

- Between 1950 and 1995, people in the "industrialized world" alone consumed more resources than all the humans who lived before them combined.[14]
- The average person in the US consumes twice as much today as we did fifty years ago, and our consumption continues to rise.[15]
- Global consumption has doubled in the last thirty years and continues to grow at 1.5 percent per year, far outpacing population growth.[16]

The enormous increases in consumption are definitely *not* across the board: the wealthiest of us consume far more. For example, the wealthiest 10 percent of us consume about *twenty times* more energy overall than the bottom 10 percent. (In other words, **the population of low-income people would have to increase twenty-fold for them to cause as much destruction in the world as wealthy people**.)[17]

The scale of destruction is often greatly magnified by our technology as well. We have pesticides and massive fishing trawlers and mountaintop removal mining and nuclear waste and clearcutting and factory farming and plastic-wrapped everything, and on and on.

It's conceivable that we could have used—and could still use—advances in technology in ways that did not generally increase our consumption and destructiveness, but that allowed us to tread more lightly on the planet. However, we consistently move in the opposite direction. We don't need to consume more just because we can, but that is what we consistently do.

THE THIRD CRITICAL factor already made a star appearance in the story of the devastation of Bhopal. It is an extremely powerful factor. It stimulates ever more consumption and stubbornly resists efforts to reduce the violence and destruction involved. At the same time, it skillfully masks that violence and prioritizes production and consumption in our society, giving them incredible inertia. I am referring to **the corporation**.

Corporations are relatively new to the world, having existed for only the blink of an eye in terms of human existence. They only appeared on the scene in their current form some 150 years ago, yet their impact on people and the planet has been unprecedented. **They are not just another player in the game; they have taken over the entire game.**[18]

- Fifty-three of the hundred largest economies in the world are corporations, meaning their economic activity exceeds that of the majority of countries worldwide. Exxon's economy alone is larger than that of more than 180 countries.[19]
- The thousand largest corporations produce about 80 percent of the world's output.[20]
- Multinational corporations control half the world's oil, gas, and coal.[21]
- They also generate half the gases responsible for climate change.[22]

- Corporations own most of the media in the US, and spent $560 billion on advertising globally in 2019, deeply shaping our sense of the world and ourselves.[23]
- Between 1998 and 2014, nineteen of the top twenty political lobbyists in the US were corporations and their trade associations, spending over $5 billion.[24] As we noted in chapter fifteen, research shows that **the preferences of the average American appear to have only a "minuscule, near-zero, statistically non-significant impact upon public policy,"** which is dominated by "economic elites and organized groups representing business interests."[25] (Studies show that these companies receive an incredible return on their investment, with the top ten companies receiving over $1,000 in federal support for every dollar they invested in lobbying and campaign contributions.[26])
- While corporations comprise only about 20 percent of US firms, with proprietorships and partnerships much more common, **corporations account for 85 percent of US business revenue.**[27]

Law professor James Gustav Speth notes that, with their vast resources, and in the absence of mechanisms that to some degree keep other major players like governments in check, corporations can "influence the world on a scale and at a speed the world has never seen before," and "they have the power to shape the economic and political frameworks within which other groups and classes must operate."[28]

IMAGINE IF WE could, like sorcerers, capture the essence of everything that so powerfully drives people to pursue money—our insecurities and fears, our concerns about status, the weight of culture and ideology. Imagine if we could distill the essence of all of these, and then, at the same time, leave out the very things that actually tend to make us happy and that could get in the way of making money—things like a deep sense of abundance and gratitude, compassion, connection with the natural world, altruism, civic responsibility. Now imagine if we could conjure a body to receive this unrestrained essence, to give it a life of its own, and unleash it in the world for the explicit purpose of making money.

This is what we have done in creating the corporation. The word actually comes from the Latin *corpus*, "body." We've even given corporations the legal status of people in some important respects.[29]

Every corporation channels the powerful material desires of the people who work there and the people who invest in it, and it focuses them on the exclusive purpose of maximizing profit, regardless of whether the corporation contributes to the public good in some way, or whether it actually undermines the public good, or if it even causes tremendous destruction and suffering.

If that seems a little overstated, consider the far more pointed comments of Leo E. Strine, Jr., the former Chief Justice of the Delaware Supreme Court and the former Chancellor of the Delaware Court of Chancery, two of the most respected and influential courts with respect to corporate law.[30] In 2012 he wrote an essay entitled "Our Continuing Struggle with the Idea that For-Profit Corporations Seek Profit." He began by writing:

> *This Essay addresses an issue that, to be candid, perplexes me. That issue is the continuing dismay evidenced in Western, capitalist nations when public corporations that pursue profit for their stockholders take actions that adversely affect the nation's economic stability, the corporation's employees, or the environment. … Corporate law requires directors, as a matter of their duty of loyalty, to pursue a good faith strategy to maximize profits for the stockholders.*[31]

Corporations used to be required to consider not only profit, but also the "public good," but those laws were overturned in the late 1800s and early 1900s.[32] There is a famous case from 1919 that played a key role in this: *Dodge v. Ford Motor Company*. For years the Ford Motor Company had paid significant dividends to shareholders, and over a period of six years it paid an additional $40 million in "special dividends." In 1916 Henry Ford, who was the majority shareholder, decided the company would not pay out a special dividend, but rather invest that money in expanding operations in order to provide cheaper and better cars, and to pay workers better wages.

Two individuals, John and Horace Dodge, who owned 10 percent of the Ford stock, sued, asking the court to order the company to resume paying the special dividends and to halt the company's planned expansion.[33]

Ford testified that he felt the company "made too much money and had an obligation to benefit the public and the firm's workers and customers." He argued that "business is a service, not a bonanza," and stated, "I do not believe that we should make such awful profits on our cars; a reasonable profit is right, but not too much."[34] (He also probably didn't like the idea that the Dodge brothers were using their profits to start up a competing company.) The Dodges, on the other hand, contended that Ford was motivated by "an improper altruism" toward his workers and customers.

The court ruled in favor of Ford with regard to the planned expansion, stating that the company's directors had the authority to decide what is in the best long-term interest of the company. However, the court ruled against Ford when it came to the special dividends, ordering the company to pay them immediately, and it rebuked the company for blatantly putting the customers and employees ahead of the shareholders:

> *A business corporation is organized and carried on primarily for the profit of the stockholders. The powers of the directors are to be employed for that end. The discretion of directors is to be exercised in the choice of means to attain that end, and does not extend to a change in the end itself.*[35]

The discretion that courts give to directors to decide how they wish to operate is quite broad, and companies actually have significant latitude to respect environmental and ethical concerns. Companies such as Dow do not have to completely disregard the well-being and lives of people and the environment—they just often do.[36] Judge Strine again:

> *The continued failure of our societies to be clear-eyed about the role of the for-profit corporation endangers the public interest. Instead of recognizing that for-profit corporations will seek profit for their stockholders using all legal means available, we imbue these corporations with a personality and assume they are moral beings capable of being "better" in the long-run than the lowest common denominator. We act as if entities in which only capital has a vote will somehow be able to deny the stockholders their desires, when a choice has to be made between profit for those who control the board's reelection prospects and positive outcomes for the employees and communities who do not.[37]*

Of course, corporations resort to illegal means all the time as well. Robert Monks and Bruce Willing, experts in corporate governance, have noted that executives simply "behave rationally" when deciding whether to comply with the law or break it. They estimate the likelihood of getting caught and convicted, as well as what the fine might be, and weigh that against how much money they stand to make. The result is that "deterrence is improbable in most cases."[38]

These entities whose entire existence is dedicated to maximizing profit, whose every action aches with "never enough" and with the idea that somehow production and consumption are what really matters—these are the institutions that dominate the US political system, that handle 85 percent of revenue in the US, whose economies are in many cases larger than those of entire countries, and that directly shape our minds and culture through their media ownership and hundreds of billions of dollars spent annually on advertising.

HUMAN ACTIVITY IS not inherently bad. We are not inherently bad.

We are of nature. *We are the fourth sister.*

That is our rightful place and our deepest calling: to live in mutually beneficial relationships with the natural world, to recognize the wonder of this world *and of ourselves*, and to live in relationships that are rooted in this miracle and connection.

We must remember that this is who we are, that this is what is possible.

Even while we are living in the midst of something very different.

Interlude

"Breathing in—I am aware of my heart;
breathing out—I am smiling at my heart."

"Why that's very nice," answers your heart. Maybe we
have never had the time to do that, and this is the first
time we have been aware of our heart. "Dear one, I know
that you are there, and I am glad about it." …

Our heart then begins to experience relief. It has been
waiting a very long time for the appearance of this
friendly attitude on our part.[39]

—Thich Nhat Hanh

Chapter 23

For the Otters, the Redwoods, the Children of Bhopal: The Reality Beyond the Picture Show

You Americans,
you've mastered the art
of living with the unacceptable.[1]

—Breyten Breytenbach

Why is it so easy to see [how violent, destructive, and doomed our society
and economy are] when you're by yourself in a cabin on a hillside,
and almost impossible to believe once you step out of the house and
join several billion folks doubling down on the status quo?[2]

—Douglas Pavlicek
(from Richard Powers' *The Overstory*)

T HE WORLD IS burning. We are laying waste to the very life support systems that gave rise to and sustain human life. We are degrading and extinguishing lives, both human and a vast breadth of others at a horrifying pace, with horrifying disregard. This economic system, this culture of materialism and consumption, is brutal and hollow. It serves neither those of us who are doing the consuming or those of us who are being consumed. Whatever successes it may have to its credit, its failures are of another order entirely, and are only growing more urgent with every day. This system is bankrupt and it is doomed. One way or another it is going down.

More than ever, what we do to each other and to the other creatures of this planet, we do to ourselves and our children and all the generations yet to come, and our consumption is at the heart of this.

Most of us understand consumption through the values of our immensely materialistic society—that consumption is responsible, important, that it is central to our happiness, our worth, and our purpose—and through the lenses that corporations hand to us via their ads, their websites, their stores—that consumption is sexy, smart, feel-good, positive—and through the brief but very real flashes of distraction, pleasure, and reassurance that consumption sometimes affords us.

As Paulo Freire noted in the opening quote to part V, "One of the greatest obstacles to the achievement of liberation is that the oppressive reality absorbs those within it," and "it engenders an entire way of life and behavior for those caught up in it."

To help those of us who are so immersed in the materialism of this culture, to help us get a better perspective on the reality that Judge Strine urged us to open our eyes to, we are going to take a reality tour of sorts. I have enlisted the help of two industries and one corporation to serve as our guides. I selected the two industries because they are deeply interwoven with most of our lives and so are relevant for most of us. As for the corporation, I let 3,800 top corporate executives, directors, and analysts pick that one for us, based on the one they have voted as their "Most Admired Company" every year for over a decade.

The industries are oil and meat, and the corporation is Apple. In the following pages we will briefly pay a visit to each of them and see what they can tell us about the impact of modern consumption in the world.

Our Daily Companion:
Oil

WE IN THE United States consume over eighteen million barrels of oil every day. Most of it goes to transportation, moving us and the products we buy.[3] Nearly a third of it, then, is used to make those products, both to power their production and as a key ingredient. Most plastics are made of oil, as are nylon and polyester, so oil is an integral part of packaging, carpets, clothing, toys, electronic devices, and much more.[4]

Oil is also central to food production. Counting the gas needed to run farm equipment and the oil used in fertilizers and pesticides, it takes about ten calories of oil to produce one calorie of food.[5] The further up we eat on the food chain the more energy required: it takes roughly thirty-five calories of fossil fuel to make one calorie of beef.[6] It takes an average of ten more calories to then process each calorie of food.[7] Ecologist David Pimentel has estimated that "if all of the world ate the way the United States eats, humanity would exhaust all known global fossil-fuel reserves in just over seven years."[8] (Organic farming uses about 50 percent less energy.[9])

When we drive somewhere, BP is our sidekick; when we sit down to a meal, Shell is sitting at the table with us; when we buy something made with plastic (bottled water, a phone, a toy), ExxonMobil is there handing it to us. Most anything we buy or produce, there is an oil company involved to some degree. This helps to explain why seven of the top twenty most profitable corporations in the world are oil companies.[10]

To live in the US today is, for almost every one of us, to live in an intimate and powerful relationship with the oil industry, and all that they do to bring us that oil.

THE LARGEST OIL spill on record took place in 1991. As part of their retreat from Kuwait, Saddam Hussein had the Iraqi military set fire to more than six hundred oil wells and release hundreds of millions of gallons of oil.[11] The Persian Gulf ecosystem was devastated, and President Bush called Hussein an "ecoterrorist."[12]

Not satisfied to let Saddam enjoy all of that infamy by himself, in 2010 the oil company BP came along and set their own record. They released over 130 million gallons of oil in the Gulf of Mexico off the coast of Louisiana, from one of their drilling rigs, the Deepwater Horizon, making it the largest ocean spill ever.[13]

Eleven people died and the gulf ecosystem was likewise devastated. More than a million birds died, about ten times as many baby dolphins washed up on shore that year, there were five times more turtle strandings, 50 percent of shrimp in the area were found to lack eyes, up to 50 percent of fish had oozing sores, and so on. On top of that, the millions of gallons of oil on the ocean floor do not seem to be degrading but simply spreading.[14]

(BP's CEO assured people that the amount of oil released was "relatively tiny" in comparison with the "very big ocean."[15])

That release was not intentional in the way the Kuwaiti spill was, with people going right over and opening the spigots, but a US district court judge ruled that BP had "intentionally" acted with "reckless disregard of the probable consequences," and that its "willful misconduct" and "gross negligence" had caused the massive oil spill. They knew about the existing smaller leaks, the false test reports, the inadequate training of workers, and more. Yet they did nothing.

Quite obvious but still worth noting, the judge determined that BP's decisions were "primarily driven by a desire to save time and money."[16]

(BP's net income from 2010–2019 averaged over $7 billion a year, and as of 2019 BP ranked seventh on *Fortune's* global 500.[17])

WHILE NOT AS singularly spectacular as the BP spill, Shell, Exxon, ChevronTexaco and other oil companies have collectively dumped more than *twice* the Deepwater Horizon amount in the Niger River Delta in Africa over the past fifty years, where they allow pipelines to fail at a rate many times higher than elsewhere in the world. Adam Nossiter, writing for *The New York Times* about the Niger River Delta in the wake of the BP disaster noted:

> *Perhaps no place on earth has been as battered by oil, experts say, leaving residents here astonished at the nonstop attention paid to the gusher half a world away in the Gulf of Mexico. It was only a few weeks ago, they say, that a burst pipe belonging to Royal Dutch Shell in the mangroves was finally shut after flowing for two months: now nothing living moves in a black-and-brown world once teeming with shrimp and crab.*
>
> *Not far away, there is still black crude on Gio Creek from an April spill, and just across the state line in Akwa Ibom the fishermen curse their oil-blackened nets, doubly useless in a barren sea buffeted by a spill from an offshore Exxon Mobil pipe in May that lasted for weeks.*

(Exxon Mobil's net profit from 2010–2019 averaged more than $26 billion a year, and as of 2019 it ranked eighth on *Fortune's* global 500.[18])

These continual spills in the Niger River Delta open a window onto a whole other nightmarish side to these oil companies, which Nossiter alludes to.

In the face of this black tide is an infrequent protest—soldiers guarding an
Exxon Mobil site beat women who were demonstrating last month, according
to witnesses—but mostly resentful resignation.[19]

Being subjected to such devastating environmental abuse, why would protests be infrequent? Why would people live with resignation? The answer is that there is a long and intense history of violence by oil companies there that includes torture and extrajudicial killings. In a famous case involving the murder of environmentalist Ken Saro-Wiwa, who denounced Shell as "waging an ecological war" against his people, Shell was charged with working closely with the military junta in Nigeria, financing and arming it, and directing it to suppress protests by locals and to keep out foreign journalists. People were tortured and killed and their crops were destroyed. Many years later, Shell agreed to pay $15.5 million to the families of Saro-Wiwa and eight other leaders who were executed.

(Shell's net profit from 2010–2019 averaged over $16.7 billion a year, and it ranked third on *Fortune's* global 500 in 2019.[20])

Not to be outdone, ChevronTexaco was sued in 1993 by thirty thousand Ecuadorians from five Amazonian tribes for dumping 18.5 *billion* gallons of toxic waste into open, unlined pits. (The BP disaster, again, involved 130 million gallons of oil.) The case was supposed to be heard in the US, but ChevronTexaco managed to have it moved to Ecuador where the company had much more influence.[21] "The company's $195 billion market valuation was more than three times the entire country's annual GDP."[22]

ChevronTexaco argued that there were no laws regulating its behavior in Ecuador and that nobody had been able to prove those 18.5 billion gallons of waste caused any real damage anyway.[23] The judge, to most people's astonishment, found ChevronTexaco's claims ludicrous and fined them $18 billion.[24]

In response, ChevronTexaco hired two thousand (yes, two thousand) lawyers to help it get out of paying the fine.[25] A Bloomberg analyst summarized, "**It's cheaper for Chevron to pay the lawyers than to pay for the lawsuit. It's a simple business case for them.**"[26] The analyst also noted that "Chevron doesn't have any refineries, storage terminals, oil wells or other properties in Ecuador that could be seized to pressure the company to pay."[27]

(ChevronTexaco's net income from 2010–2019 averaged over $14 billion a year, and as of 2019 it ranked eleventh on Fortune's global 500.[28])

THE FOURTEENTH AMENDMENT to the US Constitution was ratified in 1868 to protect the rights of former slaves, guaranteeing every "person" the "equal protection of the laws." Thirteen years later, the Southern Pacific Railroad, owned by Leland Stanford, objected to a special tax that California had placed on railroad properties. He argued that it was unconstitutional under the Fourteenth amendment because "just as the Constitution prohibited discrimination on the basis of racial identity, so did it bar discrimination against Southern Pacific on the basis of its corporate identity."

The Supreme Court decided the case on other grounds and did not address the question of whether corporations have the same Fourteenth Amendment rights as people. However, the administrator responsible for publishing the court's decision (who was also the former president of the New York and Newburgh Railroads) wrote in the summary that the court *had* ruled on that question and that it had ruled in favor of corporations.

A few years later in a different case, one of the Supreme Court justices (who had secretly provided the Southern Pacific Railroad legal team with internal Supreme Court memos during the previous case, by the way) wrote that corporations do have the same status as people in at least some regards. He based this on the summary of the Southern Pacific case, *knowing full well* that the court had not actually ruled on that question and that the summary was wrong. But his ploy worked and this concept of corporate personhood gained general legal acceptance.

In fact, by 1912 the Supreme Court ruled on only 28 Fourteenth Amendment cases that related to the rights of African Americans, but ruled on 312 cases related to the rights of corporations, using it to strike down child-labor laws, zoning laws, wage-and-hour laws, and more.[29]

In certain other critical respects, however, corporations are *not* considered people before the law. In 2002 the relatives of twelve Nigerian activists who had been reportedly tortured and executed by the Nigerian government in collaboration with Shell brought a lawsuit against Shell. They did so based on a US law that allows foreign nationals to seek remedies in US courts for human rights violations. The Supreme Court ruled that the law does not apply to corporations because they are not people.[30]

ChevronTexaco was able to dodge a lawsuit related to the Torture Victim Protection Act in the same way. The court ruled again that corporations are not people and ChevronTexaco therefore could not be sued for the torture and executions in Nigeria under consideration.[31]

The Lives That Give Us Life:
Factory Farms

Humanity's true moral test, its fundamental test,
consists of its attitude towards those who
are at its mercy: animals.[32]

—Milan Kundera

AS CLOSE AS our relationship is with oil, it cannot compare with the intimacy, even sacredness, of our relationship with food. Food is life and movement and thought. Through food we take the elements of creation and the very bodies of other plants and animals into our own bodies to become us. Through food we are intimately related to the sunshine and wind and rain that bathed those plants and animals, with the soil and land that sustained them, and with all that they experienced in their lives.

Through our food we are also intimately related to the corporations that provide us with most of the food we eat in the US today and with the decisions they make about that soil and water, and about those plants and animals and their lives and deaths.[33]

That corporations are making these decisions is awful news all the way around, but particularly for the animals. There are no federal laws and few state laws that protect farm animals from cruelty. The federal Humane Slaughter Act mandates that animals be "rendered insensible to pain" before being killed, but it excludes chickens and turkeys, which are the vast majority of the animals we eat in the US. Meanwhile, the good this act could potentially do for pigs and cows is negated by almost nonexistent enforcement.[34] This means that the way these animals live and die generally comes down to a simple question: What will make corporations the most money? Which leads to a lot of horrifying answers.

The process of shepherding animals from birth to death to market as it exists today wouldn't be recognizable to anyone from the entire ten-thousand-year history of domesticated farm animals up until about fifty years ago.[35] Prior to that, most meat in the US was produced by large numbers of small and medium-sized family farms.[36] Consolidation under corporate ownership has been so intense that a mere handful of corporations produce nearly all of the meat consumed in the US today.[37]

In their drive to maximize profits, these corporations have created an entirely new breed of "farms," referred to as "factory farms" (or more officially, "confined animal feeding operations").[38] These are generally warehouses where animals are confined for long periods of time, often their entire lives, usually with no windows, little or no room to move, and little or none of the basic things that usually occupy their time and their minds, from earth to sunshine to fresh air.[39]

In the US, we consume over ten billion animals every year, not counting fish and other sea creatures, and nearly all of them, as well as the animals we use to produce milk and eggs, are factory farmed—99 percent to be precise.[40] For egg-laying hens this means generally living their lives in a space the size of a sheet of paper. For the six million female pigs used to birth more pigs, it means living in metal crates the size of their bodies. And so on.[41]

Michael Specter, writing for *The New Yorker*, shared about his first time experiencing a chicken factory:

> *I was almost knocked to the ground by the overpowering smell of feces and ammonia. My eyes burned and so did my lungs, and I could neither see nor breathe. I put my arm across my mouth and immediately moved back toward the door, where I saw a dimmer switch. I turned it up.*
>
> *There must have been thirty thousand chickens sitting silently on the floor in front of me. They didn't move, didn't cluck. They were almost like statues of chickens, living in nearly total darkness, and they would spend every minute of their six-week lives that way. … I fled quickly.[42]*

An anonymous person who worked undercover in a factory farm put it more bluntly. "If you haven't been in a hen plant, you don't know what hell is. Chicken shit is piled six feet high … and your lungs burn like you took a torch to 'em."[43] (The ammonia burns and blinds many animals' eyes, and about 65 percent of pigs develop lesions in their lungs.[44])

Animals kept in these conditions tend to go mad, so egg-laying hens have their beaks seared off so they don't peck each other to death. (Yes, a lot of nerves in those beaks). And most pigs have their tails cut off so they don't bite each other's tails, which tends to lead to cannibalism. (Again, yes, lots of nerves).[45] This is all done without anesthesia.[46]

The conditions are so unhygienic **that antibiotics have to be laced into the animals' food to prevent massive die-offs. In fact, nearly 80 percent of all antibiotics in the US go to farm animals,** which has driven the development of antibiotic-resistant bacteria.[47] Julie Beck, writing for *The Atlantic*, notes:

> *[There has been] an explosion of antibiotic-resistant bacteria, both in the US and around the world. Deaths from resistant infections are currently at about 700,000 per year, and are estimated to rise to 10 million per year by 2050. If nothing changes, the World Health Organization predicts the future will look a lot like the past—where people die from minor injuries that become infected. "The problem is so serious that it threatens the achievements of modern medicine," the WHO wrote in a recent report.[48]*

Despite these measures, hundreds of millions of animals die painful deaths each year from diseases virtually unknown outside factory farms, like "sudden death syndrome," and a condition in which excess fluids fill the body cavity.[49] An additional thirty million

155

chickens and one hundred thousand pigs die every year due to heat or cold or injuries sustained while being transported to slaughter.[50]

Regardless of how they die, the bottom line is that every single year ten billion creatures die having no idea that the world can be beautiful, or that life can be a delight.

ANOTHER ASPECT OF factory farming that would have been mind-boggling to anyone who raised animals up until recently is the animals themselves. They have been bred to grow so fast and produce so much that in some ways they are entirely different creatures.

- Factory-farmed turkeys cannot reproduce naturally; they have to be artificially inseminated, and "not a single turkey you can buy in a supermarket could walk normally, much less jump, or fly" as they do naturally.[51]
- Broiler chickens have been bred to grow four hundred times faster than is natural and their breasts are seven times larger than they were even twenty-five years ago.[52] The result is that they hardly move by the end of their lives because they are in so much pain, and a quarter of them suffer stress fractures.[53]
- Dairy cows are artificially impregnated about every ten months to keep them producing milk, about twenty thousand pounds of milk each year to be specific, ten times more than normal.[54]
- Egg-laying hens are generally killed when they are about a year old because their productivity drops after that. It's cheaper to kill them and bring in new ones rather than keep them for their fifteen to twenty-year life span.[55] They produce about 320 eggs in that year, compared to 100 or less just a hundred years ago.[56]
- Every year 250 million male chicks that are born to laying hens are killed and disposed of because chickens of this variety have been bred to produce large numbers of eggs, not to grow fast, so it is not profitable to raise the males.[57]

If I misuse a corporation's logo, I could potentially be put in jail;
if a corporation abuses a billion birds, the law will protect not the birds,
but the corporation's right to do what it wants.[58]

—Anonymous investigator

MOST PEOPLE IN the US strongly object to these conditions, so corporations go to great lengths to keep us from seeing and hearing about them.[59] The corporations rarely allow outsiders, especially journalists, to visit their facilities.[60] Most of the pictures and videos we have from these factories and slaughterhouses come to us through committed people who go undercover on behalf of organizations like Mercy for Animals, the Humane Society of the United States, and People for the Ethical Treatment of Animals.[61]

So the corporations have, in turn, sought to create severe penalties for doing that kind of investigative work. The Animal Enterprise Terrorism Act of 1996 equated undercover investigations with terrorism. "Ag-gag" laws, making it a crime to conduct undercover investigations of factory farms and slaughterhouses, have been enacted or introduced in dozens of states.[62]

Corporations have thrown their legal weight around in other ways too. A 2009 California law required that animals too sick or injured to move be euthanized and removed from the food supply. The industry took that case all the way to the Supreme Court and in 2012 the law was struck down. Now the only "downed" animals that cannot be sold for human consumption are cows, because they have a higher chance of harboring illnesses such as mad cow disease.[63]

When chickens' bellies are mechanically cut to remove their guts, the meat is often contaminated with feces. USDA inspectors used to have to condemn this meat as contaminated, but the industry had feces reclassified as a "cosmetic blemish."[64]

And in 2008, Californians voted on a state proposition with the modest requirement that "farm animals be able to turn around, stand up, lie down and extend their limbs." It was approved by 63 percent of voters and was passed into law, but not before the industry spent $10 million trying to defeat the initiative, and even then it spent seven years appealing the law in three different courts.[65]

The "Best of the Best":
Apple

AS OF 2021, Apple has been selected the Most Admired Company in the world for fourteen years in a row by *Fortune*.[66] That was in no small part because of Apple's astronomical profit. But the thousands of corporate executives, directors, and analysts who were polled by *Fortune* ranked Apple number one across all nine categories surveyed, which went beyond profit.

Judge Strine warned us that we should expect corporations to do things that are harmful to the economy, their employees, and the environment. How does that apply when we're talking about the "best of the best"? It was my decision to focus on oil companies and factory farms. What can we learn about consumption and corporate culture when we let the corporate world choose our subject?

The Economy

In 2011 Apple sought approval from the City of Cupertino to build a new headquarters. One of the council members suggested that Apple could provide free wireless Internet to the city, something another company had done nearby. Apple's CEO at the time, Steve Jobs replied (with a straight face), "See, I'm a simpleton; I've always had this view that we pay taxes, and the city should do those things. Now, if we can get out of paying taxes, I'll be glad to put up Wi-Fi."[67]

What Jobs did not mention was that that very year **Apple skipped out on $2.4 billion in taxes in the US** by diverting money through subsidiaries in Ireland, the Netherlands, and the Caribbean.[68] Five years later, European authorities ruled that Apple owed $14.6 billion for underpaying taxes there as well. (*The Washington Post* noted that even if Apple's inevitable endless appeals ultimately failed, the $14.6 billion in back taxes was "just a slice of Apple's cash stockpile" of "more than $200 billion."[69])

Adding to the gross cynicism of Jobs's comment is the fact that Apple had received more than $840 million dollars in government subsidies.[70] And Foxconn, the company that makes the iPhone for Apple, has since then received a $4.5 billion subsidy from the state of Wisconsin.[71]

Indeed, economist Mariana Mazzucato has documented that Apple owes much of its success, arguably its very existence, to government support. In its early days, Apple received funding from the federal Small Business Investment Company, and it was only after that that Apple was able to secure private funding. More importantly, touch screens, GPS, and even Apple's voice recognition program were all products of publicly funded research.[72]

Apple's economic exploits also include false advertising, price-fixing, copyright infringement, allowing children to make unauthorized purchases, anticompetitive deals with wholesalers, and on and on. In one case their offense was deemed so harmful that the corporation was fined a record $1.2 billion dollars.[73]

Employees

Human rights and labor abuses have been common at Apple for decades. Among the most egregious violations are those suffered by children in places like the Congo and Indonesia, where, along with adults, they mine the metals used in phones and computers manufactured by Apple and other corporations.

Siddharth Kara, an expert on modern slavery and human trafficking, has documented the bleak lives of people working the cobalt mines in the Congo, sometimes literally walled off from the world in an attempt to keep their plight invisible. They brave tunnels that are sometimes miles deep and unpredictable, and they breathe toxic dust, while often earning less than a dollar a day.[74] As of 2019, Apple was facing a lawsuit filed by a human rights firm on behalf of families of children who suffered serious injuries or death working in such mines in Apple's supply chain.[75] A BBC investigation documented similar abuses of children who dig tin by hand in mines in Indonesia that supply Apple.[76]

Meanwhile, as of 2020, at least seven of Apple's suppliers have been documented using workers from camps in a predominantly Muslim region of China where more than one million people have been imprisoned and where there have been reports of torture and forced sterilization. Apple has used its weight to lobby against a bill in the US that would make the use of such labor illegal.[77]

Abuses of workers among Apple's suppliers in China are nothing new. There have been almost annual reports of such abuses going back decades. They include pervasive use of child labor, dangerous working conditions, extremely long hours, and extremely low wages.[78] A high-profile investigation in 2006 prompted Apple to promise changes.[79] Then when fourteen people committed suicide at one of its manufacturers in 2010 there was another round of public hand-wringing by Apple.[80] In the following years, child labor actually increased, workers were discovered not being given a single day off from work, over a hundred workers were poisoned by a chemical used to clean screens, and so on.[81]

Apple publicly condemns these situations, yet former Apple employees and internal documents make it clear that **Apple has not only been aware of these abuses, but it has also known that the demands it places on its suppliers actually in many cases** *necessitate* **them violating labor laws.** They indicate that Apple would have to offer more flexibility and higher pay in its contracts for suppliers to be able to comply with such laws.[82]

The Environment

In 2014, David Price, writing for *MacWorld*, noted that Apple's CEO, Steve Jobs, "thought laws didn't apply to him. Why would the principles of sustainable industry be any different?"[83] For years Apple was rated one of the worst tech companies in regard to the environment. Its legacy includes toxic air pollution, heavy metals in waterways, and tremendous carbon emissions.[84]

The US Public Interest Research Group has estimated that **if everyone held on to their smartphones for just one extra year, it would save as much in carbon emissions as taking six hundred thousand cars off the road**.[85] However, in 2020 Apple was fined more than $100 million dollars for sending software updates to previous versions of iPhones that intentionally slowed their performance, compelling people to buy new batteries or new phones.[86] At the same time, Apple will not provide even basic repair documentation to customers, it charges substantial fees for repairs, it relentlessly pursues legal action against unlicensed repair shops, and it has aggressively lobbied against right-to-repair legislation in dozens of states.[87]

It is generally not profitable for Apple to refurbish or recycle phones that are traded in, so most of the material ends up in the landfill.[88] Indeed, Apple has a strict "full-destruction" policy for the phones it directs to be dumped, whether they still work or not. **It sued one company for refurbishing** *more than a hundred thousand* **phones still in working order**.[89]

Apple has tried to change its image and, among other things, it has pledged that its entire business, including its supply chain, will be carbon neutral by 2030.[90] That would be an incredibly positive accomplishment not just for Apple but for the ripples it would cast throughout the business world. Their pledge must, however, be taken with some large grains of salt. As of 2021, Apple's products have been showing *increased* carbon emissions.[91] Meanwhile, Apple claims that some of its operations are already carbon neutral, even though they depend heavily on coal and natural gas. It justifies this claim by paying "other energy users who derive a fraction of their energy usage from renewable energy to 'credit' their renewable consumption to Apple." This juggling of numbers yields a net gain of *zero* for the environment.[92]

What We Should Expect to Find

Did you know … Dow is responsible for the birth of the modern environmental movement? The 1962 book Silent Spring, about the side-effects of DDT, a Dow product, led to the birth of many of today's environmental action groups.

—From DowEthics.com
(a Yes Men spoof of Dow's website)

AGAIN, THE 3,800 corporate leaders who ranked Apple the Most Admired corporation in the world did so not just based on profit. They ranked Apple first in all nine categories surveyed. One of those categories was actually social responsibility. Another was people management.

Whether the other corporations that were considered are actually even worse than Apple (forced sterilization? children laboring in toxic mines? billions of dollars funneled to offshore tax shelters?), or these kinds of abuses were simply not considered noteworthy by the corporate leaders who were polled—either one of these speaks volumes about the state of the world, the nature of corporations, and the impact of consumption.

To be clear, not all consumption is as violent or destructive as this. Many products and services offer important value and are, in turns, low in energy and chemical use, are local and transparent and accountable, are rigorously certified, are low in packaging and processing, their workers are paid fair wages and are treated with respect, and so on.

Similarly, not every corporation is as destructive and callous as Dow, Shell, Apple, and Tyson (the largest meat company in the US). Many are rather innocuous. And a vast majority of the people who work at corporations are wonderful people. Nonetheless, as Danny Schechter, the producer of 20/20, notes, "There's a logic to these corporations. Which means that certain values get emphasized while others get de-emphasized. And the ones that get emphasized are what's going to bring the bottom line up."[93] Law professor Joel Bakan, author of The Corporation, writes:

> *The people who run corporations are, for the most part, good people, moral people. … Despite their personal qualities and ambitions, however, their duty as corporate executives is clear: they must always put their corporation's best interests first. … The money they manage and invest is not theirs. They can no sooner use it to heal the sick, save the environment, or feed the poor than they can use it to buy themselves a villa in Tuscany.*[94]

Ultimately, a vast portion of our consumption really is as violent and destructive as what we have just documented, and a great many corporations really are this awful. The words of Judge Strine bear repeating:

> *Instead of recognizing that for-profit corporations will seek profit for their stockholders using all legal means available, we imbue these corporations with a personality and assume they are moral beings capable of being "better" in the long-run than the lowest common denominator.*[95]

In other words, the examples we just explored are not warnings about what might sometimes happen if we're not careful; **they are warnings about what we should expect to find woven into our daily consumption.**

Drink much?

In 2018, Americans bought over 70 *billion* plastic bottles of water, using over 36 million barrels of oil just to produce the bottles, to say nothing of shipping them.[96] That plastic never actually decomposes; it simply breaks down into smaller and smaller pieces, which at this point can found everywhere worldwide—including our drinking water.[97] (Poetic justice: there are about twice as many microplastics in bottled water as tap water.[98]) **Fiji Water gets special mention for helping prop up a dictatorship by on an island (Fiji) where half the population itself doesn't have access to clean drinking water.**[99]

Eat much?

Only 5 percent of food in the US is organic.[100] Eating food with pesticides and herbicides increases cancer rates by approximately 30 percent.[101] **Pesticide exposure hits farmworkers especially hard, causing neurological damage, infertility, respiratory problems, and cancer in them and their family members.**[102] Pesticides also play a central role in the dramatic decline in bee and other populations.[103]

Wear clothes much?

By 2050 the clothing industry could account for over 25 percent of global carbon emissions.[104] Synthetic textiles are not biodegradable, and nearly three times as much CO_2 is emitted to produce a polyester T-shirt as a cotton one.[105] However, cotton is hugely water-intensive, requiring 1,800 gallons of water to make a pair of jeans. And unless it is grown organically, cotton is also hugely pesticide-intensive; **some twenty thousand people die of cancer and suffer miscarriages each year as a result of chemicals sprayed on conventional cotton.**[106]

Spend time indoors much?

In 2019, nearly 16 billion pounds of vinyl (PVC) were produced in the US.[107] Most of that went into buildings as siding, windows, flooring, and more.[108] In every stage of its life, vinyl releases chemicals linked with birth defects, neurological damage, and cancer. Some of these chemicals have, like plastic fibers, spread to every point on the globe. **Every one of us has measurable levels of them in our bodies.**[109]

Drive much?

Driving accounts for one-fifth of global warming emissions in the US.[110] And the global warming impact of *producing* a car actually rivals the impact of an entire lifetime of driving the car.[111] Scientists working for car manufacturers and big oil companies warned about the consequences of global warming as early as the 1960s, but those warnings were kept secret—even while those corporations spent countless sums of money trying to sow doubt about global warming, and lobbying against climate action.[112] (Journalist Kate Arnoff writes, "**At this point, you have to wonder which heavily polluting industries** *didn't* **know about climate change half a century ago.**"[113])

Bank much?

The massive fraud that banks committed leading to the 2008 mortgage crisis caused trillions of dollars to disappear overnight, followed by the loss of eight million jobs and the biggest recession since the Great Depression.[114] The major banks were fined tens of billions of dollars for this, with much of that to be paid in the form of relief for homeowners.[115] Almost before the ink was dry, Bank of America violated the terms, raising rates on homeowners and even implementing a policy "punishing any bank employee who spent more than 10 minutes helping a victim get a loan modification."[116] Not to be outdone, Chase responded by committing more fraud, creating false documents showing that they still owned hundreds of millions of dollars of the toxic loans they had actually sold years before as part of the original fraud, and then cancelling them.[117] (**Meanwhile, as of 2015, taxpayers had provided $4.6** *trillion* **in assistance to the major banks to help keep them afloat.**[118]) HSBC Bank gets special mention for laundering billions of dollars for drug cartels during this same time.[119]

Take medicine much?

All the major pharmaceutical companies have paid criminal fines of tens of billions of dollars for an incredible range of violations:

- Hiding serious side effects of their drugs, including heart failure
- Promoting their drugs for uses that have not been approved and that are often known to be ineffective or even dangerous
- Aggressively marketing drugs to non-approved populations like elderly people with dementia and children
- Paying kickbacks to doctors for prescribing their drugs
- And on and on[120]

The former senior editor of *The New England Journal of Medicine* writes that, due to the overwhelming influence of money, "**It is simply no longer possible to believe much of the clinical research that is published, or to rely on the judgment of trusted physicians or authoritative medical guidelines.** I take no pleasure in this conclusion, which I reached slowly and reluctantly over my two decades as an editor of *The New England Journal of Medicine*."[121] The editor of the British medical journal *The Lancet* concurs, suggesting that half of the scientific literature "may simply be untrue."[122]

Is nothing sacred? The sweet refuge of chocolate?

Cocoa manufacturers use so much child and slave labor that in the early 2000s lawmakers pressed for labels on chocolate indicating whether it uses such labor or not. The industry fought the initiative and agreed instead to voluntarily eliminate the "worst forms" of child labor by 2005. When they failed to meet that deadline, they created another. Then another. Most recently they said they would try to reduce the worst forms of child labor 70 percent by 2020, but as of 2020 none of the goals had been met.[123]

(In 2011, a lawsuit was brought against Nestle, Archer Daniels Midland, and Cargill on behalf of children who had been trafficked from their home country of Mali into Cote d'Ivoire. The three companies knowingly bought cocoa from the plantations where the children had "worked 12 to 14-hour days with no pay, were routinely beaten and whipped, and were threatened with guns." A US district court ruled that because corporations are not people, they cannot be held liable for violations of international law.[124])

THIS IS THE world we live in: a world where people are routinely poisoned and left to suffer while corporations spend billions of dollars protecting themselves from takeovers; where toxic substances can be found in the bodies of every human worldwide; where ethnic minorities are rounded up and forced to create devices for the world's most admired company.

All so that people who already have all the money they need to be as happy as they can be—more money, indeed, than can do them any good—can have even more money.

A world where most if not all drinking water contains plastic fibers and most of our food contains carcinogens, people who live by terrorizing others can just go to one of the major banks to handle their money transfers, and billions of beautiful creatures live their entire lives in hells created to the specifications of profit.

So that we can live swimming in more possessions than our ancestors could have imagined, and can still be less happy than we were before people had indoor plumbing.

A world where children are enslaved to produce cocoa for chocolate conglomerates; where the companies that produce our medicines routinely trick people into taking drugs they don't need or that are actually harmful to them; where as many as one-third of the plants and animals we share the world with today will be gone forever by 2070, and the very survival of our own species is uncertain.[125]

So that we can work longer hours, have higher rates of depression and suicide, have fewer friends and feel more alone.

The world is burning. We are degrading and extinguishing lives, both human and other, at a horrifying pace, with horrifying disregard.

But this is not who we are. It is not who we have been for nearly all of human history, and it is not who we need to be.

Interlude

Breathing in,
I am aware that I am breathing in.

Breathing out,
I am aware that I am breathing out.

In …

Out …

Breathing in,
I notice that my in-breath has become deep.

Breathing out,
I notice that my out-breath has become slow.

Deep …

Slow …

Enjoy.

Breathing in,
I calm my body and my mind.

Breathing out,
I ease everything.

Calm …

Ease …

Breathing in,
I smile.

Nothing is as important as my peace, my joy.
I smile through everything,
even through my suffering, my difficulties.

Breathing in,
I establish myself in the present moment.

Breathing out,
I realize it is a wonderful moment

Present moment …

Wonderful moment …

Enjoy.

Breathing out,
I release, I let go.

This is a practice of freedom

Smile …

Release …

—Thich Nhat Hanh[126]

Chapter 24

We Are the Machine,
We Are Blossoms:
Exit Strategies

A LETTER FROM the future:

December 3, 2084

To you, our collective grandparents,

One hundred years ago today, more than eight thousand people were killed, and more than twenty thousand followed in the days and months after the Dow Chemical Disaster in Bhopal. It still stands as the largest industrial disaster of all time.

In a sense, the Dow Disaster was just a sudden and dramatic representation of what was playing out all around the world in different ways day after day, year after year. Nonetheless, the Dow Disaster is still used today as a case study to help people understand what the world was like back then, particularly the power and viciousness of corporations, and the profound materialism and disconnect that had captured so many people.

There is definitely still tremendous violence today. We are still very much in the middle of what has come to be known as the Great Turning. But in many important ways our lives are much better, and it seems we are through the hardest part.

The Global Climate Disaster (referred to inoffensively in the early days as "global warming" or "climate change") was the catalyst for much of what we've accomplished. Large numbers of everyday people had already been making changes in their personal lives—consuming less, using renewable energy, driving less, connecting more deeply with their local communities. Then increasing numbers of more prominent people—scientists, celebrities, young activists—began making public commitments to not fly, to not eat meat, to live simply. This helped draw attention to both the urgency and the possibility of the moment.

As people in the US began to experience firsthand some of the early effects of the Global Climate Disaster—unprecedented wildfires, hurricanes, and flooding, recurring cycles of new diseases, disruptions to food supplies, economic turmoil—they were motivated to pass legislation that made extensive investments in solar and wind power, and that reduced the massive subsidies to fossil fuel corporations, and that curtailed drilling and fracking.

Still, many corporations and politicians doggedly opposed the necessary changes, even in the face of the rising horror. Eventually this was their undoing. There was a tremendous backlash, and "the Great Turning" gained significant momentum. With a dedication of purpose not seen since the Great Depression and World War II, people forced politicians to pass a handful of laws and amendments that strictly limited the influence of corporations and big money in politics, that strengthened protections for people, animals, and the environment from corporate abuses, and that made corporations accountable for the entire life cycle of their products. Insurance companies, banks, and investors began to abandon fossil fuel companies and other particularly destructive businesses as too costly and too risky.

The dislodging of corporations from the center of the political system, fossil fuel companies in particular, created an opening for more expansive changes. The shift to renewable energy was completed for most sectors of society; there were huge investments in alternative transportation and local food and energy sources. Recycling was scaled up dramatically, and the use of plastic was strictly limited. Subsidies to the meat and dairy industries were ended, factory farms were abolished, and renewed support was given to protect and expand the world's forests.

Beyond that, not as explicitly linked to the Climate Disaster, but fundamental to our well-being as a society, work weeks were restricted to twenty-seven hours, and a universal basic income was established; extreme disparities in wealth, health care and education were corrected; African Americans and Native Americans were paid reparations, and land was returned to both of them.

To be clear, despite all of the progress that was made, the Global Climate Disaster kept rolling on. The amount of carbon that had been put into the atmosphere—more since 1990 than in all of previous human history—had a delayed effect, and temperatures kept climbing. The rising oceans entirely sank many island and coastal nations, and much of the East Coast of the US had to move several miles inward. The ongoing drought and desertification forced huge numbers of Africans, Asians and Central Americans to migrate. In many places temperatures stayed over 100°F even at night, and outdoor work became impossible for months at a time.

There has been incredible political strain worldwide. More than a billion people became climate refugees, and vast numbers of people have died. Countless species are gone forever. **It is excruciating that so much of this could have been avoided if the necessary changes—many of which were inevitable, were going to have to be made one way or another—had just been made sooner.**

And still ...

Despite all of this, even in the midst of so much upheaval and horror ...

What many people did not anticipate was that, thanks to the tremendous changes we have made as part of the Great Turning, for the first time since at least the 1940s, **there have been marked increases in how happy people generally are in the US and other industrialized countries where similar changes have been made.** People report feeling a greater sense of purpose as we come together to face these challenges rather than ignore them or contribute to them; we are working less and enjoying life more, including spending more time with friends and family; there is a new vitality to our local communities; we are forging deeper connections with the places we live and the people who live there with us.

Woven through all of this, there is a shift in how we feel in our lives. Like seeing the night sky in the desert, fantastical and alive with stars, after seeing it only through the haze of city lights, or like seeing trees in the late afternoon sunshine, shimmering and radiant, after endless days of clouds and rain: there is a sense that we are returning from a self-imposed exile that served none of us. **There is a sense that we are on a journey of returning to wonder and reverence, to love and belonging, to the world and to ourselves.**

Many of us in the United States and other industrialized countries resisted for so long the changes that were so desperately and obviously needed. Who knew that all along we had so much to gain?

Please know, we have so much admiration and gratitude for all you are doing and all you are going to do.

With heart,
Your collective grandchildren

[With great thanks to Bill McKibben, environmentalist and cofounder of the climate change organization 350.org, whose article "Hello From the Year 2050" inspired this reflection, and from which many of these ideas and statistics are drawn.[1] With gratitude also to ecologist and activist Joanna Macy, who developed and popularized the concept of "the Great Turning."[2]]

IF YOU FIND this letter inspiring, you're not alone. I still remember when I first read Bill McKibben's "Hello From the Year 2050," on which this letter is based. It was exhilarating to step out of our current situation, which can feel so overwhelming, and to taste the tremendous possibility before us, to experience through story what it would be like for us to really step up and do what is necessary to meet the essential challenges of our time.

If you find this letter somewhat unbelievable, I'm with you on that one, too. Even a brief glimpse at the world around us offers countless, very sound reasons to be utterly pessimistic. There are immense forces invested in perpetuating the status quo, and vast numbers of people caught to varying degrees in denial, a lack of awareness, overwhelm, and resignation.

In the end, though, what really matters is not whether this letter from the future strikes you as realistic or as improbable, but the fact that it is possible, in whole or in part. We are living in the most critical era in human history—and the most critical era in the history of millions of other plants and creatures here with us, at our mercy. **What this moment requires is not that we resign ourselves to what seems likely or realistic, but that we rise to the necessary challenge of possibility**, that we see that there is indeed light ahead and we aim straight for it, despite the storm raging all around us.

Pulitzer Prize-winning author Thomas Friedman has written, "Pessimists are usually right and optimists are usually wrong, but all the great changes in history have been accomplished by optimists."[3]

Just as important though, is to acknowledge that many of the greatest changes have also come from people who did not see it as a choice to be either optimistic or pessimistic, but for whom pursuing whatever possibility existed was the only acceptable option.

Hopelessness and resignation—these are only affordable to those of us who continue to be shielded to some degree from the direct impact of the incredible violence and destruction rising in the world. But that will not last; none of us can avoid this growing storm.

Ultimately, resignation is simply a variation on denial. It may look different and have an air of intelligence or realism to it, but it is just another way of sticking our heads in the sand and doing nothing. It is just another form of callous disregard, and it is just as great a failure.

To not actively commit to possibility—to not actively commit to making some kind of humane, livable future a reality—is to help ensure that the future is not humane or livable. To abdicate to hopelessness and resignation or "realism" is to be part of the violence and destruction.

And it is actually, in the end, to be *un*realistic. It is to deny the extraordinary things people throughout history have accomplished against the fiercest odds. It is to deny the countless times people have overturned institutions and norms that seemed unshakable, that were rooted in immense economic and political power. By all reasonable political calculations, these efforts were lost causes. Yet the people prevailed.

I am thinking of the ending of slavery in the US, where enslaved people were America's largest financial asset and were pivotal to America's most exported commodity, cotton, and were central to the amassed fortunes of so many of the elite in both the South and the North.[4]

I am thinking of the formerly enslaved people of the island nation of Haiti, winning their freedom and independence after three hundred years of colonial rule, and despite the French and British sending at least *five times* as many troops against them as Britain sent against the American colonies.[5]

I am thinking of India likewise overthrowing the brutal ninety-year reign of the British, in significant part through nonviolent civil disobedience in which millions of Indians participated.

I am thinking of England in 1940, facing the stunning military advances of Nazi Germany, which was on the brink of conquering France and capturing some three hundred thousand British and allied troops trapped there, prompting many key British figures—realistically, pragmatically—to press for England to acquiesce to a Nazi Europe. Yet England chose to fight, and ultimately prevailed.

I am thinking of the millions of labor activists in the US, many of them beaten, jailed, and killed by police, state militias, and hired hands, but who in the 1930s succeeded in securing the forty-hour workweek (down from seventy), an end to child labor (more than one in twenty of the lowest-paid workers were children), a national minimum wage, worker's compensation, and the right to organize.[6]

I am thinking of, more recently, the unprecedented shift in attitudes in the US toward gay marriage, with only a quarter of people approving of it in 1996, and it not being legal in a single state, but just twenty years later 60 percent of people approving of gay marriage and it being legal in all fifty states.[7]

In all these cases, people rejected the "obvious" and "realistic" future being handed to them—and they changed the world.

The horrifying abuses of oil companies, factory farms, banks, the pharmaceutical industry, chemical companies, on and on—these are not just the way the world is or the way it needs to be. This economic system is vast and powerful and deeply entrenched. And yet it has existed for only a flash of human history, *and it is fast nearing its end*.

As the letter from the future mentioned, the sooner we embrace the changes that our times demand—many of which we are going to have to make one way or another, regardless—the more violence and destruction we can spare ourselves, our children, and countless other species.

And, as the letter notes, *we have so much to gain.* The immense crisis we are facing can be a catalyst for achieving many significant opportunities for us to live better lives.

TO REALIZE THE possibility of this moment, to rise to the challenges and the opportunities before us, involves walking three paths of change: personal, political, and cultural. The next ten pages—the remainder of this chapter and part V—are dedicated to exploring these three paths. They are equally relevant to addressing any of the broad range of abuses we have explored in the last few chapters, but we will focus on global warming, for the reasons identified in the letter from the future: the climate disaster is indeed urgent and overwhelming, and it can indeed serve as a catalyst for change that positively affects many of these other situations as well.

We will start by looking at the first path, that of personal change.

The world carbon economy is a machine. The machine is our enemy; we are the machine. The machine does not eat unless we feed it; and we feed it every day. It lives because of us; we need to stop feeding it. If we don't stop feeding the machine, Mother Earth will remind us not to feed it.[8]

—Chase Iron Eyes
(from his blog on the website, *Last Real Indians*)

A 2017 STUDY ranked 148 actions people can take to reduce their personal contribution to global warming.[9] Here is a countdown of the top five (after I have combined three of them that relate to cars):

5) Eat a plant-based diet. If cattle were their own country, they would be the third-largest emitter of greenhouse gasses (after the US and China).[10] Raising animals for meat accounts for roughly 20 percent of greenhouse gasses worldwide, due to the impact of feed production and processing, along with the belching of methane by cows, which is a far more powerful greenhouse gas than carbon dioxide.[11]

A third of Americans are eating less meat than we did just a few years ago.[12] Simply cutting a third of the beef we eat reduces our food-related emissions by almost 15 percent.[13] **Eating zero meat has the same impact as actions six through nine from the list** *combined.* (Those are: always washing clothes in cold water, rigorously recycling, hanging clothes to dry, and using LED light bulbs).

4) Buy green energy. Residential energy accounts for roughly 20 percent of greenhouse gas emissions in the US, but the share of electricity we use in our homes that comes from renewable sources is increasing rapidly, recently topping 20 percent. Many of us have the option to specify, with a single phone call, that our electricity come from renewable sources.[14] **This is arguably the first thing all of us should do because it has such a large impact and is so simple.** (It still generally costs a little more, about $12 a year for a typical home, but soon it will actually be cheaper.[15])

3) **Fly less.** There is an increasingly popular worldwide movement of people voluntarily reducing or eliminating how much they fly.[16] It got a big boost when youth activist Greta Thunberg of Sweden took a boat across the Atlantic in order to speak at the UN in New York.[17] **Airlines are responsible for about 5 percent of global warming, yet very few people actually fly: a mere 1 percent of us account for half of airline emissions, and Americans fly far more than people in other countries.**[18] That means those of us who do fly are contributing an inordinate amount to the violence and destruction related to flying. It means that we also have a tremendously oversized potential to reduce this important source of emissions. (One study found that among people who do fly, they themselves rated nearly half of their flights as not particularly important, meaning there is an opportunity for us to significantly reduce flying without even experiencing any real loss.[19])

2) **Go car-free.** This would reduce the average American's carbon footprint by 25 percent.[20] Driving more efficient cars and driving less also have an important impact, and they are more realistic for many people.[21] Going car-free, though, is a whole other level of impact because **as much as half of a car's overall contribution to global warming is from production.** So just having a car in the first place has a huge effect. (This also means the world is generally better off when we keep inefficient cars going and only replace them with new, more efficient cars when they no longer work—or, better yet, when we are ready and able to just stop driving.[22])

1) **And, number one: have fewer children.** Given the enormous environmental impact the average American has, having one fewer child in the US has about *forty times* the impact of going car-free, and has ten times the impact of the other top twelve items *combined*. **People in the nonindustrialized world consume so little per person that, to match the overall consumption of the US, their population would have to increase from roughly 6.5 billion to roughly 70 billion.**[23]

ALL OF THIS said, a single person making these changes arguably has very little impact. Nigel Savage, founder of the Jewish environmental group Hazon, writes, "Every single thing we do to help create a more sustainable world is arithmetically close to meaningless."

And yet, he notes, to do these things—"**to take any action which is individually insignificant, but which in aggregate, as part of the wider whole, is vitally necessary—is to attest, almost theologically, to our ultimate significance, and to the moral force of choosing to be part of that larger whole.**"[24]

Savage compares it to voting, something that on an individual level makes almost no difference. It is extremely rare that an election ever comes down to one vote. Yet it is only by each person individually acting that our collective power can emerge.

Looked at in reverse, when we do *not* make any of these personal changes, we contribute to the prevailing ideas perpetuating the climate disaster: that global warming isn't really that urgent, that we can get through this without having to really change very much, that other people will take care of this, that the change has to happen "out there."

But really, who else is going to lead on addressing the climate disaster if not each one of us taking a stand and saying, "This is vitally, urgently important, and we each need to make these changes now, regardless of whether the majority of people are making them, regardless of whether the people and corporations most responsible for this horror are making them"? We need to be part of creating that majority, creating a situation where those people and corporations have to make these changes.[25] **We need to be the leaders**.

What is also true is that *we have a responsibility to make these changes regardless of any larger-scale impact*. Moral philosopher Peter Singer notes that **failing to cut our emissions as individuals is like taking a bulldozer and plowing under the crops of a subsistence farmer in Africa**.[26] It's horrifying and it's wrong and we need to stop, regardless of whether it contributes to larger political change. That is our personal responsibility.

And the difference it makes matters, even if it seems small in the big picture.

One person going vegetarian will not end factory farming, but it will spare some one hundred animals a year from the horror of factory farms.[27]

One person buying a fair trade chocolate bar every week instead of non-fair trade will not end slavery in the world, but over twenty years it can mean one less child sold into slavery, and a fund for medical services for a fair trade cooperative.[28]

One person reducing their driving by half (or doubling their fuel efficiency) will not bring oil companies to their knees, but over a lifetime of driving, there will be some $40,000 less available to those companies to bribe paramilitary forces in Nigeria or pay for lobbyists to fight climate reforms.[29]

The role of an activist is not to navigate systems of oppressive power with as much integrity as possible, but rather to confront and take down those systems.[30]

—Derrick Jensen

THAT TOP-FIVE list is focused specifically on reducing our personal contributions to global warming. But the more important question is: *What are the top ways we can each make the biggest difference in ending global warming, period?*

173

Number one on that list would have to be: **support political/systemic change**. Lifestyle changes are simply "insufficient to stop this culture from killing the planet," as author/activist Derrick Jensen writes. "We need organized political resistance." We need to lobby, and protest, and campaign, and vote, and participate in direct actions.

To offer one simple but elegant example of the power of leveraged change, let's consider number four from the previous list: "Buy green energy." A study was conducted where two Swiss energy suppliers switched a large number of people's default energy source to renewables. People still had a choice, but instead of having to reach out and proactively change to renewables, it was reversed. Despite the fact that renewables at the time cost more money, the vast majority stuck with the renewables. With this one simple action, the number of people who were using green electricity in their homes and businesses jumped from 3 percent to 80 percent overnight.[31]

Bill McKibben writes:

> *I'm fully aware that we're embedded in the world that fossil fuel has made, that from the moment I wake up, almost every action I take somehow burns coal and gas and oil. I've done my best, at my house, to curtail it: we've got solar electricity, and solar hot water, and my new car runs on electricity—I can plug it into the roof and thus into the sun. But I try not to confuse myself into thinking that's helping all that much: it took energy to make the car, and to make everything else that streams into my life. I'm still using far more than any responsible share of the world's vital stuff.*
>
> *And in a sense that's the point. If those of us who are trying really hard are still fully enmeshed in the fossil fuel system, it makes it even clearer that what needs to change are not individuals but precisely that system. We simply can't move fast enough, one by one, to make any real difference in how the atmosphere comes out. Here's the math, obviously imprecise: maybe 10 percent of the population cares enough to make strenuous efforts to change—maybe 15 percent. If they all do all they can, in their homes and offices and so forth, then, well … nothing much shifts. The trajectory of our climate change horror stays the same.*
>
> *But if 10 percent of people, once they've changed their light bulbs, work all-out to change the system? That's enough. That's more than enough. It would be enough to match the power of the fossil fuel industry, enough to convince our legislators to put a price on carbon.*[32]

Our strategy cannot be to wait for most people (or, God forbid, most corporations) to voluntarily make individual changes, or to simply make changes in our own lives without some broader engagement as well. The power dynamics simply won't allow it. "Worrying about your carbon footprint is exactly what big oil wants you to do," writes Auden Schendler in a *New York Times* op-ed.[33] Lacking forceful political action, the fossil fuel companies are powerful enough to just keep doing what they've been doing—to hell with the death and destruction raining down around us.

We have to use our collective power to literally make them stop, to not leave them any choice.

We need to act collectively in order to:

- End subsidies to the fossil fuel industry (currently $20 billion a year).[34]
- End drilling and mining on federal lands (which corporations pay cents on the dollar to use, with the public picking up the environmental and health costs).[35]
- Ban new fossil fuel projects and fund a rapid transition to renewable energy (solar panels pay back the energy required to make them in a mere four years, after that it's all free energy).[36]
- Tax carbon: have people pay a tax on all products based on how much that product contributes to global warming, thereby incentivizing people to reduce their contribution to global warming. (More than 3,600 leading economists agree that this is "the most cost-effective lever to reduce carbon emissions at the scale and speed that is necessary."[37])
- Allow secondary forests to grow back (and absorb one-third of the excess carbon we're producing).[38]
- End subsidies to the meat and dairy industry (tens of billions of dollars a year), create a moratorium on new factory farms, and legislate significantly improved standards of living for farmed animals.[39]
- Limit big money and corporate involvement in political campaigns.
- Implement four-day workweeks (reducing emissions by 30 percent).[40]

Derrick Jensen writes, "Another 120 species went extinct today; they were my kin. I am not going to sit back and wait for every last piece of this living world to be dismembered. I'm going to fight like hell for those kin who remain—and I want everyone who cares to join me."[41]

Responding to environmental destruction requires not only the overcoming of corporate evildoers but "self-overcoming." ... A more adequate response to our true problems requires that we cease to be a society that believes that wealth is the accumulation of money (no matter how much of it we're planning on "giving back" to nature), and begin to be a society that understands that "there is no wealth but life."[42]

—Curtis White

THE THIRD CRITICAL path we must walk relates to cultural change. A brief tale of coal and Model Ts and a so-called "ludicrous idea" offers an important introduction to this.

In 1850s Britain, coal was being consumed at a massive rate, rapidly eating away at the country's reserves. Some believed that more efficient technology would reduce consumption, allowing the reserves to last longer. Economist William Jevons observed that significant advances in the efficiency of the steam engine had actually led to *increased* consumption of coal, and he predicted the trend would continue. "It is a confusion of ideas to suppose that the economical use of fuel is equivalent to diminished consumption. The very contrary is the truth."[43]

It turns out he was right, and it turns out the same is true for energy consumption in general, as well as for the consumption of many kinds of natural resources.[44] As paradoxical as that may seem, it's fairly straightforward: as we achieve greater efficiency in producing certain goods or services, their price goes down; at a lower price, demand tends to go up— more people can afford it and they can afford more of it.

Now, if that were the only factor at play, research suggests the gains in efficiency would still generally outweigh the increased consumption. And sometimes they actually do. However, increased efficiency in one area tends to ripple through the economy. For example, when motors become more efficient, everything that involves a motor becomes cheaper—which is really important, because motors are fundamental to most production and transportation. People not only can afford to buy more products with motors and use them more, but *they can also use the money they save to buy totally unrelated products*, consuming further energy and raw materials. So the savings ripple through the economy, actually leading to faster economic growth and greater overall fuel consumption.[45]

This is known as the *rebound effect*.[46] **In short, we take the gains in efficiency and instead of saying, "Hey, life is good, we have enough now, let's use these gains to protect the earth and give ourselves more time for the things that really matter," we say, "Hey, look at this other stuff I can have now," and we feed those gains right back into the machine.**

Not everyone is convinced that efficiencies really work this way.[47] A prominent efficiency advocate, Amory Lovins, argues that if that were the case then "we should mandate inefficient equipment to save energy."[48] Which would obviously be ludicrous.

Or not.

As David Owen points out in *The New Yorker*, "If the only motor vehicle available today were a 1920 Model T, how many miles do you think you'd drive each year, and how far do you think you'd live from where you work?"[49] How much energy and how many raw materials would be invested in roads? How much stuff would we have trucked all over the country?

Instead, between 1960 and 2006, while the US population increased 35 percent, the "vehicle miles" we traveled increased more than 400 percent, the use of motor fuel increased 300 percent, and we have fifty million more registered vehicles than licensed drivers.[50] This is not an isolated example.

- The energy efficiency of the US economy doubled between 1984 and 2005, yet electricity consumption grew 66 percent (far outpacing population growth).[51]

- The average person in the US consumes twice as much today, in terms of energy and raw materials, as we did fifty years ago, and our consumption continues to rise.[52]

- Global consumption has likewise doubled in the last thirty years and continues to grow at 1.5 percent a year, again far outpacing population growth.[53]

- Per-person energy consumption worldwide more than doubled between 1910 and 2010.[54]

What if somewhere along the line we had, in fact, decided enough is enough, and used the gains in efficiency to give the planet and ourselves a break? If we in the US had done that in 2000, then as of 2020, even after factoring in the increase in population, **we could have been using approximately 40 percent less energy and working eighteen-hour weeks.**[55]

And that is based simply on the natural progression of efficiency and productivity. We could do far better than that by intentionally shifting to more efficient technologies that are already available. For example, between 2010 and 2020 the price of solar panels dropped 90 percent and the price of a wind farm dropped 71 percent, making solar and wind the cheapest source of electricity throughout most of the world.[56]

And still … the actual impact of these strategies depends entirely on what we do with them. All of that incredible potential could actually make zero difference if we just turn around and use the time, money, energy, and raw materials we save to simply produce and consume more. **Indeed,** *it could make things worse*, **accelerating the destruction.**

(Over a sixteen-year period the Dow Chemical Company invested $1 billion in energy efficiency measures, which resulted in over $9 billion in savings.[57] Perhaps I'm not the only one who shudders to imagine Dow Chemical having an extra $9 billion to throw around in the world. [I'll save you time looking it up: no, they did not use that money to provide medical care to the people of Bhopal or to clean the poisons out of the soil and water there.] Perhaps I am also not the only one who is confident that Dow's savings didn't actually do anything to help the environment either. As if the emissions they spared the planet through those efficiencies didn't just get blown out other pipes in other places through the increased production and consumption that they and their directors and lobbyists and advertisers and shareholders put those $9 billion dollars toward.)

This is all wrong. I shouldn't be up here [addressing world leaders at the UN]. I should be back in school on the other side of the ocean. Yet you all come to us young people for hope. How dare you!

You have stolen my dreams and my childhood with your empty words. And yet I'm one of the lucky ones. People are suffering. People are dying. Entire ecosystems are collapsing. We are in the beginning of a mass extinction, and all you can talk about is money and fairy tales of eternal economic growth. How dare you! … How dare you pretend that this can be solved with just "business as usual" and some technical solutions?[58]

—Greta Thunberg

AS THE STORY of the rebound effect illustrates, political changes alone are, like personal changes, inadequate to the challenge before us. We must commit to a deeper cultural shift. Political scientist Jan Dutkiewicz, writes very pointedly:

> *All too frequently, activists, politicians, and scientists reduce the all-consuming crisis of global warming to a question of greenhouse gas emissions: what drives them up, and how best to bring them down. … [This] climate pragmatism offers a frustratingly narrow vision of reform. Rather than challenging the instrumental view of nature that led us to this pre-apocalyptic moment, it asks us to imagine a world much the same as our current one, minus the climate change.*[59]

This represents a horrifying failure of vision—and of compassion and love—both for the plants and creatures who share the planet with us, but also for ourselves, spinning out our lives in this shallow and cruel system. It not only acquiesces to "the capitalist mindset that has brought us here," as Dutkiewicz puts it, *it actually reinforces it*. It accepts the premises of the current system about what progress and a good life look like, and it turns a blind eye to the vast death and destruction this system is causing every day entirely independent of the climate disaster. **It conveys the message that "Everything is fine over here, just a little climate problem we need to get under control and then everything will be alright."**

This "climate pragmatism" is actually a failure of pragmatism as well, because without attending to the underlying cultural norms fueling climate change and so much other violence and destruction, *it actually offers little chance of making any significant or sustained change at the political level.* Author and environmentalist Wendell Barry notes that specific goals like soil conservation, wilderness preservation, sustainable agriculture, community health, and so on, simply "cannot be achieved alone."

[These efforts] are too specialized, they are not comprehensive enough, they are not radical enough, they virtually predict their own failure by implying that we can remedy or control effects while leaving causes in place. Ultimately, I think, they are insincere; they propose that the trouble is caused by other people; they would like to change policy but not behavior.[60]

Of course, it would be ridiculous to wait to engage in political change until some day when we arbitrarily decide that our society has undergone a "sufficient" cultural shift. It would also be to miss the role that political change itself plays in shaping culture, the way politicians and policies shape norms and affect how we think about what is important and right.[61]

The same is true for personal changes—they also play an important role in shifting culture. Climatologist Michael Mann (who has made many important contributions to addressing climate change) unwittingly offers a powerful example of this. He has written that "a single scientist, or even hundreds of scientists, choosing to never fly again is not going to change the system."[62] Which is, again, obviously true in an immediate arithmetical sense. And yet my own experience is that *when I read that sentence the mere idea of what he described changed me!*

I imagined hundreds of leading scientists making a united public commitment to not fly, and I felt emboldened and inspired. I felt our culture and public perception shift right beneath my feet, and I could imagine the shudder of urgency that would ripple through the collective consciousness. "Oh, right, this is real, these things do matter, change is possible and it's already happening. Let's do this!" **The idea for me was so powerful that that was the day I decided I am going to try to never fly again.**

Still, personal changes and political fixes can only accomplish so much if they are not part of a broad and fundamental cultural shift. The climate crisis is just the latest and most epic manifestation of the underlying violence and destructiveness of this culture. As Derrick Jensen writes, **"So long as we find it not only acceptable but right and just to convert the lives of others and the life-support system of the entire planet itself into fodder for us, there is little hope for life on the planet."**[63]

Fleshing that out a bit, we could add:

So long as production and consumption remain the primary measures of our worth and purpose;

So long as we feel utterly dependent on them for our well-being and happiness, for approval, and for keeping our sense of isolation, inadequacy, and fear at bay;

So long as our default orientation is toward bigger, better, newer, instead of abundance and gratitude;

And so long as we continue to be so epically detached from our hearts, and from the wonder of the world, and from the miracle of ourselves;

Then we will continue to feed this violent and destructive machine. **Regardless of any changes that are made, we will constantly rearrange ourselves and the pieces of the machine to keep grinding forward to meet what we falsely perceive as essential needs.**

This is the critical challenge of our time.

It is also the critical opportunity of our time.

> *The most remarkable feature of this historical moment is not that we are on the way to destroying our world—we've actually been on the way for quite a while. It is that we are starting to wake up, as from a millennia-long sleep, to a whole new relationship to our world, ourselves, and each other. This is the great and necessary adventure of our time.*[64]

—Joanna Macy

THE SIGNIFICANT CHANGES we are being called to make can evoke a sense of sacrifice or loss, or anxiety or fear. For those of us who have access to the many things that must be scaled back or left off entirely—for those of us who have a choice about taking flights, or buying large cars or homes, or eating meat, and so forth (which is many of us in the US but very few of us worldwide)—we are being called to release things we are used to, things we may even feel dependent on for our well-being and happiness, for the kinds of lifestyles we expect for ourselves or that others expect of us.

Yet what lies before us is an invitation to actually live better lives, to experience greater freedom and joy and purpose.

- We know that money accounts for only 2 to 4 percent of our happiness.
- We know that "materialism is toxic for happiness," and that there is actually no amount of money that can make up for the happiness we lose when we are materialistic.
- We know that the more money and possessions we have, the less we are able to "savor positive emotions and experiences," and "the more fleeting that joy, contentment, and gratitude" are.
- We know that the more materialistic we are, the less likely we are to actualize ourselves, the lower quality our relationships tend to be, and the more fragile our self-esteem.
- And we know that we in the US suffer extreme rates of depression, suicide, loneliness, and isolation.

This is what we are being asked to give up.

These are the traps we are being invited to step out of.

Saint Teresa of Avila wrote, **"Thank God for the things that I do not own."**[65]

Every time we shed one of the things in our lives that does nothing for our happiness, health, or security, we shed one more layer of this gauze, this deception. We shed one more distraction that pulls our time and attention away from the vast abundance we are held in.

These calls for us to change are a gift. Each time you hear a call to release something, to limit your consumption, you can use that as an opportunity to shift your orientation from "the things you do not have" and "what you have to go without" to *so much that you do have.* (Again, Thich Nhat Hanh: "We are surrounded by miracles, but we have to recognize them; otherwise there is no life."[66])

It is natural that we feel hesitant or anxious in the face of the changes we must make. But at a deeper level this shift that our society is undergoing is actually an immense blossoming. It is again what Joanna Macy refers to so beautifully as "the Great Turning."[67] **It is a profound blessing that so much of what is required of us is precisely what best serves us, what most deeply nourishes happiness, health, connection, and love.**

These calls to reduce our consumption aren't a guilt trip; *they're an intimate, heartfelt invitation from the universe, and a reminder from your own deepest self.*

They aren't about sacrifice or loss; they're about liberation.

As the letter from the future put it:

> *Like seeing the night sky in the desert, fantastical and alive with stars,*
> *after seeing it only through the haze of city lights,*
> *or like seeing trees in late afternoon sunshine, shimmering and radiant,*
> *after endless days of clouds and rain:*
> *there is a sense that we are on a journey of returning to wonder and reverence,*
> *to love and belonging,*
> *to the world and to ourselves.*

Part VI

Coming Home: Abundance, Belonging, and Love

*Don't remind the world
that it is sick and troubled.
Remind it that it is
beautiful and free.*[1]

—Mooji

Chapter 25

Another World Is Possible

The logic of the new science [of happiness] is breathtaking.
If it is right, it requires us to rethink some of our most basic assumptions
about how we live, how we work, and what we are trying to achieve.
In short, the science of happiness may provide us with a new definition
of what we mean by human progress.[2]

—Mark Easton

HUMANS FIRST LEFT Africa in significant numbers some eighty thousand years ago.[3] The first of them arrived in North America about fifteen thousand years ago, having crossed over from Asia.[4] Approximately four hundred years ago, the first people arrived in North America from the other direction, primarily Europe, and began to settle on the eastern shore. Their population grew at a furious pace, as did the area they settled and roamed. In an abrupt, apocalyptic period of just a few hundred years, the vast majority of Native people were killed off as a result of warfare, famine, and epidemics. Of those who survived, most were driven from their homelands by the newcomers, under extreme deprivation and threat of death. Hundreds of thousands were enslaved.[5]

There was an assumption by many of the Europeans who came to this land, and their descendants, that what was done to the Native Americans was not only acceptable, but good, inevitable even, "the design of Providence," as Benjamin Franklin suggested.[6] President Andrew Jackson stated, "Established in the midst of another and a superior race, and without appreciating the causes of their inferiority or seeking to control them, [the Indian tribes] must necessarily yield to the force of circumstances and ere long disappear." President Theodore Roosevelt concurred, "The settler and pioneer have at bottom had justice on their side; this great continent could not have been kept as nothing but a game preserve for squalid savages."[7]

But there was something happening in their midst that defied this vain and brutal logic, and even flipped it on its head. Benjamin Franklin wrote in 1753:

When an Indian child has been brought up among us, taught our language and habituated to our customs … if he goes to see his relations and make one Indian ramble with them, there is no persuading him ever to return.

This was not the case with white children who had been raised among Indians and later returned to white society, or even with many white people captured as adults, who would "take the first good opportunity of escaping again into the woodlands." Sometimes the Native people would try to forcibly return white people in prisoner swaps. They often refused, or had to be tied up and dragged. Sometimes they managed to escape and return anyway.

Journalist Sebastian Junger writes, "As early as 1612, Spanish authorities noted in amazement that forty or fifty Virginians had married into Indian tribes." These settlers had only been there a few years, having previously lived their entire lives in England. "These were not rough frontiersmen who were sneaking off to join the savages; these were the sons and daughters of Europe." Some of the colonies implemented severe penalties for white people who sought to live with Indians.

More than 150 years later Hector de Crèvecoeur observed, "Thousands of Europeans are Indians, and we have no examples of even one of those aborigines having from choice become European."[8]

I first read about this history several months ago in Sebastian Junger's excellent book "Tribe." It has haunted me since. It raises the possibility that our culture is built on some fundamental error about what makes people happy and fulfilled.[9]

—David Brooks

PEOPLE HAVE A tendency to see themselves as living at the pinnacle of history. Evolutionary biologist Richard Dawkins refers to this as the "vanity of the present," the idea of "the past as aimed at our own time, as though the characters in history's play had nothing better to do with their lives than foreshadow us."[10]

In the same vein, Cornell historian Morris Bishop pointedly wrote:

The Middle Ages is an unfortunate term. It was not invented until the age was long past. The dwellers in the Middle Ages would not have recognized it. They did not know that they were living in the middle; they thought, quite rightly, that they were time's latest achievement. … As our modern age ceases to be modern and becomes an episode of history, our times may be classified as the Late Middle Ages, for while we say time marches forward, all things in time move backward toward the middle and eventually to the beginnings of history. We are too vain, we think we are the summit of history.[11]

Whether we think of our times as the pinnacle of history, or the Late Middle Ages, or something entirely different, there is often a tendency to imagine that we are somehow more advanced or better off than the people who lived before us. This can be particularly tempting here in the United States where we place a premium on looking forward, and abandoning or even disdaining the old, and where we have brought about unprecedented change in terms of productivity, technology, and material wealth. Often when people in the US refer to "progress," they actually mean quite simply "economic progress," and by that they often mean quite simply "more."

Jeremy Caradonna, writing for *The Atlantic*, notes, "This narrative remains today an ingrained operating principle that propels us in a seemingly unstoppable way toward more growth and more technology, because the assumption is that these things are ultimately beneficial for humanity."[12]

As we have noted, this narrative minimizes the overwhelming violence and destruction our economic growth has entailed, and it hides from us the ways that even those of us who are supposedly the greatest beneficiaries of that growth are in many ways actually worse off for it—not to mention those who quite clearly are not the beneficiaries.

We began chapter one by noting that as late as the 1940s most Americans lived in conditions that few Americans today would consider acceptable: A third of homes didn't have indoor toilets or running water, more than half lacked central heating, almost none had air conditioning, and so on. And yet they were generally happier than Americans are today.

More dramatic still, then, is the comparison with Native Americans, who lived with extremely few material possessions, far fewer than the average 1940s American, and yet whose lives, if we are to judge by the significant one-way migration between them, were markedly better than those of the more recent arrivals.

These facts suggest an intriguing possibility, one that many Americans today might consider simply ludicrous because of our extreme "vanity of the present": that **most people throughout human history have been better off than we in the US are today.**

Consider it: happiness in the US has been in a constant decline for at least eighty-some years. And the Native Americans seem to have been markedly happier than the European Americans prior to that. And the Native Americans lived very similarly to the way most humans have lived for most of human history.

This isn't actually a new idea. Anthropologist Marshall Sahlins published a landmark essay in 1968 in which he argued that our "foraging" ancestors were, in fact, "The Original Affluent Society" because of the relative ease with which they satisfied their material desires and how much leisure time they enjoyed.[13] (They are often referred to as "hunter-gatherers," but they didn't actually get much food from hunting, so the term "foragers" has been suggested as more accurate.[14])

As we know, leisure time is a consistent predictor of well-being, and Sahlins and others have calculated that our foraging ancestors spent a mere three to five hours a day securing food, and those were fairly leisurely hours as well, with frequent pauses, socializing, days on and off.[15] Many foragers today, the Kalahari Bushmen and the Hadza of Tanzania among them, still spend that much time or less collecting food.[16] Junger writes about a European American woman who was captured by the Seneca at age fifteen and later hid from a search party trying to "rescue" her. She wrote, "We could work as leisurely as we pleased. No people can live more happily than the Indians did in times of peace … their lives were a continual round of pleasures."[17]

By contrast, Americans today spend an average of nearly seven and a half hours working and commuting on weekdays, and for those of us who work full-time the average is ten hours a day.[18] And, as we noted earlier, that time is, for many of us, definitely *not* a continual round of pleasures.

More than five decades on, allowing for certain minor revisions, Sahlins's core ideas have been generally supported by subsequent research and have been largely embraced by anthropologists.[19] **But the affluence of our ancestors goes far beyond the leisure time they enjoyed. It seems to span every one of the factors we have noted are central to human well-being.**

Consider our **social lives**. Junger writes that the Indigenous people did "almost everything in the company of others," that "they would have almost never been alone," something that is true in most foraging communities. He notes, by contrast, that the economic system and values of the new Americans led many of them to accumulate personal property and make "more and more individualistic choices," which diminished their reliance on each other and their presence in each other's lives. **In a sense, to choose between living with the Native Americans and the European Americans was to choose between people and possessions, and people consistently won out.**[20] Americans today socialize with friends and family for an average of only forty-five minutes a day.[21]

Some of the new Americans who chose to live with the Native Americans also noted that women had greater freedom and respect in the Native communities, that attitudes toward sex were more relaxed and positive, and that the communities were far more egalitarian.[22]

Consider also **physical activity**. Foraging lifestyles tend to be very physically active, and it has been documented that when people transition away from foraging they lose aerobic fitness and muscle strength.[23] A full 80 percent of Americans today do not get the recommended minimum of two and a half hours a week of moderate exercise.[24]

Consider **sleep**. Prior to electrification, Americans slept an average of nine hours a night, and evidence suggests that in the US and certain other parts of the world, we have traditionally slept eight hours a night in two blocks, with two to three hours of relaxing in between, for a total of ten or eleven hours of sleep and relaxation. There are Indigenous communities that still sleep that way today, while others sleep multiple times throughout the day and night.[25] Americans today, on average, sleep less than the minimum seven hours a night that are recommended, and 40 percent of us regularly sleep six hours or less.[26]

Consider also that **Americans today live in a state of chronic stress**, "extreme stress" for a quarter of us. The top stressors? Money, work, and the economy, followed by job stability and housing costs, most of which were entirely unknown to our ancestors.[27]

But didn't our ancestors live with their own chronic stress? Didn't they live in constant fear of diseases and food shortages and predators? The seventeenth-century philosopher Thomas Hobbes famously referred to their lives as "solitary, poor, nasty, brutish, and short."

He didn't actually have any evidence for this, though; he thought it was just obvious, or "manifest." That first one on his list, "solitary," gives you a clue as to how little Hobbes actually knew about the lives of our ancestors, since the opposite is much closer to the truth.[28] He was just as wrong about the "short" part. Going back at least thirty thousand years, our foraging ancestors who survived childhood could expect to live over seventy years, not far from the life expectancy of seventy-nine years in the US today, and beyond what it is throughout much of the world today.[29]

That note about "surviving childhood" is important, though. There was, in fact, vastly higher infant and child mortality. The mortality rate was perhaps as high as 40 percent before age fifteen, compared to 4.3 percent by age five worldwide today, and less than 1 percent in the US.[30] So we are far better off in that regard.

Other than that, though, not so much. Considering health more broadly, researchers did an extensive study of the skeletons of over twelve thousand people in North and South America spanning seven thousand years ago to the early 1900s. They found that of all the Native Americans, African Americans, and European Americans they studied, **the healthiest of all were generally the Native people living prior to 1,500 years ago.** The expansion of agriculture and urban centers after that led to less varied diets and promoted the spread of disease and social inequality, all of which led to marked declines in health.[31] Similar declines have been documented following the rise of agriculture and urbanism in other parts of the world, as well.[32] A notable exception to their findings were the "equestrian nomads of the Great Plains of North America" who, in the 1800s, "were not fenced in to farms or cities," and not yet fenced into resettlement camps. They "seemed to enjoy excellent health, near the top of the index."[33]

Our foraging ancestors also seem to have enjoyed much greater food security than many of us today. Pulitzer Prize-winning author Jared Diamond writes that it is "almost inconceivable that Bushmen," modern-day foragers in southern Africa, "who eat 75 or so wild plants, could die of starvation the way hundreds of thousands of Irish farmers and their families did during the potato famine of the 1840s."[34] Likewise, the diet of the Hadza in Tanzania "remains even today more stable and varied than that of most of the world's citizens," even though they have exclusive rights to only a fraction of their original homeland and live on briny soil with scarce fresh water. In fact, they "have no known history of famine; rather, there is evidence of people from a farming group coming to live with them during a time of crop failure!"[35]

A Kalahari Bushman, when asked why his tribe hadn't taken to farming, famously replied, "Why should we, when there are so many mongongo nuts in the world?"[36] By contrast, each day some forty million Americans, 10 percent of us, go hungry.[37]

Humans have dragged a body with a long hominid history into an overfed, malnourished, sedentary, sunlight-deficient, sleep-deprived, competitive, inequitable, and socially-isolating environment with dire consequences.[38]

—Journal of Affective Disorders

IT DOESN'T SERVE anyone to romanticize the lives of Indigenous people, either our ancestors or Indigenous people today. Pick any group of Indigenous (or other) people and you will find things that were/are less than ideal. The tremendous violence that many Native Americans practiced against each other in warfare (well before the Europeans arrived), for example, was far from idyllic.[39] As were the famous human sacrifices of the Aztecs and Maya, which were also practiced by the people of the massive city of Cahokia in what is today the American Midwest.[40] Not as headline grabbing but still poignant: a group of modern-day Yanomami people were forced to leave their foraging lifestyle in the Amazon to live in a town in Brazil. Several of them mentioned appreciating that their feet no longer hurt all the time, as they had when they lived barefoot in the rainforest. (Unfortunately, their physical health declined markedly, though.)[41]

What we find, then, is that, with the important exception of infant mortality and the slight exception of life expectancy, in all of these areas that are most vital to human well-being, our ancestors weren't just slightly better off than we are today, they were *significantly* better off. (Foragers also "consume less energy per capita than any other group of human beings."[42]) We are talking about our ancestors from (at least) the dawn of *Homo sapiens* over 300,000 years ago up until we began to transition to agriculture and become more urban—10,000 years ago for some in the Middle East, 1,500 years ago for some in the Americas, right through to the present for many Indigenous people today.[43]

And we haven't even touched on what might be the most fundamental difference between our ancestors and us, namely *the way we experience the world and ourselves.* As we know from part III, the subjective, internal factors related to well-being are often far more important than the objective facts of our lives.

It's about how we understand our relationship with the cliffs and ferns and flowers and muskrats. It's about our place amidst the stars and mosses and water buffalo and butterflies. It's about our connection with the world and each other and our own hearts. **It** is ultimately a question of belonging. And in this sense yet again, the way we are living is in many ways deeply impoverished, cruel even.

The remainder of our journey is in large part dedicated to reclaiming the profound, inherent, and immutable belonging of every one of us, of you.

For now, what is important though, is not whether we can definitively state that most people throughout time have been happier than Americans today, or even that it seems quite likely. It is enough to acknowledge that it is merely *possible*, that there is actually a robust and valid conversation to be had on the matter. **What is important is that we loosen the grip of the "vanity of the present" and the common assumption that ours is the way, that we are a pinnacle of human evolution, and that we just need to keep plowing forward in this same general direction, despite so many obvious and devastating failures.**

It is not to suggest that we should just return to a foraging lifestyle, something that would be impossible for most of us anyway. Nor is it that we should abandon everything we and our non-foraging ancestors have learned and created, much of which is wonderful and tremendously helpful. It is simply to acknowledge that in many profound ways *we are not being served by the current state of things—and that it does not have to be this way.*

The isolation, lack of time, chronic stress, and depression that characterize so many of our lives are an extreme aberration in the history of humanity, one that most people who have come before us—and hopefully most who come after us—would not recognize or understand.

Another world is possible. Many other worlds are possible, worlds that are much more deeply aligned with our real needs, that nourish us in the ways that really matter, worlds that are ultimately truer to the sacredness of each of us and our vital place in the universe.

Chapter 26

From Owning to Belonging

*Maybe we've all been banished to lonely corners by our obsession
with private property. We've accepted banishment even from ourselves
when we spend our beautiful, utterly singular lives on making more money,
to buy more things that feed but never satisfy ... believing that belongings
will fill our hunger, when it is belonging we crave.*[1]

—Robin Wall Kimmerer

OWNERSHIP IS A concept that has profound implications. To a large degree it defines
how we think about and treat other people, creatures, and the natural world. It also
affects how we feel about the world. It is powerfully related to our sense of belonging—or
a lack thereof.

As it is generally understood in the US and throughout much of the world today,
ownership is about control—who gets to decide what is done with something. As the
influential English judge Sir William Blackstone put it in the 1700s, it is "that sole and
despotic dominion which one man claims and exercises over the external things of the
world, in total exclusion of the right of any other individual in the universe."[2]

Which sounds pretty definitive, not to mention epic. Ownership in real life is, of
course, more complicated than that (which Blackstone did also acknowledge). Our "sole
and despotic dominion" can dissolve in an instant: an object can be engulfed in flames, a
company can go under, or things can just be taken from us, by another person, a
corporation, an army, the government. We also can't just do *anything* we want with the
things we own; there are laws and customs intended to protect us from people doing things
like driving their car autobahn-style through a park, or running a chainsaw in a
neighborhood at three in the morning, or building a casino as an add-on to their garage.

Still, our basic ideas of ownership aren't so far from what Blackstone described three
hundred years ago. Most things are owned by somebody, and whoever owns them gets to
largely do with them what they want, and the rest of us don't. We have vast and deeply set

legal and economic systems that enshrine this approach to ownership. These systems can, in fact, create the impression that this approach is the *only* way to really do things, that it is the natural and obvious way of relating to the world and the things in it.

But what is ownership, really? If we step outside our particular cultural and legal context? If we consider it in the bigger picture—*the really bigger picture*? What is ownership within the deeper reality of the unfolding of our lives and the universe?

To help us in thinking about this, I offer a rather simple and immediate example: my shirt. By the norms of this society, I own this shirt that I am wearing as much as anyone can own anything. But what does that really mean? What is the relationship between this shirt and me in the larger story of existence?

That story begins some fourteen billion years ago when all matter came into being. It continues as many of the earliest atoms were transformed into heavier elements in the hearts of stars, which lived out their lives and then cast them into space. After about nine and a half billion years, the atoms that make up my shirt and me were part of a very exclusive club of elements that were in close enough proximity that they coalesced to form this planet (or to later join the earth as meteors).

Four billion years ago some of those elements became organized in patterns that we identify as living things. Those patterns have diversified and evolved until a mere three hundred thousand years ago one of them took the form that we call human beings. Then just seven thousand years ago another one of the patterns took the form we call cotton. Some fifty years ago some of those elements were drawn together in the human pattern to create me (with atoms being continually swapped in and out ever since). Then just a few years ago another bunch of elements were drawn together in the cotton pattern, and then into the pattern of a shirt (extra-large, plaid). Soon after that I gave someone something we call money, that person let me walk out with the shirt, and here we are—I "own" it.

But only for an instant.

In just the blink of an eye our paths will separate again. Soon the elements in this shirt will dissolve out of the forms of shirt and cotton, and the elements in me will dissolve out of the form of human. In fact, it won't be very long in the big picture before the very patterns of cotton and human will cease to exist entirely, whether they evolve into other patterns or just come to an end. Such is the nature of time, evolution, extinction, impermanence. The elements will go on to be part of many other patterns over the lifetime of the earth, and beyond. This shirt and I, though, in these particular forms, are not long for this world or each other.

Before our paths separate, though, *for this brief wondrous moment*, this shirt and I are in an intimate and special relationship. In all of time and space, we are here together, now. Vast interstellar journeys brought us together, intimate companions in this tiny corner of the universe, sharing our days and moving through the world together, each of us complementing the other's form, intermingling our identities and experiences, transferring and absorbing elements from each other, each helping sustain the other.

We are, my shirt and I, both evidence of the wonder of creation. We are two miracles in the presence of each other. We exist in a very intimate and mutual companionship.

Meanwhile, unbeknownst to the shirt or the cotton or the elements in the cotton: I "own" them. Thanks to the human imagination and a blend of expediency and hubris, they are my property.

It is an important concept, ownership. I appreciate the nuance it lends to my relationship with this shirt and the other things I "own," knowing that they will generally be around for me to make use of in the foreseeable future, and not go walking off with someone else. Practically speaking, there is a lot to be said for private property, at least in degrees. But, at the same time, this way of thinking about ownership, this one-way relationship of power and control, is also rather silly—it is so hollow and deceptive, so devoid of the awe and respect and mutuality that are the true hallmarks of our relationships with the things we "own."

And if the idea of ownership is so limited and misleading when it comes to a shirt, how much more deceptive and meager is it when we consider not a shirt but a tree, or a field, or a pond, or other living beings?

FOR MOST OF time, nothing has "owned" anything else. This was true for billions of years before humans emerged, and it has been true for most of the three hundred thousand years since then as well.

There is a territoriality that exists among many animals, and at some point we humans extended this sense to include specific objects we had made. Archeologist Steven Mithen has noted, "You can see notches and marks on various items" from around forty thousand years ago that clearly exhibit some notion of connection or ownership.[3]

These ideas do not seem to have extended beyond certain personal objects though. Political economist Henry George has noted, "Wherever we can trace the early history of society … all members of the community had equal rights to the use and enjoyment of the land of the community."[4] This is a concept that still holds true for many around the world, notably Indigenous societies.[5]

The idea of owning the earth and the legal norms related to land ownership as we know them today were born in the crucible of sixteenth-century Europe and the rise of the "nation-state." Then, in an abrupt and violent flash, "Between 1520 and 1900, colonial powers put billions of acres in private hands," writes historian Louis Warren. "They achieved this mostly by wresting lands from indigenous owners at great cost in blood (most of it indigenous), and transferring titles to colonists who operated in a market economy."[6]

Classical economist Ludwig von Mises pointed out that all private property came into existence "when people with their own power and by their own authority appropriated to themselves what had previously not been anybody's property." He noted, "Virtually every owner is the direct or indirect legal successor of people who acquired ownership either by arbitrary appropriation of ownerless things or by violent spoilation of their predecessor."[7]

Fast forward to today, and the range of originally "ownerless things" that people have appropriated is vast. Indeed, there are few if any things that are not owned at least somewhere in the world: water, seeds, soil, images, sounds, ideas, even other people and creatures. For each of these there are people somewhere in the world who claim the authority to control them, and this "ownership" is generally viewed, at least by those immediately around them, as legitimate, beneficial, natural even.

In their essay "The Myth of Ownership," Liam Murphy and Thomas Nagle, professors of law and philosophy, write, "We are all born into an elaborately structured legal system governing the acquisition, exchange, and transmission of property rights, and ownership comes to seem the most natural thing in the world." Yet it is, in truth, immensely *un*natural. It is entirely a human invention, a shared imagining, and it is subject to an endless range of varieties and preferences and perceptions.[8]

Classical philosopher John Stuart Mill noted:

> *The Distribution of Wealth depends on the laws and customs of society. The rules by which it is determined are what the opinions and feelings of the ruling portion of the community make them, and are very different in different ages and countries; and might be still more different, if [people] so chose.*[9]

IN 2017, THE third-largest river in New Zealand was granted the status of a living being and given the same legal rights as a person. This was not some environmentalist strategy for how to get the river more rigorous legal protection, somehow convincing people to just extend the legal rights of people to a river. This decision was nothing less than the legal recognition and institutionalization of the worldview of the Whanganui tribe of Maori people.

The Whanganui perceive the world and nature as sacred entities in vital and vibrant relationship with humans, not with humans sitting on top of all creation, but humans being in mutual relationship with all creation, "equal to the mountains, the rivers and the sea." Gerrard Albert, the lead negotiator for the Whanganui, noted:

> *We have fought to find an approximation in law so that all others can understand that from our perspective, treating the river as a living entity is the correct way to approach it, as in indivisible whole, instead of the traditional model for the last 100 years of treating it from a perspective of ownership and management.*
>
> *We can trace our genealogy to the origins of the universe. And therefore, rather than us being masters of the natural world, we are part of it. We want to live*

like that as our starting point. And that is not an anti-development, or anti-economic use of the river, but to begin with the view that it is a living being, and then consider its future from that central belief.

We consider the river an ancestor and always have.

This legal decision reflected an unprecedented 140-year long battle in court, and "hundreds of tribal representatives wept with joy when their bid to have their kin awarded legal status as a living entity was passed into law."[10]

For some in the US it may seem strange to grant a river the legal status of a person. Of course, for many Maori people it may seem strange to grant corporations the legal status of people, something we have done in the US to some degree since the 1800s.[11] The status accorded to these two entities offers a window onto the different worldviews of these two people. Americans do not actually experience corporations as living beings, as the Whanganui do the river, but we have granted them the highest legal status and privilege available—acknowledged them as one of us. For the Whanganui, according this status to the river reflects the essence of mutuality and reverence they experience in relationship with the natural world. For Americans, according this status to corporations reflects the essence of utility and separateness we experience in relation to the natural world, and the primacy we give to profit, production, and consumption.

The difference between these two worldviews has been described by some as the difference between owning as *property* and owning as *belonging*. Owning as property invokes a sense of humans as distinct from "the external things of the world" and a sense of human superiority and "dominion" over the rest of creation. Owning as belonging involves a sense of inseparability of humans and creation, a sense that nature, the world, the universe, are sacred, living, that they are ancestors, the source of life and people.[12]

The oneness of people and the land is embedded in the very language of many Indigenous cultures.[13] Pulitzer Prize-winning poet and language activist Natalie Diaz offers her own native language from the American southwest as an example:

In Mojave thinking, body and land are the same. The words are separated only by letters: 'iimat for body, 'amat for land. In conversation, we often use a shortened form for each: mat-. Unless you know the context of a conversation, you might not know if we are speaking about our body or our land. You might not know which has been injured, which is remembering, which is alive, which was dreamed, which needs care, which has vanished.[14]

The Potawatomi language, indigenous to the Great Lakes region, practically glistens with the aliveness of the world. Robin Wall Kimmerer, author of the profound book *Braiding Sweetgrass: Indigenous Wisdom, Scientific Knowledge, and the Teachings of Plants,* writes about this and the significance of the fact that most of the words in Potawatomi are verbs—70 percent of them compared to 30 percent in English:

> *A bay is a noun only if water is dead. When bay is a noun, it is defined by humans, trapped between its shores and contained by the word. But the verb wiikwegamaa—to be a bay—releases the water from bondage and lets it live. "To be a bay" holds the wonder that, for this moment, the living water has decided to shelter itself between these shores, conversing with cedar roots and a flock of baby mergansers. Because it could do otherwise—become a stream or an ocean or a waterfall, and there are verbs for that, too. To be a hill, to be a sandy beach, to be a Saturday, all are possible verbs in a world where everything is alive.*[15]

Speaking directly to the idea of ownership, anthropologist Bradford Keeney writes of the profound mutuality that the Kalahari Bushmen, indigenous to southern Africa, experience:

> *When Bushmen say they own something, it means not only that they own the feeling for it, but also that the feeling has transmitted its essence, its complex nexus of relationships, into their very being. We become the other—whether a friend, butterfly, redwood forest, giraffe, or seahorse—through our intensely felt union with it.*[16]

THE WORD *OWN* derives from the root *aik*, meaning "to be the master of," and *possess* derives from the root *poti*, meaning "powerful; lord."[17] The way these words are used today is very true to their origins. They are both very much about "power over." Neither has any flavor of the temporary, imagined, and arbitrary nature of ownership, nor do they convey any sense of wonder or mutuality.

Maybe to ever truly step outside the worldview and assumptions of this culture, and to deeply experience ourselves and the world in this truer, more miraculous way, requires stepping outside this language, transcending the blinders embedded there—an honoring of earth-centered languages as liberation.

Alternately, perhaps each of us can contribute to rebirthing the English language into greater truth and reverence through our personal choice of words. For example, when we want to convey some of the arbitrariness and imagined nature of ownership, perhaps, instead of own and possess, we could use the words *appropriate* and *claim*. For example, instead of "One percent of the population owns ...," we could say:

One percent of the population appropriates to itself …
Or, *One percent of the population claims ownership over …*
Or simply, *One percent of the population controls …*

When we want to evoke some of the reverence and mutuality of our relationships with objects and the earth, maybe the word belong can be helpful. Instead of "I own this shirt," for example, we could say:

This shirt belongs to me.

Or maybe we can stretch our language further and say:

This shirt and I belong to each other.
Or, *This shirt is part of my belonging.*

Or, with gratitude to Potawatomi:

This being-a-shirt is part of my belonging.

Robin Wall Kimmerer has proposed a beautiful shift in language that goes much further than any of this. She notes that in English we never refer to a person as "it," that that would be disrespectful, reducing them to a thing. Similarly, in Potawatomi and most other Indigenous languages, the same words are used to refer to the living world as for people.

And so, having consulted her elders, she proposes a bit of "reverse linguistic imperialism":

> *Fluent speaker and spiritual teacher Stewart King … suggested that the proper Anishinaabe word for beings of the living Earth would be Bemaadiziiaaki. I wanted to run through the woods calling it out, so grateful that this word exists. But I also recognized that this beautiful word would not easily find its way to take the place of "it." We need a simple new English word to carry the meaning offered by the indigenous one. Inspired by the grammar of animacy and with full recognition of its Anishinaabe roots, might we hear the new pronoun at the end of Bemaadiziiaaki, nestled in the part of the word that means land?*
>
> *"Ki" to signify a being of the living Earth. Not "he" or "she," but "ki." So that when we speak of Sugar Maple, we say, "Oh, that beautiful tree, ki is giving us sap again this spring." And we'll need a plural pronoun, too, for those Earth beings. Let's make that new pronoun "kin." So we can now refer to birds and trees not as things, but as our earthly relatives. On a crisp October morning*

we can look up at the geese and say, "Look, kin are flying south for the winter. Come back soon."[18]

These ideas strike me as very beautiful and very important. I wonder, if I use "ki" and "kin" for the rest of this book, will it be distracting and take away from the reading? Or will it elevate the reading? I find myself wanting to try it out. And so I write:

At this moment, when I turn my head toward the window, I can see the sky. Ki (the sky) is bright with morning light, and ki fills me with a sense of spaciousness.

I find that in writing that brief sample, the act of using the word "ki" to refer to the sky absolutely fills my heart. I notice myself looking forward to being outside this afternoon beneath the sky, and getting to spend time with ki, like meeting up with a friend. And as I look out the window again, I can actually feel ki's gratitude, and my own, for the gift of this word, "ki."

And so I have just gone through the rest of the book and replaced "it" with "ki" whenever referring to the living world.

[Several months later, reflecting on this change, I find that I actually used "it" infrequently enough that, now, every time I come across the word "ki" instead, it *does* trip me up for a moment. It's not like the word is everywhere and I just get used to it. But personally, I have found myself appreciating that momentary distraction. It invites me to bring attention to whatever is being referred to, to ki's aliveness and vitality. It warms me and grounds me. I hope it is a positive experience for you as well.]

I FIND MYSELF humbled by the idea of one day touching the Whanganui River or entering ki. I am moved by the spirit of reverence and deep relation the Whanganui people feel for ki. Through them I see that body of water differently, feel ki differently.

Through them I now see other bodies of water differently as well. The grasses, flowers, soil, and mountains also. Through the Whanganui I have changed; they have awakened something within me that was dormant—a reverence, a connectedness, a belonging. Even as someone who has spent most of my life loving being in nature and helping try to protect ki, this is different.

I recognize how much of my own sense of the natural world has been conditioned and impoverished by the worldview that is embedded in the culture and legal system around me, in my language. The Whanganui people's worldview, as it is now inscribed in law, to a certain degree, invites us to a much deeper and richer place of connection with nature and ourselves. I am grateful to them for their vision, their courage, and their tenacity. It is a gift not just to the river or themselves, but to me and to all of us.

THIS ISN'T A treatise against private property, and the goal isn't that we just abandon ownership.

Philosopher and revolutionary Jean-Jacques Rousseau once wrote, "The first man who, having enclosed a plot of land, took it into his head to say this is mine, and found people simple enough to believe him, was the true founder of civil society."[19] In the spirit of Rousseau, what is vitally important is that—in the face of people claiming something is theirs (quite often that *so very much* is theirs), and in the face of elaborate systems built around perpetuating this idea—we are not simple enough to believe them. *At least not in any kind of fundamental or absolute way.* As Rousseau went on to warn, "**You are lost if you forget that the fruits of the earth belong to all and the earth to no one!**"[20]

Just as important, we must take care that our current system of ownership, and the language and values it constantly presses on us, do not diminish our sense of the world and ourselves. That the power and separateness and utility that define ownership within this system not blind us to the much more wondrous and reverent truth of our relationship with the world.

Chapter 27

Infinite Gratitude
and the Gift of Everything

*[Conscience demands] that we pay back to others the help they have given us
in making it possible for us to do our work and secure ourselves in our
possessions. … In other words … we are entitled to claim as absolutely our
own, the product of our own labor, after we have paid back what we owe to
others.*

*But that one condition alters the whole stand-point. When can we ever say
that we have fully settled that account? … If we admit that we owe
something to all those living and dead, the product of whose work has aided
us in doing our work, what is there left as the fruit of our labor of which we
can individually say, "I alone have made it, and hence it is mine"?* [1]

—W.L. Sheldon (from "What Justifies Private Property?")

S OME 90 PERCENT of Americans believe that hard work and ambition are the most
important factors affecting people's financial success, with more than 80 percent
citing education as a close runner-up.[2] In other words, a vast majority of us believe that
how financially wealthy people are is a reflection of how hard they work, how ambitious
they are, and how educated. Which is to say, we believe that how wealthy (or not wealthy)
people are is generally an accurate reflection of how wealthy they deserve to be.

This can be a very affirming belief for those of us who are well-off financially: our
wealth is a kind of signal to ourselves and others that we are worthy of admiration and
respect, that we are of the skilled, intelligent, hard-working set, that we are deserving,
favored even. Research shows that the more money we have, the more strongly we hold to
this belief, with financially wealthy people more likely to agree with statements like, "I
honestly feel I'm just more deserving than other people."[3]

A similar phenomenon was documented in a study where people were asked to play
Monopoly, but one person was given twice as much money to start, plus twice as much
money every time they rounded the board, plus was allowed to roll two dice versus one.

Not surprisingly, the people who were given these big advantages invariably won. A little more surprising, though, is the fact that those people also consistently felt they *deserved* to win. The advantages were huge and they were obvious to everyone, yet the winners still attributed their success to their own skill, and became "less attuned to all of the other things that contributed to" their winning.[4] (It's like that expression, "Born on third base, thinks he hit a home run.")

Which probably helps to explain the fact that the wealthier we are, the more we also tend to agree that "greed is justified, beneficial, and morally defensible"—it just feels like we deserve to have more.[5]

For those of us who struggle and can't ever seem to get ahead, though, these beliefs can be an added weight. What are we not doing right? Is there something wrong with us? What do others think of us? The belief that people generally get what they deserve is so strong in the US that we suffer a greater sense of shame when we are unemployed than people in other countries. Social scientist Vicki Smith notes that when we fail to recognize structural problems in the economy or that the "deck is stacked in certain ways," it can lead to depression and a lack of motivation.[6]

Wherever you fall on the socioeconomic spectrum, I have wonderful news: this belief that hard work, skill, and intelligence are the main ingredients in success is wrong. Super-duper totally wrong. **It turns out that hard work and skill have virtually *nothing* to do with a person's financial standing.**

Preposterous, I know, but true nonetheless. Branko Milanovic, a former lead economist at the World Bank, has done extensive research on this topic and he has calculated that the single greatest factor in how wealthy we are financially is—*by far*— where we were born. No matter how hard someone works or how smart or skilled they are, they will still, almost every single one, never make any more or less money than the general range of incomes for the country where they live. And those ranges vary dramatically. For example, the poorest 5 percent of Americans are, on average, wealthier than 68 percent of the rest of the people on the planet (after adjusting for cost of living). The poorest 5 percent of Americans are in fact, on average, wealthier than the average of the *richest 5 percent* of people in India. Only the top 3 percent of Indians have higher incomes than the poorest 1 percent of Americans.[7]

Warren Buffett, the third wealthiest person in the world as of this writing, speaks to this powerfully when he notes, "Society is responsible for a very significant percentage of what I've [received]. If you stick me down in the middle of Bangladesh or Peru or someplace, you'll find out how much this talent is going to produce in the wrong kind of soil. I will be struggling thirty years later."[8] Expanding on this, he writes:

Both my children and I won what I call the ovarian lottery. For starters, the odds against my 1930 birth taking place in the US were at least 30 to 1. My being male and white also removed huge obstacles that a majority of Americans then faced. My luck was accentuated by my living in a market system that sometimes produces distorted results, though overall it serves our country well. I've worked in an economy that rewards someone who saves the lives of others on a battlefield with a medal, rewards a great teacher with thank-you notes from parents, but rewards those who can detect the mispricing of securities with sums reaching into the billions. In short, fate's distribution of long straws is wildly capricious.[9]

Milanovic has calculated that more than **60 percent of the variation in people's incomes worldwide can be accounted for exclusively by where they live.** As he puts it, "Citizenship is fate."[10]

But there's more. Because family is fate also. In some countries, like China and Peru, there is almost no opportunity for someone whose parents are financially poor to end up rich, and vice versa. In other places like Norway, Canada, and Denmark, there is significant opportunity for movement in both directions. The US is somewhere in the lower third, just ahead of Argentina, just behind Pakistan and Singapore.[11]

Economist Kathryn Wilson notes, "The best developed country in which to be born rich is the United States, because the likelihood is that you will remain rich." But it's the worst if you're poor, "because you're most likely to stay poor."[12] More specifically, "Children from low-income families have only a 1 percent chance of reaching the top 5 percent of the income distribution, versus children of the rich who have about a 22 percent chance."[13] This is illustrated by the list of the four hundred richest people in the US. One hundred of them were born into the upper class, another one hundred inherited at least $1 million, and one hundred of them inherited enough to automatically get them on the list—$1.1 *billion*.[14]

John Cassidy, writing for *The New Yorker*, puts it pointedly, "For all too many working-class Americans … US society is less of a launchpad than a glue trap. With their feet stuck to the ground, they have little prospect of ascending very far."[15]

Milanovic calculates that **how wealthy a person's parents are accounts for fully 20 percent of how wealthy the person is.**[16] So more than 80 percent of a person's wealth is based on geography and family alone. And we're not done yet.

As Buffett noted, **race and gender** play a significant role as well. Take three people whose *parents* all have the same income—one Black person, one Latinx, and one White—and the White person can expect to receive 33 to 50 percent more than the Latinx or Black people.[17] Men can expect to receive 27 percent more than women with the same job, hours, experience, and education, each of which is also affected by a person's gender, magnifying the gap even further.[18]

Milanovic concludes that once you add race, gender, and a range of other factors that he lumps under the umbrella of "luck," the portion of income that can be explained by actual effort is "very, very small."[19] Of course, ultimately, none of us did anything to actually earn (or be burdened with) how industrious, clever, or ambitious we are anyway. Some blend of nature and nurture gets the credit for these as well.

Still, there are two more factors, more significant than any of these others, that Milanovic doesn't even mention. These two require expanding our vision to a whole other scale, beyond even the question of geography. The first is *when* **we were born.** For most of the three-hundred-thousand-year history of *Homo sapiens*, humans have lived with incredibly few material possessions and couldn't have even fathomed the number and variety of things most of us in the US have today.

The second is **the fact that we were born as humans.** There are some seven million other kinds of animals out there, each of them graced with their own variety of beauty and intelligence, but possessions and income—not so much.[20]

None of this is to ignore the difference that hard work and skill do often play in our lives. *Many of us have absolutely made tremendous sacrifices and worked immensely hard to secure what we have, and that deserves to be honored.* It is simply to acknowledge that the influence of work and skill can only be measured on the minutest of scales, that they are indetectable when considered alongside these other factors. The bottom line is that, ultimately, **the money and possessions in your pocket and in your name are there, and not in the pocket or name of somebody else in the world, only by complete and utter chance. That someone else's extreme poverty is not yours is likewise entirely serendipity.**

GOING ONE STEP further, it is important to note that there is often actually an inverse relationship between how much money a person has and how deserving they are of that money. Many people have accumulated great wealth by doing truly horrible things. As we noted in part IV, for many people and corporations it is simply the norm to regularly cause profound harm to other people, other creatures, and the planet in the course of making money.

But it goes beyond just obvious and direct acts of violence. It's also about the ways violence and destruction so often *indirectly* contribute to financial wealth. It's not just about what any of us as individuals have done; it's about the things our families have done, our employers, the businesses we buy from or invest in, the government, and so on, perhaps totally unbeknownst to us, and sometimes at quite a distance. Virtually all of us in the US are touched by this to some degree. Scholar Michelle Alexander notes, "**But for slavery, genocide and colonization, we would not be the wealthiest, most powerful nation in the world—in fact, our nation would not even exist.**"[21] Financial prosperity is often a reflection of how much the spoils of these and other atrocities, large and small, have flowed to us, both directly and indirectly, by having funded colleges and research and roads and cheap goods and services, and so on.

Let us acknowledge, also, that the flip side of this is all too often true as well. Many of the people who have done the most to benefit the world live in poverty or with modest financial wealth. They often don't get paid at all for their time, effort, brilliance, love. Indeed, many of the things that matter most in the world don't generally pay anything, and even require foregoing paid work—parenting, caring for an elderly relative, protesting, walking in the woods, and so on.

All of this challenges or even flips on its head the idea that many people hold, that when a person is financially wealthy it is an indication that they (or their family or industry or country) are worthy of admiration and respect, that they are uniquely clever or hard-working, or that they have made important contributions to society—that they *deserve* that wealth—and that people who are poor somehow deserve their poverty.

It also completely upends one of the fundamental justifications for some people having immense financial wealth and others suffering and dying in profound poverty.

IN THE EARLY 1900s the US was replacing Great Britain as the "world's wealthiest nation." In 1915 economist Wilford King conducted a study with the goal of reassuring the public that the wealth was being shared by all Americans. Instead, he was distressed to find that "the richest 1 percent possessed about 15 percent of the nation's income." He noted, "It is easy to find a [person] in almost any line of employment who is twice as efficient as another employee, but it is very rare to find one who is ten times as efficient. It is common, however, to see one [person] possessing not ten times but a thousand times the wealth of [their] neighbor."

By comparison, today the richest 1 percent of us possess 24 percent of the nation's income.

And the contrast in wealth is even more stark. Over the past forty-five years roughly $50 trillion has been redistributed from the poorest 90 percent of us to the richest 1 percent. The result is that all of us in the 90 percent put together today own less wealth than those of us in the 1 percent.[22]

I WANT TO offer another suggestion around language: that we try to avoid using the words *earn* and *make* when referring to income. For example, we could simply say:

> *People with incomes of $1,000,000 a year ...*
> Or, *People who receive $10,000 a year in income ...*

This is not to take away from the incredible effort many people put into their work, regardless of their level of income, from the lowest-paid subsistence worker to billionaires—that there is indeed usually some aspect of "earning" the money and material things we receive. It is vitally important, though, that we not perpetuate the idea that those of us with high incomes did anything particularly to our credit to merit them, or that those of us with little or no income are getting what we deserve.

TO THE DEGREE that our sense of identity or worth is tied to our financial status, or to the degree that we believe our happiness is tied to having a lot of material wealth, we may find all of this threatening—the fact that our financial status ultimately has nothing to do with merit. **Yet, in the end, for all of us, these truths are liberating.**

How and why this is so is highlighted by one final factor affecting how materially wealthy we each are. And despite the enormity of the other factors we've already considered, they all pale in comparison with this final one. I am referring to nothing less than the fact that the world exists, and that we exist—*that anything exists at all.*

This is perhaps obvious, like it goes without saying. But in not saying it, it can also be invisible, and so perhaps is not actually so obvious.

It is a profound miracle and gift that we were born at all, and that we were born into air and water and soil and light, into intelligence, skill, and movement, into all the things that make our lives and endeavors possible in the first place. This, **the gift of existence, is the foundation for all other gifts.**

To paraphrase Trappist monk Thomas Merton: to live in gratitude is to recognize the love in everything we have been given—and we have been given *everything.*[23] **We did nothing to earn any of it, yet here it all is, somehow, stunningly, here, flowing all around us and through us.**

To recognize that everything is a gift, that fundamentally none of us have done anything to earn any of it, is itself an immense gift:

- It invites us to shed any judgments or fears we have that money and possessions say anything about our, or other people's, worth.
- It reminds us that we do nothing on our own, that the soil, the sun, the waters are our constant companions, making our lives and every effort possible.
- It lays bare the immense hubris and ingratitude in the way we often claim so much for ourselves and cling so tightly to it.
- It allows us to perceive the extreme disparities in the world with a renewed heartache, and outrage, and to more powerfully challenge them.

Recognizing the gift nature of everything is, then, a gift of love, belonging, gratitude, and conscience.

And it is also a gift of possibility. It further frees us from the traps of "that's just the way things are," and "we're all better off this way." It further underscores that our ideas and systems of ownership are simply human imaginings—often deeply ill-imagined—and that they are subject to change when and how we wish. And it frees us to think much more expansively about what is possible, about how we wish to relate to the world, and about what we want for ourselves and for all beings.

Chapter 28

Feel the Wind,
Drink the Stars,
and Take Only What You Need

*In order to live, I must consume. That's the way the world works, the
exchange of a life for a life, the endless cycling between my body and the body
of the world. …*

*If we are fully awake, a moral question arises as we extinguish the other
lives around us on behalf of our own. Whether we are digging wild leeks or
are going to the mall, how do we consume in a way that does justice to the
lives that we take?*[1]

—Robin Wall Kimmerer

I N HER BOOK *Braiding Sweetgrass*, Robin Wall Kimmerer tells the story of an
Algonquin ecologist who sought funding from her tribal council to attend a conference
on "sustainable development." The members of the council asked her what sustainable
development meant. She explained that it is basically about managing natural resources in
a way that ensures the ability of people to meet their economic goals both for present and
future generations.

After considering that for a moment, an elder replied, "This sustainable development
sounds to me like they just want to be able to keep on taking like they always have. It's
always about taking. You go there and tell them that in our way, our first thoughts are not
'What can we take?' but 'What can we give to Mother Earth?' That's how it's supposed to
be."[2]

The idea of *sustainable development* arose in the 1970s and it represented an important
departure from previous economic thought. Until then mainstream economists generally
believed (and many still do today) that technology, human ingenuity, and the free market

will always be able to meet any challenges related to limited natural resources or environmental problems.

In the 1960s and 70s, though, the scale of environmental destruction taking place gained increasing attention from the general public. Air and water pollution, strip mining, nuclear waste, the population explosion, and the widespread use of toxic chemicals like DDT were the subjects of nightly news programs and influential books like Rachel Carson's *Silent Spring*. In 1972 a group of leading economists and scientists published a report titled *The Limits to Growth* that included this warning:

> *If the present growth trends in world population, industrialization, pollution, food production, and resource depletion continue unchanged, the limits to growth on this planet will be reached sometime within the next one hundred years. The most probable result will be a rather sudden and uncontrollable decline in both population and industrial capacity.*[3]

The idea that economics might have to take the environment into account—that there might be limits to what we can take from and do to the environment—was new and radical for many, and it is still debated by many economists. Even among nontraditional economists, economics had always concerned itself with people and products and services, not the environment.

Nonetheless, this idea—that we must be careful that our economic activity does not impact the environment in a way that undermines the ability of future generations to meet their own economic goals—has gained tremendous credibility. There have been two global sustainable development forums attended by most of the world's leaders; one of the eight UN Millennium Development Goals focused on sustainable development; there are institutes, schools, and government offices dedicated to promoting sustainable development; and environmental economics has become institutionalized as a subfield of economics.[4]

And still, at the end of the day, the Algonquin elder nailed it. Sustainable development is indeed, in many critical ways, simply more of the same, and it is, ultimately, about taking. It implies that we've basically got a good thing going here, we just need to figure out what the maximum is we can get away with taking and somehow stay under that. In other words, it's about keeping this same general paradigm of "development," but just getting it to be "sustainable."

By perpetuating the idea that material wealth makes a significant contribution to our well-being even after our basic needs are met, and the idea that economic growth is a vital, central human pursuit, and that more is better (just so long as we don't go *too* far with it), sustainable development actually *contributes* to the very violence it seeks to prevent. **It perpetuates a bleak and false story about who we are, about what our place is in the world, and about what our deepest callings might be in how we live our lives.** Even as it has contributed to some important shifts in how economists and others think about the environment, ultimately it also contributes to keeping us all trapped.[5]

By so many measures—screaming, flashing, urgent measures—this economic paradigm is rigged against all of us. We need to forget the rules for how to be a "winner" or how to be "green" or "sustainable" inside this box, all of us crammed in here, lonely and disconnected, crushing and killing each other as we produce and consume more, as it careens down the hill, off ever-larger cliffs.

We need to step back from all of that, step out of the box and into the vast world, and

feel the wind,

drink the stars,

listen to the trees,

and remember:

that they are sacred, alive,

that they are our ancestors

that none of us earned any of this, yet here it is, flowing all around us and through us,

that we are all of us and all of nature, miracles.

Then—not trying to cobble together adjustments to the current model, but building anew from the foundation of this truth—we can ask ourselves the question Kimmerer posed at the beginning of this chapter:

"How do we consume in a way that does justice to the lives that we take?"

And, we might add:

How do we consume in a way that also does justice to ourselves and to the whole?

In other words:

What does consumption look like when it is rooted in wonder and sacredness, in reverence and love?

KIMMERER NOTES, "OUR ancestors, who had so few material possessions, devoted a great deal of attention to this question, while we who are drowning in possessions scarcely give it a thought."[6] The answers they came up with varied to some degree from culture to culture, but there are also some fundamental principles that are still nearly universal among

Indigenous people.[7] They are often referred to as the "Honorable Harvest," and this is how Kimmerer summarizes them:

The Honorable Harvest

Know the ways of the ones who take care of you, so that you may take care of them.

Introduce yourself. Be accountable as the one who comes asking for life.

Ask permission before taking. Abide by the answer.

Never take the first. Never take the last.

Take only what you need.

Take only that which is given.

Never take more than half. Leave some for others.

Harvest in a way that minimizes harm.

Use it respectfully. Never waste what you've taken.

Share.

Give thanks for what you have been given.

Give a gift, in reciprocity for what you have taken.

Sustain the ones who sustain you and the earth will last forever.[8]

These principles stand in stark contrast to the laws and norms that generally govern our society, which serve primarily as limitations, where we are generally encouraged to take and do everything we want right up to the legal limit (and to not necessarily let that stop us either). The principles of the Honorable Harvest, instead, evoke an entire way of relating to the natural world that is honoring and rich. *They are lush invitations as much as they are guidelines.*

These principles reflect the wisdom of people who have a very close relationship with the land and the sources of their food and clothing and shelter, people who understand that their own thriving and the thriving of the plants and creatures around them are intimately linked. They also reflect an appreciation for these plants and creatures as deserving of our reverence regardless of their value to people.

And these principles also communicate that we are precious, that we are also deserving of reverence. They acknowledge the importance of meeting our needs, but they also make it clear that consumption is just one facet of our beings. In fact, **they make it clear that** *how* **we consume can be even more important than** *whether* **we consume.**

Kimmerer writes about the early colonists arriving in the Great Lakes region, and of their awe at the extraordinary abundance of wild rice. In just a few days the Native people could fill a canoe with enough rice to last the year. But many of the settlers were bothered by how much rice was left "to waste." After four days of harvesting, the Native people could not be compelled to collect any more, leaving it for "the Thunders."

Kimmerer notes that for many settlers this was an example of what they perceived as a general laziness and lack of industriousness on the part of the Native people. Those settlers seem to not have appreciated that these practices were actually related to the extraordinary abundance around them. Or that the quality of life the Native people enjoyed was rooted in the respectful relationship they had with nature, and in their appreciation of "enough," and not needing to accumulate more.[9] Those settlers seem to not have appreciated that the Honorable Harvest represented an advanced way of living that had served those people in good stead for millennia.

IN HIS BOOK *Eating Animals*, author Jonathan Safran Foer shares about a conversation he had with his grandmother about some of what she suffered in World War II. She told him:

> *During the war it was hell on earth, and I had nothing. I left my family, you know. I was always running, day and night, because the Germans were always right behind me. If you stopped, you died. There was never enough food. I became sicker and sicker from not eating, and I'm not just talking about being skin and bones. I had sores all over my body. It became difficult to move. I wasn't too good to eat from a garbage can. I ate the parts others wouldn't eat. If you helped yourself, you could survive. I took whatever I could find. I ate things I wouldn't tell you about.*

That led to the following exchange between Foer and his grandmother about the vital importance of *how* we consume, in this case about respecting the traditional Jewish laws around eating.

"The worst it got was near the end. A lot of people died right at the end, and I didn't know if I could make it another day. A farmer, a Russian, God bless him, he saw my condition, and he went into his house and came out with a piece of meat for me."

"He saved your life."

"I didn't eat it."

"You didn't eat it?"

"It was pork. I wouldn't eat pork."

"Why?"

"What do you mean why?"

"What, because it wasn't kosher?"

"Of course."

"But not even to save your life?"

"If nothing matters, there's nothing to save."[10]

IT'S NOT AN easy thing, to root our consumption in reverence and wonder. This is true even when we have the collective wisdom of our ancestors and Indigenous people worldwide distilled for us in thirteen principles.

Some of the principles of the Honorable Harvest can be applied fairly easily and immediately, no matter our circumstances: *Use what you take respectfully, never waste; Share; Give thanks.* Others are not so easy. Most of us are vastly removed from the sources of the material things in our lives. How do we ask permission of a plant at the other end of the food we buy? How do we sense whether there is a willingness from the land on the other end of the clothes we buy? How can we be sure we are not participating in taking more than half of something?

Robin Wall Kimmerer tells about a time she went to a mall to buy paper and pens, and tried to do so in the spirit of the Honorable Harvest. Despite her best efforts she couldn't feel any life in anything there, the pens and their petrochemical origins especially, but even the trees in the paper. They were all so removed from their origins and so processed.[11]

With the principles of the Honorable Harvest—as with most of the ideas throughout this book—it's more about leaning into them than it is about a precise action or a specific outcome. Kimmerer writes, "I think my elders would counsel that there is no one path, that each of us must find our own way."[12] It's about letting these principles and ideas guide our decisions about consumption as much as we can, to let them feed greater peace and justice in the world—even though we cannot do all we would like.

And it is about letting these principles feed *us* as well, it's about **letting the reverence and interconnectedness inherent in them inspire us and fill us, regardless of limitations in how we apply them.**

PERHAPS *ALL* OF the principles of the Honorable Harvest speak to you. Or perhaps one or two of them stand out. There is a way in which any one of the principles can, by itself, lead us in the direction of the others. For example, *Give thanks for what you have been given.* To live in a spirit of gratitude simply wouldn't be consistent with wasting what we take,

211

or with hoarding it. Likewise, it may not be easy to *Be accountable as the one who comes asking for life*, but embracing the spirit behind this principle would likewise just not be consistent with harvesting in a way that is destructive, or being disrespectful with it, and so on.

One of the principles of the Honorable Harvest can be particularly helpful in this regard. It can, to a certain degree, act as a stand-in for the others—a general rule that can help us honor all of the principles even when they are elusive on their own. **It is a principle that has emerged as a central theme throughout this book, crying out to us chapter after chapter:**

Take only what you need.

Without having worded it this way, we already know that *Take only what you need*—prioritizing money and possessions only to the point where we can meet our basic needs—is an important guidepost for living a rich and joyful life. We also know that this principle is central to addressing the pain and destruction in the world related to our consumption. And now we find that it can also go a long way toward helping us root our consumption in reverence. By taking no more than we need, we minimize how much we might be taking in any uncertain or harmful ways—when it is difficult to ask permission and hear an answer, for example, or to know if something was harvested in a way that minimizes harm.

The following are some of the ways we might apply this principle:

Buy no more food, clothes, appliances, devices, and other things than you need.

Buy no more new things than you need, when second-hand or recycled are available.

Buy no more disposable things than you need.

Choose those things that consume the least materials and energy, and that involve the least violence (that are Fair Trade or Forest Stewardship Council certified, for example).

Live in a home no larger than you need. Heat and cool your home no more than you need.

Drive no more than you need, and use a car that consumes no more fuel than needed. Fly no more than you need.

Use no more water than you need.

Have no more children than you need.

Support politicians, policies, and organizations that support "take only what you need."

As WE KNOW, "need" can be a slippery concept. We tend to always think we need more than we have, no matter how much that is. More than that, though, we *live in a culture that is profoundly committed to the exact opposite of "Take only what you need."* It is a core assumption of this culture that "money and possessions are central to happiness and success," while "take as much as your income/power/circumstances will allow" is just the way things are done. So this principle, *Take only what you need*, and that extended list of examples (buy no more new things than you need, and so forth), may evoke a sense of imposition or sacrifice.

As we noted in chapter twenty-four, for those of us who have a choice about consuming more than we need—which is many of us in the US, but very few of us worldwide—the call to limit our consumption does indeed involve a significant shift from the kinds of lifestyles we are often used to. And that is often challenging and uncomfortable.

But what we have before us is a beautiful invitation to actually *elevate* our existence, and to live much better lives.

- We know that materialism is "toxic" for happiness, that it undermines our relationships and self-esteem, and our ability to experience joy, contentment, and gratitude.
- We know that we in the US suffer extreme rates of depression, suicide, loneliness, and isolation.
- And we know that the way we are doing things now is causing extraordinary pain and suffering in the world, on an unprecedented and catastrophic scale.

This is what we are being invited to give up—what Mother Teresa, referring specifically to the United States, called the deepest "poverty of the soul" she had experienced anywhere in the world.

We also know that:

- Money accounts for only 2 to 4 percent of our happiness.
- And that time spent socializing contributes much more to our happiness than income.
- And that caring for our physical health contributes much more to our happiness than income.
- As does working less.
- As does living with presence and gratitude.
- As does having a sense of purpose and making a difference.
- As does having a good relationship with ourselves.

We know that *all* of these are generally far more important to our happiness than income.

Remember the study where a significant number of people voluntarily made a lifestyle change that involved receiving significantly less money? And 85 percent were happier for it? And another bunch of people had to make similar changes involuntarily and a quarter of them described it as a "blessing in disguise"? Remember also the people who in 1930

213

had ten hours cut from their workweek and they were markedly happier? ("I wouldn't go back for anything. I wouldn't have time to do anything but work and eat.") *Research consistently shows that the vast majority of us are happier when we voluntarily limit our consumption.*[13]

This is what *Take only what you need* means. *This* is what we are being invited to step into: *the fullness of a joyful life.*

TO TAKE ONLY what we need, to leave something be, rather than consume more of it, is to affirm that the people, creatures, and places on the other end of our consumption—everyone bearing the brunt of global warming, child and slave labor, oil spills and toxic waste, and on and on—*that they matter.* It is to affirm that the forests and waters and bison and otters are so much more than resources, that they are each of them a face of creation, that they deserve our reverence.

Every time we decline to take something we do not need, it is like a quiet, intimate moment, us whispering to these people, plants, and creatures, most of whom we will never know, "I see you, I am grateful for you."

To take only what we need is to affirm that *we* matter as well. For if the lives of other people and creatures are less important than production and consumption, then so are our own lives. If the world does not really matter, then *we* do not really matter.

As we noted in chapter twenty-four, every time we find ourselves considering whether to consume something we don't really need, we can use that as an opportunity to shift our orientation from "the things we don't have" and "what we have to go without" to so much that we do have. We can take a moment and, with a brief gesture or thought, we can show appreciation for the things in our lives that we were perhaps not noticing before.

Thank you, blanket.

I'm glad you are here, birds.

I see you, bowls, books, roof.

I am blessed to be here with you, moon.

I am grateful to you, photograph of someone I love.

I honor you, food, entering me to become me.

What a miracle that you and I are alive here together, grasses.

At some point it's no longer even a question of whether a particular action or some additional consumption would actually do any harm. Living in reverence and connection is its own reward.

Chapter 29

The Foundation of the Universe: Being Abundance

B UCKMINSTER FULLER HAS been called "the twentieth century's Leonardo da Vinci." He was a philosopher and inventor, a futurist and poet, an architect and more.[1] In Lynne Twist's landmark book *The Soul of Money: Transforming Your Relationship with Money and Life*, she writes that the first time she heard Fuller speak in the 1970s was a turning point in her life. She describes his vision that so affected her in the chapter "Scarcity: The Great Lie."[2]

> *Buckminster Fuller … said that for centuries, perhaps thousands of years, we have lived in the belief that there's not enough to go around, and that we need to fight and compete to garner those resources for ourselves. Perhaps it had been a valid perception at some time, or perhaps it hadn't been, he said, but at this point in history—in the 1970s—we were able to do so much more with so much less that as a human family we clearly had reached a point where there actually was enough for everyone everywhere to meet or even surpass their needs to live a reasonably healthy, productive life. This moment represented a dramatic breakthrough in the evolution of civilization and humankind, he said.*
>
> *Whether it was a recognition of something already true or a moment of transformation in the status of civilizations, he said, either way it could be the most significant turning point in our evolution. …*
>
> *The new threshold completely changes the game, and it would take fifty years, he predicted, for us to make the necessary adjustments in our world so we could move from a you-or-me paradigm to a you-and-me paradigm, a paradigm that says the world can work for everyone, with no one and nothing left out. He said that our money system, our financial resources system, would need to adjust itself to reflect that reality and it would take decades for us to make the*

adjustment, but if and when we did, we would enter an age, a time, and a world in which the very fundamental ways we perceive and think about ourselves and the world we live in would be so transformed that it would be unrecognizable.

Jesus said to his disciples …
Sell your possessions
and give to the poor.

—Luke 12:24, 12:33, NIV

IN LEVITICUS, THE third book of the Jewish Torah and the Christian Bible, it is required that all land be returned to the original owners every fifty years, the Jubilee year. People can buy and sell the land, people can fall on misfortune or make bad decisions, an entire generation can benefit from or suffer the consequences. But every fifty years the means of providing for our basic needs and producing wealth is to be equalized, and our relationships with our ancestral homes are to be reestablished. If a person has the resources to pay a fair price for the land and structures on ki (the land), they are to pay it; if they do not, ki is to be returned anyway.

Through all of this, it is understood that nobody can ever actually own the land in some kind of fundamental way. Ultimately, the land is bigger than that, ki is of something else entirely. "The Lord said to Moses at Mount Sinai: 'The land must not be sold permanently, because the land is mine and you reside in my land as travelers and sojourners.'"[3]

Theologian Ron Sider, in his book *Rich Christians in an Age of Hunger: Moving from Affluence to Generosity*, notes that the mandate laid out in Leviticus "prescribes justice in a way that haphazard handouts by wealthy philanthropists never will. The year of Jubilee was an institutionalized structure that affected all Israelites automatically. It was the poor family's right to recover their inherited land at the Jubilee. Returning land was not a charitable courtesy that the wealthy might extend if they pleased."[4]

Obviously, it would be a bit complicated to meet these requirements in today's world. But consider what it would look like if we took them at face value: dividing the financial wealth in the United States ($98 trillion in 2019) by the adult population (290 million), every adult would get nearly $340,000 in debt-free assets.[5] Divide up the US gross domestic product ($21.43 trillion) and we would every one of us receive more than $70,000 a year on top of that.[6]

That means every person in the US would receive roughly seven times the abundance point ($10,000 per person), and more than twice the saturation point ($30,000), every year. There is so much financial wealth in this country that if we honored the requirements of Leviticus, we would all still be living vastly beyond (and in contradiction of) the ethic of the Honorable Harvest.

Sharing the abundance in our lives is a central tenet of cultures and religions worldwide. It is honored as a way to help everyone meet their basic needs and weave together the members of a community, and as a way for us to purify ourselves, to nourish our gratitude for the gifts we have received, and to honor the sources of life and sustenance.[7]

The prescriptions around sharing that are laid out in these traditions, like those of Leviticus, can often seem radical to us in the US, given how extreme our own norms are around ownership and the accumulation of material wealth. Indigenous cultures have some of the most generous sharing traditions, consistent with their fundamental view of "ownership as belonging" and the principles of the Honorable Harvest. Many things are "owned" communally in the first place, and then on top of that, wealth has often been measured in Native communities not by how much a person has in material terms, but by how much they give away. Potlatch ceremonies in the Pacific Northwest, which involve significant gift giving, were outlawed for nearly fifty years in the US, and seventy years in Canada, because the degree of sharing offended "the white man's sense of frugality and respect for private property."[8]

The Sun, each second, transforms four million tons of itself into light ...
giving itself over to become energy that we, with every meal, partake of. ...
For millions of years, humans have been feasting on the Sun's energy stored
in the form of wheat or reindeer, as each day the Sun dies as Sun and is
reborn as the vitality of Earth. ...

Human generosity is possible only because at the center of the solar system a
magnificent stellar generosity pours forth free energy day and night without
stop and without complaint and without the slightest hesitation. This is the
way of the universe. This is the way of life.[9]

—Brian Swimme

IF YOU DONATE $20 today, in a matter of months forty children who currently have worms will be free of them.[10] If you donate $50 today, in several months someone who is currently blind will see again.[11] If you donate as little as $200, in one year someone will be alive who will otherwise be dead.[12]

Some life-saving treatments and interventions cost as little as a dollar per person, but not every one of them definitely saves a life. Not everyone who gets a mosquito net, for example, even in areas with a high incidence of malaria, would otherwise contract malaria and die. Also, there's a cost to delivering those nets. However, once we factor in these costs and variables, for somewhere between $200 and $2,000 (estimates vary) it is statistically reliable that if we provide mosquito nets to enough people in areas where malaria is endemic, someone's life will be saved.[13]

That is the power of you, and that is the power of allowing ourselves to be part of the flow of energy and resources that is the way of the universe, the way of life.

That is also how profoundly unjust the current distribution of abundance is in the world—that amounts of money that for many of us in the US can seem very small, are so inaccessible to many other people that they can make a vast difference in those people's lives, even the difference between life and death.

And sharing is only to our benefit. Maya Angelou writes:

> *I have found that among its other benefits, giving liberates the soul of the giver … and that intangible but very real psychic force of good in the world is increased. … The gift is upholding the foundation of the universe … strengthening the pillars of the world.*[14]

Indeed, **we are *all of us* vastly better off when we share the abundance of the world— both when we are giving and when we are receiving**. We know that:

- Money does little for our happiness beyond our basic needs.
- Donating money or spending it on others boosts our own happiness—as much as doubling our income.
- Fundamentally, none of us has done anything to earn the material wealth in our lives anyway. We are no more deserving of it than anyone else.

And for relatively small amounts of money, people's lives can be changed in dramatic, joyful, liberating ways. Their very lives can be saved, and in a very meaningful sense, so can those of the people sharing.

In other words, we should not just *take only what we need*, we should also (as the Honorable Harvest also notes) *share all that we do not need*.

In his book *The Life You Can Save*, philosopher Peter Singer puts it this way: if we are able to prevent the death and suffering of others without sacrificing something nearly as important, we should do so. Thus, we have a moral obligation to donate whatever money we do not really need, in order to prevent the death and suffering of others. (He makes this clear statement without even adding the very important fact that the person on the sharing end benefits as well.)

This may seem straightforward and obvious, but, as Singer notes, "If we were to take it seriously, our lives would be changed dramatically."

> *What many of us consider acceptable behavior must be viewed in a new, more ominous light. When we spend our surplus on concerts or fashionable shoes, on*

fine dining and good wines, or on holidays in faraway lands, we are doing something wrong.[15]

To put a finer point on it, **when we withhold from others a portion of the abundance of the world that they need to meet their basic needs or survive—when we ourselves don't really need it, don't even really benefit from it, have no fundamental right to it anyway— we are participating in their suffering and we are complicit in their deaths,** *even as we are diminishing our own well-being.*

This flips on its head the kind of spending that otherwise tends to represent not just the norm in our society, but even success and status. Any money we spend on more home than we need, more car than we need, more products than we need, more travel than we need, and so forth, literally means casting aside other people's lives. It represents a callousness we usually associate with kings and queens sitting in a castle on a hill: "Let them eat cake." Yet it is simply the norm here in the US and many other places. This is most obviously and glaringly relevant to those of us who are millionaires, not to mention billionaires. (A message circulated on social media succinctly noted, "Being a billionaire is a moral failing."[16]) But it is relevant to many of us in the United States, no matter how far we are from being millionaires. Some 130 million of us (40 percent) live beyond the abundance point ($10,000 per person), meaning we generally already have all we need financially to maximize our happiness. Another 100 million of us (30 percent) live beyond the saturation point ($30,000 per person), meaning we have more money coming in than contributes *anything* additional to our happiness.

Again, these are averages and generalizations. Still, in a broad sense, that means that 230 million of us (70 percent) are living beyond the wildest dreams of billions of people on the planet. Continuing the metaphor—a bit simplistic, but also in many ways true: we are like a privileged aristocracy living in our castle on the hill, with walls that few people on the outside can ever scale, and to a large degree we can do whatever we like with the riches in our lives, without ever needing to consider the needs of the people on the other side. We will not generally be challenged on this by the people inside the castle with us. Quite the opposite, we will generally be encouraged, even celebrated by them for making more, spending more, accumulating more.

To be fair, each of us is just making our way in the world as best we can, given the culture we were born into and have internalized. In that sense, all of this is arguably more a failure of our society than a failure of each of us as individuals, though it plays out, of course, person by person. And that is our individual responsibility—and opportunity.

Generosity links us, beyond time and place, to people of conscience and action everywhere who have made our world freer, kinder, and more just.[17]

—Alfre Woodard

MANY OF US do, definitely, share some of the money that comes into our lives, though for those of us who are financially wealthy, the amounts are rather ridiculous. Those of us with lower incomes consistently share more than those of us with higher incomes, as a percentage of income. Those of us with incomes of less than $20,000 a year (and not receiving government assistance) donate an average of 10 percent of our income. Nobody in any income range gives that much except the four hundred individuals with the highest incomes, who donate 11 percent. *Yes, the four hundred highest income people in the US donate 1 percent more of their incomes than people with incomes of less than $20,000 a year.*

Those of us with incomes of less than $20,000 who *do* receive government assistance still donate nearly 5 percent of our income. Other than people with low incomes who are not receiving government assistance, nobody in any income bracket matches that level of giving up to those of us with incomes over *$10,000,000*, who give nearly 6 percent. Yes, it takes going from $20,000 a year to over $10 million for us to bump up the amount we share by just over 1 percent.

We also tend to give more to people who are close to us geographically and to programs that serve people like us. Only $15 of every $100 in donations in the US go to addressing basic needs, and less than a third of that $15 goes to support people living in extreme poverty abroad, where our money has a vastly greater impact (and where the US has all-too-often had a hand in sowing some of that poverty).[18]

There are important exceptions to these numbers. Warren Buffett, one of the richest people in the world, is a wonderful example. He has pledged to give away 99 percent of his fortune. He notes, "Were we [Buffett and his family] to use more than 1 percent of my claim checks on ourselves, neither our happiness nor our well-being would be enhanced."[19] He also very thoughtfully acknowledges that his own generosity still does not compare in certain respects with that of so many lower-income Americans:

> *Millions of people who regularly contribute to churches, schools, and other organizations thereby relinquish the use of funds that would otherwise benefit their own families. The dollars these people drop into a collection plate or give to United Way mean forgone movies, dinners out, or other personal pleasures. In contrast, my family and I will give up nothing we need or want by fulfilling this 99 percent pledge.*[20]

In 2010, Warren Buffett, Melinda French Gates, and Bill Gates created the "Giving Pledge," urging the financially richest people in the world to join them in committing to donate at least 50 percent of their wealth over their lifetimes or upon their deaths. As of 2022, more than two hundred people have signed.[21]

(As of this writing, the person who has appropriated the most money in the world for himself, Elon Musk, has signed the pledge, though he hasn't actually donated very much.[22] Jeff Bezos, the second-most appropriating person, has *not* signed the pledge. Notably, though, his ex-wife, MacKenzie Scott, has signed the pledge and started giving away significant amounts of money immediately after their divorce.[23] To be clear, Musk, Bezos, Buffett, and the Gates control more wealth than 50 percent of all Americans combined.[24])

Peter Singer actually urges us *not* to aim for *share all you do not need*. He believes it is too high a bar, that it is simply unrealistic and will actually result in us sharing *less* than we would otherwise. He recommends, instead, a graduated scale where we give more as our incomes increase, topping out at 33 percent for incomes over $10 million.[25]

If everyone shared at Singer's graduated amounts, it would yield $471 billion dollars a year in the US alone. That is more than twice what it would take to meet the UN's eight "Millennium Development Goals," including eliminating extreme poverty and hunger worldwide, reversing the spread of HIV, malaria, and other diseases, providing a primary education for every child worldwide, and making significant strides toward "environmental sustainability."[26] **In other words, for *less than half* of what we would generate in the US alone using Singer's graduated scale (topping out at 33 percent), we could invite everybody, every person on the planet, into the vibrant human community Buckminster Fuller described, where we all have enough to live healthy and meaningful lives. (*If and when we so choose ...*)**

Even then, Singer ultimately suggests that we not worry too much about scales and percentages, but that we simply share an amount that is "significantly more" than we have in the past. That way we can try on the experience, and see how it feels to stretch in that way to help other people, creatures, and places.[27] And this is relevant for all of us. There is love and power in all sharing, small portions as well as large.

My heart is moved by all I cannot save:
So much has been destroyed.
I have to cast my lot with those
Who age after age, perversely,
With no extraordinary power,
Reconstitute the world.[28]

—Adrienne Rich

IT IS WORTH acknowledging that the wonderful projects and nonprofits we might want to support in the world are run by human beings—and all that that implies. We should expect differences of opinion about the best way to make a difference. And we should expect some degree of inefficiencies and mistakes and people who put themselves first. These are reasons to be thoughtful about how we share the resources in our lives; they are not reasons to *not* share.

"The old saying 'Everything takes longer and costs more than you expect' holds as true when you are trying to repair the world as it does when you are engaged in home repairs," writes Thomas Tierney in *Give Smart: Philanthropy that Gets Results*.[29] There are organizations that can help point you to other organizations reliably doing important work. Give Well, the Open Philanthropy Project, The Life You Can Save, Thousand Currents, Women's Funding Network, and Grassroots International are just a few of the many that can be found with an Internet search as of this writing.

As The Life You Can Save notes: *Real lives are saved every day. People with real names whose families weep with joy to see them still alive.*[30]

It looks like giving, but actually it's participating in creation. In the world here it looks like giving, but spiritually it is becoming more equal to the creative force of the universe.[31]

—Thomas Hübl

YOU ARE HELD in a flowing river of abundance and generosity.

It is the earth and sun and sky.
It is your bones and blood and hair.
It is whales and gorges and moths.
It is the miraculous, somehow-true existence of everything—all of it our larger body, all of it a gift.
It is all of creation gushing all around you.

And it is all of creation gushing *through* you, and back out into the world, nourishing and delighting you, and then going on to nourish and delight other places and other beings.

Because that is the only way to really receive this abundance, to experience the truth of it, is not to dam it up or try to grab for more of it or cling to it, but to let it flow freely through you, *to be of it.*

It is to be the sun, transforming ourselves into light, pouring forth.

It is to be the earth, both receiving and giving on incomprehensible scales, providing enough for everyone.

It is to be the river, entirely in and of the flow.

Take only what you need. Share all that you do not need.
Be the sun and earth and river. Be abundance.

Chapter 30

What If?

WHAT IF THIS is what success actually looks like? *Take only what you need; Share all you do not need.*

What if—contrary to the messages we are continually drenched in about consumption—this is what the "good life," accomplishment, "made it," *really* looks like, is what we should all be aiming for and celebrating? **Pursue only those material things you need in order to thrive, and then get on with the thriving.**

What if this is what doing our part and carrying our weight looks like? Not so much contributing to the economy and being "productive," but taking time to be still, to be social, to delight.

What if one of the most essential covenants we need to keep with the stars and the trees and each other, *and with ourselves*, is to uncompromisingly make time to do the things that bring us joy and the things that deepen our presence and self-love?

What if some of the most worthy of undertakings don't add anything to our income or the gross national product, and sometimes actually involve less profit and production and consumption, a reduction in GNP?

What if more money and stuff wasn't the default, obviously better choice? What if taking more than we need—bigger cars, bigger houses, more flights, more dollars—wasn't a signal of intelligence, skill, worthiness, but was a signal of ignorance and selfishness, of immaturity and violence?

What if living with "less" (but enough to meet our needs) reflected *positively* on us? What if it was a signal of maturity, wisdom, care?

What if this is what a life that is admirable, just, and rich looks like?

To ask the earth and other living beings what they want and need, and to honor and celebrate their answer. To hear them crying out ever more loudly, in every way that is possible, in response to our constant ask for ever more: no.

To likewise ask the deepest parts of ourselves what we want and need, and to honor and celebrate the exact same response to that same ask for ever more: no.

What if that is the kind of life we held up and pointed out to children and said, "Now *that*, kids, is what I hope you will do when you grow up: sit on the stoop/laugh with your friends/meditate/pray/make love/attend the rally/sip a coffee/eat lunch with your family/meet with your therapist/play an instrument/nap on a blanket in the park—at *least* as much as you spend 'on the job.'"

What if we said to them:

> *Children, listen to your hearts, and listen to the hearts and voices of the world. Abide by what you hear.*

What if we said:

> *Children—everyone—let your life be a song of love and gratitude for all you have been given. Dwell in the sacredness of you and the world. Take only what you need; Share all that you do not.*

Chapter 31

Awakening from a Dream
Into the Universe

The fear for me is that the world has been turned inside out. … Indulgent self-interest that our people once held to be monstrous is now celebrated as success. We are asked to admire what our people once viewed as unforgiveable. The consumption-driven mind-set masquerades as "quality of life" but eats us from within. It is as if we've been invited to a feast, but the table is laid with food that nourishes only emptiness, the black hole of the stomach that never fills. We have unleashed a monster.[1]

—Robin Wall Kimmerer

T HIS SYSTEM DOES **not exist to support your liberation.**
This system exists to maximize profit.

It is not about maximizing well-being. It is not even about maximizing utility. It hardly matters if anybody even uses what is produced, not to mention if it is good for them, so long as it makes money. In fact, it is generally fine, routine even, if the product harms people or other creatures or the environment. Unless the harm is really serious and obvious, in which case it may require doing a cost-benefit analysis and running some ads with dolphins or kids or a celebrity, or paying for some studies that show it's not really a problem.

There are countless people, institutions, practices, and many other resources—*and knowings within you*—that exist within and alongside this system that absolutely *are* committed to your liberation. But, echoing the words of Judge Strine in chapter twenty-two, our failure to be clear that profit is the only true end of this system—and that any means that might produce greater profit *will* be pursued—keeps us trapped, and feeds the tremendous peril of our time.

Look at what this system has done to anyone and anything that got in the way of ever more: Indigenous people, enslaved people, children, animals, rivers, mountains, ancient forests, sacred sites, all of the beautiful people who have tried to stand up against it who have been marginalized, jailed, killed.

This system that celebrates a handful of people appropriating more to themselves than most of the world combined, while tens of thousands of people die every day for lack …

This system that rewards the ravaging of the very earth systems that birthed us, that are necessary for our survival, and the survival of millions of species here with us …

This system that saturates life on earth with thousands of chemicals, hazardous waste, and highly radioactive waste that will be dangerous to life for a million years …

This system that withholds medicines from hundreds of millions of people who need them, while encouraging hundreds of millions of others to take medicines they do not need, that are sometimes even harmful to them …

This system that condemns billions of creatures every year to live their precious lives as cogs in factory farms …

This system that generally considers all of this just normal—either not worthy of consideration or just part of the unfortunate but necessary price of making a profit and growing the economy …

This system cannot ever ultimately serve your joy, wonder, health, your very life, because it cannot see past money and possessions as the ultimate measures of well-being and success, purpose even.

Some of us are clearly, significantly better off than others under this system. Many of us clearly benefit in certain ways from it. **But ultimately *all* of us would be far better off** living in a society that didn't measure our worth and purpose in material terms, one that didn't prioritize material things above all else, one that didn't train us to close our hearts to our feelings and our intimate connection with the world, one not so comfortable with violence and destruction and misery.

Again:

One in twenty-five of us seriously considers suicide every year.

Drug overdoses are the number one cause of death under age fifty.

One in fifteen of us abuses alcohol.

Depression rates in the US are up ten-fold since 1940,
and we consume two-thirds of the world's anti-depressants.

More than one in twenty teens are on psychiatric medication.

We are already in the grips of the sixth great extinction.

This is *all* of us. Indeed, you may recall that rates of depression are highest among the financially wealthy, and are twice the national average among corporate executives. And that wealthy children have lower rates of happiness, higher rates of depression and anxiety, and higher rates of substance abuse (of every kind of substance).

We have been trained to try to make up for the lack of things that really do matter in our lives—joy, connection, belonging, time, purpose—not by stepping outside the system, but by always hooking back into it, using work and money and consumption to try to fill these holes, to distract and comfort ourselves in the face of our fears and insecurities and shames.

Meanwhile, the system creates layers upon layers of distance and feel-good illusion between us and the violence on the other end of so much of our consumption—mountains of websites and packaging and ads and stores—and the people in those ads and stores, all of them, at turns, happy, caring, sexy, accomplished, knowledgeable, beautiful, neighborly—all of them conveying the same underlying message: "Stick with us and everything will be fine." All of them collectively reassuring us that this is the natural and obvious way of being, that it is the best of possible worlds and should be celebrated, supported.

This is the make-believe world the system wraps us in, has woven into our economy and politics, into our language and our stories, into the way we think about ourselves and the way we feel in our lives, even as—in the real world—people, cultures, knowledge, plants, animals, the earth's life support systems, are being killed off. Even as—in the real world—our spirits and bodies are being diminished, denied, sacrificed.

AND YET …

[Take a deep breath.]

If we pause for a moment …

[Another breath.]

If we hop off that thundering, obsessed (doomed) train …

[Another deep breath in and out].

If we take a step back—into nature, into our hearts …

[See if you can feel your heart.]

227

When the dust clears a little and the train disappears in the distance,

    when we take a nice long breath and look around,

        and see the trees and mountains,

            the expanse of the night sky,

When we start to hear the sounds of the wind and birds, the crickets and creeks,

    when there is space for us to feel again,

        to hear our selves,

When we notice the miracle of the world vibrating all around us

    and within us,

        the depth and richness of everything,

            as if awakening from a dream into the universe …

There it is, right there before us: the possibility of a different way of living and seeing the world—not a throwback to the past but our own unique journey, reflecting the needs and opportunities of the times we live in, yet rooted in the same reverence and connection that have nourished countless generations of our ancestors.

If we listen to the earth and the other living beings here with us, we will hear them not only saying, "*No,*" to our constant ask of ever more, but also:

Return.

Feel the length of the days, the preciousness of water and light, earth and green. Feel your sacred place among us,

If we listen to the depths of ourselves, we will likewise hear them not only saying "No" to the same ask of ever more, but also:

Come home.

Feel the vastness and beauty of you and the world… Rest in the oceans of love and belonging that are inherent within you, that can be forgotten but never lost.

Return. Come home.

Chapter 32

A Love Song to You and the World

Now is the time to know
That all that you do is sacred.[1]

—Hafiz

If Only We Would Drink

MY FATHER DIED when I was six. In the tremendous shock and pain of that experience, I managed to carry on in large part by cutting him off. What I mean is that I hardly thought about him or talked about him again after he died. His loss was a profound part of me, but losing him cut so deep I couldn't go anywhere near it in terms of touching those memories and feelings.

It wasn't a conscious decision; I was six, it was just what I did, based partly on what was modeled for me by the people and culture around me, and partly on what some part of felt was the only option in order to get through it. The lesson was deeply set in me that sadness and pain can swallow you up whole and block everything else, and that it is unbearable. And so I should always look forward. Do not look back and do not open the door to those feelings or you will be swallowed up.

As I got older, whenever I told anyone about my dad dying, it was like a story I told about someone else. There was very little feeling attached to it. In my thirties and forties, though, I came to realize just how much of my loss of my dad was about what happened inside me and not what happened in the world around me. What cystic fibrosis didn't take from me, I myself had cut away—the memories, the feelings, the connection.

An important part of my journey as an adult has been trying to go back and do some of the grieving I was too overwhelmed to do at the time, to put myself back in that story and to welcome my dad back into my life. I have two kids of my own and that's been an important part of my process. I love them so much and I'm all the time showering them with love in different ways, often when they're not even aware, like when they're asleep or not even with me. Often it's in ways they don't necessarily feel or understand, sometimes in ways they don't particularly appreciate, like when it involves limits related to safety or cookies.

I think about the degree to which they know I love them and can feel it, and also about the degree to which they absolutely do *not* know how much I love them. And I think about what would happen if I were to die while they are still young, as my father did.

It may sound obvious, but at one point well into adulthood it occurred to me that my father must have loved me very much. Perhaps (likely) as much as I love my kids. This realization rocked me. Something long locked away inside me cracked open a bit, the part of me that once experienced the world as someone who had a father, someone loved by their father—feelings long forgotten and walled off.

I thought about what it would be like for me to be in my dad's situation, to know I was going to die soon and that I wouldn't get to be there to experience or support my kids in their lives. With each of these reflections I continued to crack and open further. I still remember how powerfully it affected me when—not until some point in my forties—it occurred to me, and I knew with absolute certainty, that the last time I visited my dad in the hospital he must have thought or written, "I love you." He couldn't actually speak then, so it would have had to be a thought or a note. But even though I had no memory of it, I knew it was true. And I could actually feel his love. It moves me to tears right now just thinking about it. I can open to it, and across the expanse of more than forty years his love can reach me. I can feel it, wonderful and deep and real.

Again I think of my own kids. I love them so much that I know they will be swimming in oceans of my love for the entirety of their lives. My wish for them is that even long after I die they will be able to feel how much I love them, they will be able to dip down into those oceans and their known feelings of my love and be held in it and nourished by it. In fact, how could there not be moments when they feel my love but are not even conscious of it, do not understand why a certain song or word or place evokes that feeling? I hope that in those moments, despite their not necessarily being able to explain it, they will accept it, embrace it. That they will let that younger self within them who knows that connection—who lived and delighted in that connection—be with it and soak up my love, that they will draw on that well of comfort, strength, assurance.

Please let yourselves feel my love. It is wonderful and deep and real. I love you.

THE WORD LOVE is a pretty big umbrella and there are many different experiences and feelings we associate with it. A lot of them have to do with other people—love as something we give to, and receive from, others. These include:

- Companionship and connection
- Tenderness and care
- Affection and sex
- Communion, or dissolving of the self

We don't *need* other people to experience these. To different degrees we can experience them on our own, or with nature, spirit, ancestors, the universe, God. However, there is a quality to how we experience them with other people that can be uniquely powerful and special.

There are other feelings and experiences often associated with love, though, that particularly lend themselves to being experienced on our own, even if there is much to be enjoyed about them with other people as well. Among these are:

- Affirmation that we are special, attractive, that we matter
- A sense of belonging and purpose
- Intimacy, and a space for us to more deeply experience our hearts and emotions
- A sense of wonder, aliveness, joy
- Security

- Escape or distraction from troubles and insecurities

To some degree, each of these may not just simply *lend* themselves to being experienced on their own; they may actually *require* being nurtured independent of other people in order to be fully realized. As Robert Holden noted in chapter twelve, the quality of our relationships with ourselves determines the quality of our relationships with everyone and everything, from family and friends to God and the taxman, from money and time, to happiness and love.

Our ability to both give and receive love is strongly related to where we are in terms of loving ourselves. Looking at some of the things on that list, **how much can we receive the affirmations of others if deep down we don't believe them? How much can we experience our emotions and hearts with others if we are not comfortable with them ourselves? How much can we experience safety or belonging if we feel that another person is all that stands between us and *in*security or *not* belonging?** And so on.

Other people's affirmations and support can be very important. It was, after all, other people that originally muted these important feelings and pressed these negative ideas onto us. Having other people to help us undo that can be very powerful. And still, **a yearning for love, if we can notice more specifically what is behind that yearning, whether it's something on that list or something else entirely, may be a signal not that we should seek these things in other people, but that we should seek them within ourselves, or in other aspects of the world around us,** at least to some degree.

In her book *All About Love*, bell hooks writes:

> One of the best guides to how to be self-loving is to give ourselves the love we are often dreaming about receiving from others. … It is silly, isn't it, that I would dream of someone else offering to me the acceptance and affirmation I was withholding from myself.[2]

Brené Brown writes similarly about our sense of belonging, noting that it "can never be greater than our level of self-acceptance."[3]

> Belonging is the innate human desire to be part of something larger than us. Because this yearning is so primal, we often try to acquire it by fitting in and by seeking approval, which are not only hollow substitutes for belonging, but often barriers to it. … True belonging is the spiritual practice of believing in and belonging to yourself so deeply that you can share your most authentic self with the world and find sacredness in both being a part of something and standing alone in the wilderness.[4]

Toko-Pa Turner, in her moving book *Belonging: Remembering Our Way Home*, writes eloquently in a similar vein:

> Being at home in our lives and embracing our sacred place in all things, is to live into the fullness of life, and at times that may feel like "not belonging" in

some ways, feeling alone or disconnected or disjointed. This will be true even while a knowing, a belonging deep within us is still alive and breathing, saying, "Yes," to this moment. ...

Can I be with my longing? Can I allow the emptiness of what is missing from me to remain without trying to fill it with stand-ins or facsimiles? Can I be longing without the expectation that it be soothed? Can I, in living with absence, become the presence that it is so hungry for?[25]

CLEARLY WE CAN feel love when someone tells us they love us, or when they demonstrate their love for us in some way, with a hug or a kiss, with attention, affirmation, and so on. Very clearly we can also *not* feel love even when someone gives us these things. We can be swimming in love, but **maybe we are hurt or angry or distracted. Maybe we are closed to love, out of fear or pain. Maybe we just don't believe it—don't believe this person loves us, don't believe we are worthy of being loved, don't believe love exists.**

The reverse is true as well: we can feel love even when it is *not* being actively communicated or demonstrated. We can feel someone's love when they aren't even thinking of us or when they're thousands of miles away. We can feel the love of someone even when they're long gone from this world.

These things are true because feeling love is only partly about what's going on outside of us. It is ultimately an inner experience, and it is intimately related to our inner processes. In this way love is different from, say, sunshine, where we can only feel ki (sunshine) when ki is shining on us. It is more like gratitude, not something outside us at all but a state of being that we can potentially awaken within ourselves at any time. The things that happen outside us can absolutely help us open to that state, but:

> **Love is ours to potentially enter into and savor at any time,**
> **if we orient ourselves toward it, if we open to it.**

IT PAINS ME deeply to imagine my kids ever cutting themselves off from my love after I die, or to different degrees before I die. I recognize they would be cutting themselves off from a source of tremendous comfort and support. I can imagine how unnecessarily impoverished their experience of life would be by comparison. What a loss that would be for them and the world, to have such a treasure available and to not know it or be able to ground into it.

Nonetheless, as I reflect on my father and my own life, I can see clearly that this is what I have done. I am moved by the obviousness, the fullness, of what is right here within me and around me if I just open to it, what has always been here for me. And I mourn how much comfort, belonging, and love I have missed out on because I kept it locked away. I'm not beating myself up for this, for what I perhaps needed to do; I am simply trying to acknowledge this truth so I can embrace different possibilities going forward.

At this very moment I can feel my father's love, can feel a warmth and rightness inside me, and I feel lighter. I can tell how much effort it actually takes—*has* taken for more than forty years—to *not* let myself feel that love, to keep it down, to deny myself that deeply impressed, special feeling that was once a core part of my being and my life. It's like there are vast arid terrains within me that are slowly come back to life with just a little water and care—profound, precious parts of myself.

I wonder what rivers of love are flowing through each of us.

How much love have you been held in that you didn't know about at the time, or have forgotten or closed to? How much love from your parents or others who raised you? How much love from those who cared for you, fed you, held you, even briefly? A midwife, a nurse, a family friend. How much love and dear wishes from relatives and neighbors, perhaps ones you hardly knew, perhaps never even met?

How much love have you received from friends and peers throughout your life? How much are you still receiving, even from ones lost to the years, people who remember you once in a while, just as you sometimes unexpectedly remember them? How much have you been loved by people you met only briefly, or even by complete strangers—beaming warmth and delight just seeing you walk down the street when you were little, or perhaps more recently glimpsing you through the window of a shop or in a crowded hall? What about the countless people all around the world who in their regular prayers or meditations send love and well-wishes to all beings everywhere, including you?

I wonder how much love has been showered on you by your ancestors. How much did they love you and pray for you even when you were only a glimmer in their imaginations, or a knowing in their hearts? How much are they still holding you in everything you do and experience?

How much are you being loved in every moment by the wind and trees and light and soil, by nature, creation, the universe, spirit?

I wonder how much each of us might be deeply thirsting for love and belonging
while they are constantly raining down on us
(are perhaps the very essence of what we are),

if only we would drink.

*Lost in awe at the beauty around me, I must have
slipped into a state of heightened awareness. It is
hard—impossible, really—to put into words the
moment of truth that suddenly came upon me then.
Even the mystics are unable to describe their brief
flashes of spiritual ecstasy. It seemed to me, as I
struggled afterward to recall the experience, that self
was utterly absent: I and the chimpanzees, the earth
and trees and air, seemed to merge, to become one
with the spirit power of life itself.*

*The air was filled with a feathered symphony, the
evensong of birds. I heard new frequencies in their
music and also in the singing insects' voices—notes
so high and sweet I was amazed. Never had I been
so intensely aware of the shape, the color of the
individual leaves, the varied patterns of the veins
that made each one unique. Scents were clear as
well, easily identifiable: fermenting, overripe fruit;
waterlogged earth; cold, wet bark; the damp odor of
chimpanzee hair, and yes, my own too. And the
aromatic scent of young, crushed leaves was almost
overpowering.*[6]

—Jane Goodall

Neurological Gating
and the Response of the Heart

Native scholar Greg Cajete has written that in indigenous ways of knowing, we understand a thing only when we understand it with all four aspects of our being: mind, body, emotion, and spirit. I came to understand quite sharply when I began my training as a scientist that science privileges only one, possibly two, of those ways of knowing: mind and body.[7]

—Robin Wall Kimmerer

WE ARE CONSTANTLY receiving an immense amount of information about the world through our senses. Almost all of it is filtered out by different parts of our brains so that only the information that is of the utmost importance is relayed to our conscious minds. We simply could not function if everything our senses are picking up on in every moment were sent to our consciousness for consideration.[8] Brain researcher Arash Javanbakht writes:

> *In a world where we are simultaneously bombarded with a great deal of stimulation, we learn to focus our attention on important stimuli, while filtering out (gating) less relevant stimuli. Sensory gating … is a way of habituation to repetitious and unimportant stimuli for the brain to reserve its limited resources to focus on important stimuli that need processing.*[9]

Every neural network has a series of gating channels, each of which reduces the flow of information. For the deeper parts of the brain, this is done within 50 milliseconds; for the hippocampus, where all of the pathways converge, it takes 250 milliseconds. What information is gated and what is allowed through is determined by a number of factors. These include how new the stimulus is, or how rare or intense it is, or how much it contrasts with other background stimuli. A sudden loud or unusual sound will be allowed through. The same is true for a sudden movement, or the entrance into our visual field of something possibly desirable or threatening, or if there is a significant dimming or increase in light, and so on.[10]

Our conscious minds play a role in this. We might choose to focus on what we are reading in one moment, and in another pause and consider everything we can hear or see around us. Or we can focus on a particular sound, like the wind or a cricket or traffic, or the details of a picture or a flower or our hand. If you take a moment to scan everything you can see and hear, it can help you appreciate just how much information your brain filters out before arriving at the conscious level.

Researcher Stephen Harrod Buhner, in his brilliant book *Plant Intelligence and the Imaginal Realm*, notes that sensory gating is also strongly affected by what we learn to look for or exclude, through our personal experiences or schooling or cultural habituation. He offers a pair of studies as simple examples of this. In one, participants were asked to look at words and determine whether the capital letter in the word was a vowel or a consonant (jewEl, fAble, oRacle, breaTh). In the other, they were asked whether a series of words ended in a vowel or consonant (garden, bottle, potion, leaf). It was found that after these exercises, people's ability to simply read the words for their meaning was strongly disrupted. Buhner writes, "Long-term training in one perspective or the other actually creates a long-term template that automatically gates incoming sensory data." Even stimuli that are novel and intense "will be gated if they do not conform to the nature of expected sensory inputs."[11] **In other words, we can be trained to only experience certain aspects of a thing, like immediate superficial facts, and to not even notice others, like the meaning behind them, or our felt sense of them.**

Buhner cites other studies that demonstrate how significant this can be on a larger scale. For example, it turns out that whether we perceive something as having characteristics associated with being alive (such as having feelings or being responsive to the environment), is highly dependent on whether we believe those things are alive or not in the first place. The sense of the world that has been trained into us—in this case what is alive in the world—determines whether we are able to see attributes of livingness in the world. That information can be completely gated out subconsciously.[12]

(Of course, we do this at a conscious level as well—ignore clear evidence of something we just do not believe, or do not want to see. We can see this in the centuries of horrifying experiments that have been done on animals. People often truly believe that animals writhing in pain from being cut open, or bellowing, or becoming despondent at being separated from their mother, children, or friends, that these are just robotic responses of automatons that do not really have feelings.[13])

Buhner notes that this kind of subconscious gating is relevant also to our "feeling" sense of the world, what might also be called our heart sense of the world, or our intuition, or gut feelings. This felt sense is the primary way—a deeply nuanced and intelligent way—that most living beings, including our evolutionary ancestors, have experienced and responded to the world for billions of years. These "senses" of things, these ways of knowing, are highly evolved and deeply rooted in the core of our beings. They are, as botanist Jolie Elan notes, the mother tongue to each of us.[14] This is true not just in an inherited, ancestral sense, but in our own lives as well—our felt sense of the world is how each of us experienced and responded to the world through all of our pre-language lives, each of us living in our general sense of things and our general sense of how to respond. We didn't need to have words, or even a conscious "thought," our feelings guided us.

Author and activist Audre Lorde once wrote, **"Our feelings are our most genuine paths to knowledge."**[15]

Buhner notes, however, that when the dominant paradigm holds that our feelings are not to be trusted, or that this deep felt sense of the world does not even exist, then our experience of them will be gated, sometimes entirely.

> *As we are schooled, as life has its way with us—and with our hearts … we learn not to follow our hearts, not to grow outward into the world, but to grow up … away from the world. In so doing, we grow away from something essential to our humanness, to our habitation of this world. We can no longer see or feel what is within the surface sensory inputs that we receive; we can no longer experience the luminous with which we are surrounded. We have lost, as [psychologist] James Hillman once put it,* **the response of the heart to what is presented to the senses.**[16]

But that is not the end of the story.

"Despite our cultural immersion in surfaces, our 'growing up,' and our schooling, somewhere inside each of us, those memories reside." We have both our personal memories of our pre-language experience, as well as those billions of years of ancestral knowing built into us. We are hardwired with this beautiful, essential ability, and, Buhner notes, "All of us have the capacity to free those parts—and their unique perceptual experiences of the world."[17]

> *If we should recapture the response of the heart to what is presented to the senses, go below the surface of sensory inputs to what is held inside them, touch again the "metaphysical background" that expresses them, we would begin to experience, once more,* **the world as it really is: alive, aware, interactive, communicative, filled with soul, and very, very intelligent.**[18]

*In the midst of a gentle rain ... I was suddenly
sensible of such a sweetness and beneficent society in
Nature, in the patterning of the drops, and in every
sound and sight around my house, an infinite and
unaccountable friendliness all at once like an
atmosphere sustaining me. ... Every little pine
needle expanded and swelled with sympathy and
befriended me. I was so distinctly made aware of the
presence of something kindred to me, even in scenes
which we are accustomed to call wild and dreary,
and also that the nearest of blood to me and
humannest was not a person or villager, that I
thought no place could ever be strange to me again.*[19]

—Henry David Thoreau

The Fire Behind the Equations

The intellectual history of the last few hundred years has shown that we are not the center of the universe (thanks, Galileo), not the center of God's creative nature (thanks, Darwin), and not the center of a coherent self (ditto, Freud).[20]

—James Twitchell

WE COULD ADD many other discoveries that have further stripped us of much of our sense of importance and mystery: our sun is only one of thirty billion trillion stars; humans have been around for only about .007 percent of the history of the planet (not to mention the universe); the manual on how to make a human, the human genome, has been fully mapped; how we behave and think and feel is determined to a large degree by the three billion base pairs in that genome (for example, the roughly 50 percent of the variation in happiness that is accounted for by our genes); our genes are 99 percent the same as those of chimpanzees and bonobos (and are 70 percent the same as those of sea urchins). On and on it goes. The history of science over the last five hundred years has been, in the words of science writer George Johnson, a "long string of demotions" for humans.[21]

It has likewise involved a long string of apparent demotions for the god/s of many of our ancestors. To the degree that our concept of God has to some degree been about explaining things that we could not otherwise understand, science's stunning and thorough advances in every direction, providing greater understanding and mastery, have seemingly left fewer and fewer blank spaces on the map for God to occupy.

For example, in referring to the once commonly held belief in Europe that the "celestial spheres" moved under the influence of angels, Carl Sagan writes, "The Newtonian gravitational superstructure replaced angels with GMm/r^2, which is a little more abstract. And in the course of that transformation, the gods and angels were relegated to more remote times and more distant causality skeins. The history of science in the last five centuries has done that repeatedly, a lot of walking away from divine microintervention in earthly affairs. ... So as science advances, there seems to be less and less for God to do."[22] This is true to such a degree that many people have concluded that, as Stephen Hawking put it, "science makes God unnecessary. ... The laws of physics can explain the universe without the need for a creator."[23]

Whether this line of thinking resonates with you or not, you are probably familiar with it. It has assumed a powerful place in the world, and by certain standards it has served us incredibly well. Its ability to describe and predict how the universe works in meticulous detail has fueled the stunning technological advances we've seen in the last few centuries, and that are occurring all around us at a frenzied pace. When it comes to sending a rocket to the moon or treating diseases or designing a computer, this scientific worldview offers an amazingly accurate way of understanding and interacting with the world around us.

All of this has had massive implications both for how we think about ourselves and the universe—and for how we *feel* about the ourselves and the universe. Science writer Timothy Ferris notes that for many people, **the scientific revolution challenged most fundamentally our sense of** *belonging*, and that this wasn't so much because of the basic facts that were revealed but because of an entire way of thinking that these scientific discoveries engendered.

> *Thus began an age of apprehension, in which it became fashionable among popularizers of science to wield the nasty vastness of the universe like a club. To be scientifically hip was to parade one's unblinking acceptance of the unflattering proposition that we are but slime, clinging to a speck of dirt in a galactic outback, hurtling ignorantly through a lethal vacuum dotted with uncaring stars.*[24]

In that sense, the scientific history I've just shared is not only significant, but it is also profoundly misleading—and violent—at least in the way it is often told. Because while science is stunningly useful in understanding certain things, it is wholly inadequate when it comes to others—incredibly important others. Philosopher Galen Strawson is eloquent on this point:

> *Physics is magnificent. It tells us a great many facts about the mathematically describable structure of physical reality, facts that it expresses with numbers and equations (e = mc², the inverse-square law of gravitational attraction, the periodic table and so on) and that we can use to build amazing devices. True, but it doesn't tell us anything at all about the intrinsic nature of the stuff that fleshes out this structure. Physics is silent—perfectly and forever silent—on this question.*
>
> *This point was a commonplace one 100 years ago … Stephen Hawking makes it dramatically in his book "A Brief History of Time." Physics, he says, is "just a set of rules and equations." The question is "what breathes fire into the equations and makes a universe for them to describe?"*[25]

Countless scientists have gone to great lengths to emphasize this point. Indeed, it was a point made by every one of the founders and grand theorists of modern physics, including Einstein, Schrödinger, Heisenberg, Bohr, Pauli, de Broglie, Planck, Eddington, and James Jeans.[26] Jeans put it this way: "The essential fact is simply that *all* the pictures which science now draws of nature … are *mathematical* pictures. … They are nothing more." They are not "in contact with ultimate reality" and never can be.

In Plato's Allegory of the Cave, he invites us to imagine a group of people who live in a cave and can see only the wall in front of them and the shadows cast there from the light behind them. For all that they may come to understand of the shadows, all that they can describe and predict about them, on another level they clearly have no sense of what is actually going on. James Jeans writes that no matter how brilliantly we may be able to describe and predict the universe around us using scientific methods, "We are still imprisoned in our cave, with our backs to the light, and can only watch the shadows on the wall," we cannot penetrate any further.[27] Jeans noted that this had always been true of the models that science provided of the world, but he argued that, **from a "broad philosophical standpoint, the outstanding achievement of twentieth-century physics" was not the theory of relativity, or quantum mechanics, or the dissection of the atom, but the way that modern physics made this fundamental limitation of science absolutely clear in a way it hadn't been previously.**

Arthur Eddington, a pioneer of nuclear fusion, put it this way: "The scheme of physics is now formulated in such a way as to make it almost self-evident that it is a partial aspect of something wider," and that it does not and cannot ever tell us anything about this "something wider." "However much the ramifications of [physics] may be extended by further scientific discovery, they cannot from their very nature trench on the background in which they have their being."[28]

The difference between the picture of a thing and the reality of a thing is something like the difference between describing in mathematical terms what a sunset is and seeing a sunset, or describing in scientific terms what a song is, what laughter is, or what love is, versus experiencing them. It's the difference between the map and the territory. It can be beautiful and very useful to understand love and laughter in chemical and neurological terms, and to likewise be able to explain sunshine as electromagnetic radiation and colors as different wavelengths of that radiation, and so on. But these are very limited aspects or representations of them, indeed extremely sterile and detached ways of understanding them.[29]

That detachment is part of the strength of the scientific method, but is also its profound shortcoming compared to the lived experience of things. There is so much wonder and truth in the elegant and insightful explanations science provides of the world, and yet **so much of the wonder and truth of the world lies beyond what formulas and principles and models can ever convey.** Erwin Schrödinger, one of the pioneers of quantum mechanics, wrote:

> *The scientific picture of the real world around me is very deficient. It gives us a lot of factual information, puts all of our experience in a magnificently consistent order, but it is ghastly silent about all and sundry that is really near to our heart, that really matters to us.*[30]

To the degree that we embrace the equations and models of these great figures of science but ignore their warnings about the limitations of these equations and models, we can become distanced from the deeper wonder and truth of the world and ourselves. **To the degree that we believe science defines what is real, what is important, what everything is ultimately about, we can become blind to, and skeptical of, the very existence of everything science cannot speak to, the entire "something wider," and "all that is really near to our heart."**[31]

(And just as we can cloak science and technology in a mystique of authority that extends beyond their actual domain, we can likewise cloak *ourselves and our society* in a mystique of superiority for being scientifically minded or scientifically "advanced." We can use that cloak to keep other things at a safe distance if we feel uncertain or uncomfortable with them, things like emotions, vulnerability, intuition, spirituality. We can hold tightly to science and technology to give ourselves a sense of control and reassurance in the face of complexity, insecurity, mystery, mortality.)

In short, we can presume that in knowing the scientific facts of a thing we have somehow captured the most important aspect of them, when *the opposite is often closer to the truth*. We can write off the importance and legitimacy of our own lived experience, our own feelings, insight, intuition, wisdom. We can close to much of the vibrance, love, and miracle present all around and within us. To borrow from Stephen Hawking's words, **we can find ourselves living in the equations instead of the fire behind them.**

Again Schrödinger:

> *The scientific world-picture vouchsafes a very complete understanding of all that happens—it makes it just a little too understandable. It allows you to imagine the total display as that of a mechanical clockwork which, for all that science knows, could go on just the same as it does, without there being consciousness, will, endeavor, pain and delight and responsibility connected with it—though they actually are. And the reason for this disconcerting situation is just this: that for the purpose of constructing the picture of the external world, we have used the greatly simplifying device of cutting our own personality out, removing it; hence it is gone, it has evaporated, it is ostensibly not needed. …*
>
> *Science is reticent too when it is a question of the great Unity—the One of Parmenides—of which we all somehow form part, to which we belong. Science is very usually branded as being atheistic. After what we said, this is not astonishing. If its world picture does not even contain blue, yellow, bitter, sweet—beauty, delight, and sorrow—if personality is cut out of it by agreement, how should it contain the most sublime idea that presents itself to the human mind?*[32]

Stephen Harrod Buhner writes, "**In substantial ways, the reductionistic science that has been practiced since the mid-twentieth century has programmed all of us, to varying extents, to gate the *meanings* that flow into us from the world.** For some of us, so strongly that life indeed feels meaningless."[33] ("We grow away from something essential to our humanness," "we can no longer experience the luminous," we lose "the response of the heart.")

Intellectual and humanitarian Vaclav Havel writes:

> *The relationship to the world that modern science fostered and shaped now appears to have exhausted its potential. It is increasingly clear that, strangely, the relationship is missing something. It fails to connect with the most intrinsic nature of reality, and with natural human experience. It is now more of a source of disintegration and doubt than a source of integration and meaning. … Classical modern science described only the surface of things, a single dimension of reality. And the more dogmatically science treated it as the only dimension, as the very essence of reality, the more misleading it became. Today, for instance, we may know immeasurably more about the universe than our ancestors did, yet it increasingly seems that they knew something more essential about it than we do, something that escapes us.*[34]

Critical to making the best use of science is to be clear about what it is saying—and what it can say—and all that it is not saying or cannot speak to. It is to go beyond the surface of things that science can illuminate so powerfully, and to open to the luminous other dimensions around us and within us, to live again in "the response of the heart," to reinhabit the depth of ourselves and the world.

It's not about disregarding what science can tell us about the world—reverting to an idea of the sun revolving around the earth, for example, just because that's what it seems like, or ignoring what psychology tells us about the complexities and deceptions of the mind, and so on. *It's about not making the even more dangerous mistake of swinging too far the other way.*

It is about letting science take us everywhere it can, and then, from atop those pillars leaping into the realm beyond. **It's a realm not of certainty or of uniformity; it is rather a realm of immense and dazzling possibility.**

Suddenly the key turned and the door to the universe opened. Nothing changed in my outward perceptions. There were no visions, no sprays of golden light, certainly no appearances by the Virgin Mary. The world remained as it had been. Yet everything around me, including myself, moved into meaning. Everything became part of a single Unity, a glorious symphonic resonance in which every part of the universe was a part of, and illuminated, every other part, and I knew that in some way it all worked together and was very good.

My mind dropped its shutters. I was no longer just a little local "I," Jean Houston age six, sitting on a windowsill in Brooklyn in the 1940s. I had awakened to a consciousness that spanned centuries and was on intimate terms with the universe. Everything mattered. Nothing was alien or irrelevant or distant. The farthest star was right next door and the deepest mystery was clearly seen. It seemed to me as if I knew everything, as if I was everything. Everything ... was in a state of resonance and of the most immense and ecstatic kinship. I was in a universe of friendship and fellow feeling, a companionable universe filled with interwoven presence and the dance of life.[35]

—Jean Houston

Option D:
Opening to the Mystical

Great lions can find peace in a cage,
but we should only do that
as a last resort.

So those bars I see
that restrain your wings,
I guess you won't mind
if I pry them open.[36]

—Rumi

I WAS NEVER exposed to these ideas when I was growing up, either in my family or in school or in the media. My science classes never had a unit on "The Fundamental Limitations of Science" or "The Profound Danger of Reductionist Thinking." At least implicitly I very soundly got the exact opposite message: "If you can't measure it, it doesn't exist," and "Science will continue to roll forward, eventually filling in all the blanks and leaving room for nothing else." (I actually found myself arguing just this point many years ago as a student in a seminar at the University of Costa Rica. I remember the professor looking at me with awe and saying, "Wow, you're so American!")

Unfortunately, as part of that, I did largely fall into the trap of "living in the equations instead of the fire." I have lived what Timothy Ferris noted earlier was perhaps the most significant consequence of the scientific revolution: a stark deficit of a sense of belonging— in the world, in my life, in the universe.

I can hardly give science all the credit though. An equal measure certainly goes to religion. My sense of the universe and myself as radiant and wondrous and alive has been just as powerfully squashed by a similar reductionism in this more "officially" spiritual realm. This is true for many of us.

To be clear, more than half of us in the US identify as religious, finding at least some degree of resonance in the religious traditions we are connected with. For many others, though, our experience has been more the opposite. A quarter of us describe ourselves as spiritual but not religious, and nearly 20 percent say we're neither.[37] For the half of us who do not identify with any religion, a majority of us were "raised as a member of a particular religion before shedding [that identity] in adulthood."[38]

Personally, I can relate to all three of these perspectives: religious, spiritual, and neither. When I was young I felt a vibrancy inside myself and in the world around me, and I was drawn to places where people paid attention to their inner lives and to questions of purpose and community. For me that meant religious spaces. I didn't really encounter that anywhere else, and there was a time when I made a real effort to consistently attend weekly services.

At the same time, though, in those same spaces, I found my feelings and thoughts in many ways boxed up and diminished or denied. I was asked to take everything I experienced as miraculous, wondrous, and companionable in the world and make it all fit within a ready-made concept of the divine as a particular kind of deity. It was a deity with superhuman abilities of knowing everything, being everywhere, being able to do anything, plus being invisible. In other ways it was all too human, with powerful prejudices and anger and violent tendencies. It was a concept that seemed to me pretty obviously a projection of human hopes and fears and limited understandings of the world, along with some of our own worst inclinations. This embodiment of all things miraculous and transcendent was even assigned a gender: male.

I found my relationship with the miracle and vibrancy of the universe reduced to litanies of rules that for me were not nourishing or inspiring, but were instead alienating, and often even contradictory to my own sense of justice and love, including a demand that I worship this supernatural being or things would not go well for me.

I also felt very deeply the oceans of pain and suffering in the world (powerfully impressed on me through my father's death), all of which this all-powerful being seemed to simply allow, or even encourage as just and appropriate. I felt a sense of betrayal in all of this, and it stirred in me a powerful rejection of everything related to these religious ideas. Poet David Whyte describes this rejection vividly:

> If you have a really fierce loss, the loss of someone who's close to you, the loss of a mother, a father, a brother, a sister, a friend—God forbid, a child—then human beings have every right to say, listen, God, if this is how you play the game, I'm not playing the game. I'm not playing by your rules. I'm going to manufacture my own little game, and I'm not going to come out of it. I'm going to make my own little bubble. And I'm going to draw up the rules. And I'm not coming out to this frontier again. I don't want to. I want to create insulation. I want to create distance. Many human beings do that for the rest of their lives. Many do it for just a short period and then reemerge again. But all of us are struggling to be here.[39]

And so it was that around the age of thirteen I just turned away from it all. And by "it all," I don't just mean I turned away from the particular religious views I found so alienating. I mean I just shut off any sense of the miraculous or transcendental entirely. I lacked the imagination within myself, and the guidance outside myself, to conceive of anything of the mystical beyond the limited concepts that were being presented to me in the religious realm, and the similarly limited views being presented—and being refuted or ridiculed—in the scientific realm.

I am struck by this definition of God from the Oxford Dictionary, which captures perfectly what seemed to be my options:

> 1. *(in Christianity and other monotheistic religions) the creator and ruler of the universe and source of all moral authority; the supreme being.*

> 2. *(in certain other religions) a superhuman being or spirit worshiped as having power over nature or human fortunes; a deity.*

There I had it: I could choose between a "ruler of the universe" on the one hand and a "superhuman being" that accounts for wind and fire and human fortunes on the other. I chose instead "none of the above." Or perhaps more accurately, I chose the religion of science, and came to disbelieve or distrust anything people suggested—or that I felt—that didn't accord with a strict reductionist "scientific" worldview.

There was, of course, an option D (an endless variety of option D's, in fact), a fourth way beyond these two simplistic paths and the path of just shutting it all down. And there were plenty of people—some of the most significant figures in both scientific and spiritual traditions—who had walked different varieties of that fourth path, and who had done their best to share with the rest of us some of what they could see and feel there, to bring us closer and invite us to join them. I just couldn't hear any of those voices. I don't think I actually had any contact with them, through school, the media, family, friends, or if I did, they were drowned out in an ocean of options A, B, and C.

Among those voices, Albert Einstein's was the most prominent from the scientific community. To be clear, Einstein was a powerful advocate for science, reason, and empirical study, and he was disdainful of anyone who would contradict science based on religious tradition, dogma, projection, and so forth.[40] He was scathing in his rejection of "God" as generally portrayed in mainstream society and institutional religions, referring to that concept of God as "naive" and an expression of human "fear."[41] He wrote, "It is always misleading to use anthropomorphical concepts in dealing with things outside the human sphere—childish analogies."[42]

Yet Einstein was even more scathing of those who not only threw out this common concept of God but threw out *everything* related to the mystical and divine. He wrote, "I do not share the crusading spirit of the professional atheist whose fervor is mostly due to a painful act of liberation from the fetters of religious indoctrination received in youth."[43] These people "are like slaves who are still feeling the weight of their chains which they have thrown off after hard struggle. They are creatures who—in their grudge against the traditional 'opium of the people'—cannot hear the music of the spheres."[44]

In fact, Einstein felt that believing in a traditional concept of God was preferable to a "lack of any transcendental outlook."[45]

248

The most beautiful emotion we can experience is the mystical. It is the power of all true art and science. ... To know that what is impenetrable to us really exists, manifesting itself as the highest wisdom and the most radiant beauty, which our dull faculties can comprehend only in their most primitive forms—this knowledge, this feeling, is at the center of true religiousness. In this sense, and in this sense only, I belong to the rank of devoutly religious [people].[46]

Einstein was in good company in his thinking. In fact, every one of the major founders of modern physics I referred to earlier was not only explicit about the fundamental limitations of science in understanding the universe, but *every one of them* also embraced a mystical view of the universe. These masters of the hard sciences wrote essays such as "Embracing the Rational and the Mystical" (Pauli), "The Mystic Vision" (Schrödinger), "Beyond the Veil of Physics" (Eddington), "The Mechanism Demands a Mysticism" (de Broglie), "The Mystery of Our Being" (Planck), and "If Science is Conscious of its Limits" (Heisenberg).

Just as the more expansive and mystical ideas expressed by these leading figures are generally absent from most scientific discussions (my science classes also did not feature a unit on "The Mystical Perspectives of the Founders of Modern Physics"), so too are the expansive and mystical often absent from mainstream religious discussions. The religious perspectives most of us are exposed to are extremely limited—and limiting—compared to the ideas shared by some of the deeply respected saints and sages from these very traditions. In fact, just as many of these prominent scientists urged us to recognize the limits of what science can speak to, so too some of the most prominent religious figures have urged us to recognize the limitations of mainstream religious thought. Indeed, their critiques of mainstream ideas of God have often been just as scathing as Einstein's. Consider, for example:

Renowned Catholic theologian Meister Eckhart: *I find nothing more destructive to the well-being of life than to support a god that makes you feel unworthy and in debt to it. I imagine erecting churches to such a strange god will assure endless wars that commerce loves. A god that could frighten is not a god—but an insidious idol and weapon in the hands of the insane. A god who talks of sin is worshipped by the infirm.*[47]

Saint Thomas of Aquinas: *God's compassion and light can never be limited: thus any God who would condemn is not a God at all but some disturbing image in the mind of a child. ... One may never have heard the word "Christ" but be closer to God than a priest or a nun.*[48]

Sufi mystic Hafiz: *Some gods say, "I am not the scarred yearning in the unrequited soul; I am not the blushing cheek of every star and planet." ... Some gods, the ones we need to hang, say, "Your mouth is not designed to know the love of All, was not conceived to consume the luminous realms." Dear ones, beware, beware of the tiny gods frightened men worship to bring an anesthetic control and relief to their sad days.*[49]

Saint Catherine of Siena: *All has been consecrated. The creatures in the forest know this, the earth does, the seas do, the clouds know, as does the heart full of love. Strange a priest would rob us of this knowledge, and then empower himself with the ability to make holy what already was.*[50]

Rabbi Zalman Schachter-Shalomi: *Our ancient faiths have become ververbalized and underexperienced. … Theology is the after-thought of spiritual experience, not the other way around. We are not trying to construct some top-down authoritative system, but to nourish the seeds of our own personal spiritual experience.*[51]

Saint Teresa of Avila: *How did those priests ever get so serious and preach all that gloom? I don't think God tickled them yet. Beloved—hurry.*[52]

Saint John of the Cross: *Only the beauty and light you cannot describe has a place in God's house. … Why does my sacred church not tell you: God only sees God.*[53]

Hindu saint Tukaram: *Look what the insanity of righteous knowledge can do: crusade and maim thousands in wanting to convert that which is already gold into gold.*[54]

The ideas of these and other prominent figures are so much more expansive than what many of us have been conditioned to think of that it can often be hard for us to hear what they are really saying. This is particularly true when they use the word "God," which has such specific and limited connotations for many of us. (Rabbi Zalman writes, "GOD! Oy, what a word. Which handful of letters in the history of the world has accumulated more baggage than these?"[55])

J.B. Phillips, celebrated for his translation of the *New Testament in Modern English*, wrote an influential book titled *Your God is Too Small*. In it he identified thirteen simplistic conceptions of the divine that he found common in Europe and the US. Many of them capture vividly what I kept running into when I was young, and they may sound familiar to you as well. Among them are the Resident Policeman, the Parental Hangover, the Grand Old Man, the Heavenly Bosom, God-in-a-Box, the Managing Director, the Projected Image, and the God of Bethel (who will faithfully look after our interests if we "obey certain rules").

Phillips was heartbroken and angered by these concepts, which he found to be "joyless," "manipulative," "narrow," and "reactionary." He noted, "[People] may be made in the image of God; but it is not sufficient to conceive God as nothing more than an infinitely magnified [person]."[56] He felt that these "dangerous" conceptions often get in

250

the way of our experiencing the vastly richer and more wondrous divine aspect of the universe that he experienced in the world and in his work.[57]

Unitarian Universalist Reverend Forrest Church wrote, "**Sometimes I believe that we're divided between fundamentalists of the left and fundamentalists of the right. The fundamentalists of the right set up a tiny, little God on their altar and worship it, and the fundamentalists of the left torch that God and throw it down and say there is no God.**"[58]

I find myself intrigued, stunned even, by the warnings of these people. They inspire in me a curiosity, a sense of possibility where before I was tightly closed. I have been quite satisfied to torch the "tiny gods" Phillips and Forrest refer to, and believed that was the end of the conversation. Yet here are some of the most renowned figures in these traditions torching these same tiny gods, but also saying, "There is so much more," and inviting me to look again.

I think the eighteenth-century Rabbi Levi Yitzchak put it succinctly when he wrote, "The God you don't believe in, I don't believe in either."[59] **For me, this simple expression is a potent distillation of these other warnings/invitations. It's like a magical phrase, or a radiant, beautiful doorway. "The God you don't believe in, I don't believe in either." Beyond these words I see a luminous path—an option D.**

I imagine Rabbi Yitzchak standing beside me, with these other saints and poets gathered around, me hunched down looking at a small patch of dry, cracked earth in front of me. I imagine them shaking their heads pitifully with me, acknowledging how stifling and arid so many of the ideas are that have been presented to me. And then the rabbi smiles, and with a twinkle in his eye says, "Now look up and see all that still can be, all that is." And I raise my head to see the vibrant, miraculous expanse of the earth and the universe spread before me.

To be clear, it is not that these religious figures or the scientific ones we considered before share a single vision of the divine or the mystical nature of the universe. Rather, what they are collectively offering is an invitation, an urging, that we keep our hearts and minds open to possibilities and experiences far more expansive than what is generally presented to us—what we are generally pressed to accept, *and to go no further*—in both the scientific and religious realms.

It's not about accepting some doctrine or specific vision being forced on us, it's about not fixating on that tiny patch of dry earth, it's about lifting our heads and looking up. To paraphrase Rumi, it's about prying open the bars that restrain so many of our wings. Because, again:

Great lions can find peace in a cage, but we should only do that as a last resort.

[Walking alone through a grove of redwoods in California,] I was struck by the centuries held in their stillness, imagining countless seasons of summer fog and winter storms passing through them like days. And the centuries, too, in the down hulks of fallen trees, grown over with beds of moss and shrubs, new seedlings sprouting from their bark, the fallen petals beside them.

Then something happened. My sense of time collapsed. I felt myself expand beyond the moment of my existence. Suddenly, I saw the forest itself as a blossom, delicate and ephemeral ... I sensed the world as a momentary spring in which life bloomed ... Not large and everlasting ... but small and precious in the immensity of the universe ... it brimmed with unfathomable mystery and beauty.[60]

—Tim McNulty

Glimpses of
the Source of All of Us

IN HER BOOK *The Luminous Web: Essays on Science and Religion*, theologian Barbara Brown Taylor writes:

> In Sunday school, I learned to think of God as a very old white-bearded man on a throne, who stood above creation and occasionally stirred it with a stick. When I am dreaming quantum dreams, what I see is an infinite web of relationship, flung across the vastness of space like a luminous net. It is made of energy, not thread. As I look, I can see light moving through it as a pulse moves through veins. What I see "out there" is no different from what I feel inside. There is a living hum that might be coming from my neurons but might just as well be coming from the furnace of the stars. When I look up at them there is a small commotion in my bones, as the ashes of dead stars that house my marrow rise up like metal filings toward the magnet of their living kin.
>
> Where am I in this picture? I am all over the place. I am up there, down here, inside my skin and out. I am large compared to a virus and small compared to the sun, with a life that is permeable to them both. Am I alone? How could I ever be alone? I am part of the web that is pure relationship, with energy available to me that has been around since the universe was born.
>
> Where is God in this picture? God is all over the place. God is up there, down here, inside my skin and out. God is the web, the energy, the space, the light— not captured in them, as if any of those concepts were more real than what unites them—but revealed in that singular, vast net of relationship that animates everything that is.[61]

I shared some of the warnings that a number of saints and sages offered about what they do *not* want us to get caught up in. Now I want to share some brief glimpses of what they very much *do* want us to get caught up in, their invitations to possibility and love and belonging.

All of them use a variety of words when referring to the divine, including Beloved, Creation, the Infinite, the One, Love, and so on. I believe both they and we deserve to hear them in the fullness of what they were wanting to express, and because the word God generally means something very different for us today than what it meant for them, *starting now and through the rest of this book I will use only these other words when quoting people.*

(Meister Eckhart has suggested that "All language has taken an oath to fail to describe the divine" anyway.[62] And anthropologist Bradford Keeney writes, "It matters not what you call this great mind. It needs no name. Only your mind concerns itself with naming."[63])

I have also in some cases edited lightly to avoid gendering the divine.

These selections are very brief. It could be easy to just read through the words quickly, to take in their surface meaning but miss something deeper. I urge you to take a little time with them. Perhaps take a moment after each one and notice if there was anything you felt drawn to, anything that resonated (despite, perhaps, anything that didn't resonate, or that put you off).

I urge you to consider, what if there is something these people have seen or felt that is true and important for you to hear?

Or, going one further: *What if there is something nestled in among these words that is* ***meant for you***, *something important that is trying to reach you?*

Lao Tzu:

There is something, Something mysterious.
There wasn't sky, nor earth, but this was already.
It was silent, it was alone.
It was standing alone and unchanging,
Making a circle, but still unchanging.
I could call it Mother of the World
But I do not know its name.
I could say Existence or Intelligence
Or—awkwardly—I can say the Great.
The Great: it is eternal motion
Always in motion, infinity,
And the infinite returns into itself.[64]

Saint Thomas Aquinas:

In the meadows my spirit becomes so quiet
that if I put my cheek against the earth's body
I feel the pulse of the One.[65]

"Returning" to the Beloved …
Think about that. Inherent in that word is separation,
and separation from the Divine is never really possible. …

"You cannot be what I am not,"
the Beloved once said to me.[66]

The delight a child can know
tossing a ball into the air,
my Beloved confessed to experiencing
whenever looking at you.[67]

Saint Catherine of Siena:

What then is not a sanctuary?
Where then can I not kneel
and pray at a shrine
made holy by the Divine's
presence?[68]

I remember once walking out in the winter. …

The cold can enliven thanks, my wool coat
became a sacred robe, how happy I felt to be alive.
I waited in a world of magic,
smells of good food,
the street lamps, the smoke coming from chimneys. …

Angels feasted, as I did, on existence, and the Beloved kept saying,
"Have more of what I have made."[69]

255

Hafiz:

My Beloved said,
"My name is not complete without yours."

I thought:
How could a human's worth ever be such?

And the Divine, knowing all our thoughts—and all our
thoughts are innocent steps on the path –
then addressed my heart. ...

"I am made whole by your life. Each soul,
each soul completes me." [70]

The divine lowers its cup into you to drink.[71]

Where is the door to the Infinite?
In the sound of a dog barking,
In the ring of a hammer,
In a drop of rain,
In the face of
Everyone
I see.[72]

Meister Eckhart:

All beings are words of the Divine,
the Beloved's music, art.
Sacred books we are,
for the infinite camps in our souls.
Every act reveals the Divine and expands the Beloved's Being.[73]

256

Is this not a holy trinity: the [heavens], the earth, our bodies?
And is it not an act of worship
to hold a child,
and till the soil,
and lift a cup?[74]

How long do you think
you can just flirt with the divine
before you dissolve in ecstasy?[75]

Saint John of the Cross:

If you want,
the Virgin will come walking down the road
pregnant with the holy, and say,
"I need shelter for the night, please take me inside your heart,
my time is so close." …

Each of us is the midwife of the Divine, each of us.
Yes there, under the dome of your being does creation
come into existence eternally, through your womb, dear pilgrim—
the sacred womb in your soul.[76]

Black Elk:

The first peace, which is the most important,
is that which comes within the souls of people
when they realize their relationship, their oneness,
with the universe and all its powers,
and when they realize that at the center of the universe
dwells Wakan-Tanka,
and that this center is really everywhere,
it is within each of us.[77]

Saint Teresa of Avila:

I have no seams, no walls, no laws.
My frontiers and the Divine's are the same.[78]

Just these two words the Beloved spoke changed my life,
"Enjoy Me."

What a burden I thought I was to carry—a crucifix. ...

Love once said to me, "I know a song,
would you like to hear it?"
And laughter came from every brick in the street
and from every pore in the sky.

After a night of prayer, the Divine changed my life, singing,
"Enjoy Me." [79]

Any real ecstasy is a sign you are moving in the right direction,
don't let any prude tell you otherwise.[80]

The Baal Shem Tov (quoting the Talmud):

Each person should vitally remember,
"For my sake, the entire world was created." [81]

Saint Francis of Assisi:

The Beloved's admiration for us is infinitely greater
than anything we can conjure up
for the Beloved.[82]

No one lives outside the walls of this sacred place, existence.[83]

258

Rumi:

All these miracles are about to drive me crazy:
my elbows, my ears, my nose, my wife's nagging,
and the sweet darkness of the night, and this blanket existence
around my soul,
and my heart connected to the pulse of
every creature.[84]

The wonder of water moving over that rock in the stream
justifies existence. ...

There is a wonderful problem waiting for you
that the Divine and I share:
how to keep from fainting when we
see each other.[85]

Rabia:

I saw that the divine beauty in each heart
is the root of all time and space.

I was once a sleeping ocean
and in a dream became
jealous of a pond.

A penny can be eyed in the street
and a war can break out
over it amongst the poor.

Until we know that the Divine lives in us
and we can see the Divine there,
a great poverty
we suffer.[86]

[I am deeply grateful to Daniel Ladinsky for his beautiful book
Love Poems from God: Twelve Sacred Voices from the East and West,
from which many of these selections are drawn.]

Coming Home

IN HIS FAMOUS lectures on *The Varieties of Religious Experience*, philosopher William James defined religion as the "feeling of being at home in the Universe."[87]

There is something so warm and inviting for me in this definition. Not religion as a set of doctrines that must be believed and obeyed, but **religion as the experience of profound belonging**. Religion as the experience of:

>*the infinite camped in each of our beings …*
>
>*each of us an essential part of the universe, expanding its being …*
>
>*none of us outside the walls of this sacred place, existence …*
>
>*the frontiers of the divine and each of us exactly the same.*

Physicist Michael Gilmore has offered what is for me an equally powerful perspective on science and belonging. As we noted earlier, perhaps the most fundamental challenge the scientific revolution posed was that for many people it stole from them a uniquely special and removed status they thought was theirs. But Gilmore invites us to "view these revolutions **not as demotions but as** *inclusions*."

Seen through this lens, each of these discoveries actually chipped away at a false sense of separateness. It was science inviting us to pull in closer, to feel ourselves tightly woven into the fabric of all. It is science welcoming us back:

>*to the family of living beings,*
>
>*to the family of our geography and food and water and light and air,*
>
>*to the family of all creation.*

It is science as the experience of profound belonging.

Gilmore writes, "The more you understand our origin, evolution, and place in the universe, the more you can feel a part of it. At some point you might even feel more at home. In fact, it could make you feel downright spiritual."[88]

Yesterday as I sat among the rock formations and trees of the Gila I felt a change come upon me. I could feel the soul of that place as a delicate feather touch upon all my senses and, in that moment, a light began to emanate from every stone, every plant and tree, even the soil itself. I dropped then, into a silence as deep and still as any I have known. And as I listened I began to hear ... the music that lies deep within these hills.

It seemed as if I had come home to a place I had always known, the place in which my true family resides. I felt companioned, and loved, and as I began to look deeper, I could sense, even see, the living connections between everything around me; there were golden threads moving through the world, weaving all the apparently separate organisms and objects together. I reached out and touched the nearest and a certain feeling came over me. And I suddenly realized I could follow that feeling through my enhanced sensing, home to where it originated, deep in the foundations of the world.[89]

—Stephen Harrod Buhner

The Intimacy of
You and the Divine

*I could hear the trees talking, the animals talking, people talking—the sound
of the universe singing to itself. Dancing.*

*I was everywhere at once. I was me, but I was bigger than me. I knew the
answers to all the questions. I knew why it all was the way it was. I just
knew.*

*I didn't have a name for that experience. I was awed and at peace and
scared as well. It was big. Bigger than a dream or a wish or a fantasy. It
was real. It completely changed me forever.* [90]

—Jennifer
(from a study on "The Life Impact of Transcendent Experiences")

STUDIES OF MYSTICAL and transcendental experiences reveal some frequent and
significant similarities: a powerful sense of union with other people, nature, the universe,
or "the source of being;" a heightened state of awareness and sensitivity to the senses and
reality; a feeling of being outside of time and ourselves and our normal senses; feelings of
awe, boundless love, peace, belonging. There is also often a sense that this kind of
experience is more real or true than other experiences—there is "a quality of authenticity
that puts it beyond doubt."[91]

There are often significant *differences* among mystical and transcendental experiences
as well. That is certainly true for those of the religious and scientific figures we have met
so far. As I mentioned earlier, the mystical realm they are inviting us into is not one of
certainty or uniformity but of possibility.

This diversity of experiences is sometimes viewed as a critical flaw, a sure sign that we
must be deluding ourselves—either about the nature of any kind of reality beyond the
strictly scientific or about the very existence of any such reality. If we are all having different
experiences of what this "something wider" is like, then they must simply be tricks of the
mind, projections of our hopes and fears, or chemical or hormonal imbalances. It's
considered equally damning that our experiences often mirror the beliefs of the culture
around us (you rarely hear of someone having a mystical experience involving Krishna in
Mexico or the Virgin of Guadalupe in India). So—the argument goes—these experiences
must be social constructs, visions based on our cultural conditioning. If we were having

direct experiences of something "true" about the divine, then our experiences would be consistent across cultures and across time.

I am going to suggest that (1) that is ridiculous, (2) we are not wrong for having different subjective experiences; we are, in fact, every one of us right, in our own way, and (3) it would be bizarre to expect that people not have unique experiences of the divine, experiences that deeply reflect who we are as individuals and who we are within a particular culture and time and place.

Consider the Indian parable of the blind people touching different parts of an elephant and each of them reporting back on what an elephant is like. Their descriptions are totally different, contradictory even. They are each obviously based on a very limited perspective. And even if you put all their descriptions together you would still know almost nothing about an elephant.

But none of that matters or makes them wrong. And so it is with our mystical experiences as well. Rabbi Zalman Schachter-Shalomi, one of the founders of the Jewish Renewal movement, writes, "**I realized that all forms of religion are masks that the divine wears to communicate with us. Behind all religions there's a reality, and this reality wears whatever clothes it needs to speak to a particular people.**"[92]

Author and lay theologian C. S. Lewis, reflecting on trying to make sense of divergent mystical ideas, once suggested, "Heaven will solve our problems, but not, I think, by showing us subtle reconciliations between all our apparently contradictory notions. The notions will all be knocked from under our feet. We shall see that there never was any problem."[93]

It's about humbly acknowledging how vast and incomprehensible the universe is, and life and thoughts and feelings, and that there is only so much any of us could conceivably fathom of the whole. More than that, though, it's about honoring how natural and obvious—*and beautiful*—it is that our mystical experiences are intimately related to who we are, to our unique personalities and experiences, to our individual hopes and fears and pains, and to the textures of our societies and cultures.

Let me confess that my experiences of my children are like those of the blind people. They are based on my own tremendously subjective and limited perspective. They do not match other people's experiences of children, at times they totally contradict them, and even if you put all of our experiences together you would still know almost nothing about the entirety of children. And yet **none of that takes anything away from my experiences of my children, how rich and meaningful they are, and how** *true.* The same goes for my experiences of sunshine, food, music, love, and life in general, all of which are (thank goodness) utterly inseparable from my unique personal history, the culture around me, and my humanness. To try to separate out the subjective aspects of our experiences, or to demand uniformity in them, would be to miss the point. *It would be to walk right past the shimmering jewel in search of a shimmering jewel.*

In fact, going one step further, the analogy of the blind people and the elephant is, at least in some ways, actually deeply misleading. It suggests that there is some consistent, single, objective whole that could possibly contain or describe the divine in the first place. Obviously, there are limitations to what our minds can discern and what our vocabulary can communicate about the miracle of existence and the infinite. But I am speaking of more than this. The analogy suggests that we could at least theoretically bound and objectively describe the mystical, the way we could an elephant. Yet these saints and sages are pointing to something quite different: **the divine as dynamically (beautifully, wondrously) transformed, expanded, recreated in every moment, through each of us and our experiences—through you.** (*Every act reveals the Divine and expands the Beloved's Being … Each of us is the midwife of the Divine … Yes there, under the dome of your being does creation come into existence eternally.*). The mystical is thus, at once, as intimate as our breath and our thoughts, yet also fundamentally incomprehensible, infinite, and ever-changing.

This has several profound implications. First, it means that we each have an utterly unique connection with the divine that is so intimate and appropriate to us that nobody else can ever know the divine or interact with the divine in quite the same way as us. *Nobody can ever know the face of the divine that you know.*

Second, it likewise suggests that the universe/divine can only realize and express a certain aspect of itself through each of us—through you—and can only be seen/appreciated/loved in that aspect through you, that you are an essential aspect of the divine. (*My Beloved said, "I am made whole by your life. Each soul, each soul completes me."… The divine lowers its cup into you to drink …*)

Yes, you and the divine really are that blessed; you get to have the most intimate of connections, one that reflects who you are and what the divine is within you and through you in every new moment, a connection that is utterly singular.

Third, this also suggests that for each of us to have our most meaningful personal experience of the divine—not to mention our most likely and accessible experience—we should definitely *not* reject our experiences just because they seem to reflect our personal longings (and experiences and culture and humanness). It suggests, on the contrary, that **we are best served if we allow ourselves to be drawn into mystical experiences in a form that does, in fact, most deeply resonate for us personally.** (In Jewish commentary, it is suggested that Moses heard the divine speak to him in the voice of his father, that Moses's experiences of his father's love, authority and earliest connection were thus an inseparable, vital part of his experience of the divine.[94])

And it suggests that we very much *should* lean into our longings regarding the mystical, that we should follow those golden threads, not reject them as subjective or suspect ("Oh, that's just what I *want* to believe"). Jesuit priest James Martin has said, "I believe that your deepest desires, the things that you're drawn to, the person you're called to be, are really the Divine's desires for you. I mean, how else would the Divine call us to something?"[95]

Perhaps you imagine the divine as a reflection of human form, perhaps with what are often thought of as female or male energy. Or both. Or neither. Or perhaps for you the divine evokes some aspect of nature, or perhaps light, energy, a luminous net, an infinite circle, or something entirely unknowable. Maybe you associate a certain tone or feeling with the divine. Whatever you are drawn to, be with that. Feel the "yes" within you and around you when you do. *That is not you deceiving yourself; that is you entering an intimate and powerful well of truth that exists in that exact way only for you.*

Rabia, a prominent figure in the Sufi tradition, speaks to the importance of the very name we use for the divine:

> *Would you come if someone called you*
> *by the wrong name?*
> *I wept, because for years the Divine did not enter my arms;*
> *then one night I was told a secret:*
> *Perhaps the name you call the Transcendent …*
> *is just an alias.*
> *I thought about this, and came up with a pet name*
> *for my Beloved I never mention to others.*
> *All I can say is—it works.*[96]

Rabbi Zalman invites each of us to use a name for the divine that evokes what we are most deeply seeking, what most deeply strengthens our sense of connection with the divine. We can draw from the language of religious traditions, or from the writings of people outside those traditions, or from something else entirely, perhaps something that just emerges within us spontaneously. Zalman encourages, "Repeat it over and over to yourself. Say it out loud if you can. Try it out on your tongue. Savor it."[97]

Anthropologist Bradford Keeney writes that the Bushmen actually give the divine a different name each week.

> *Once, after a thunderous Kalahari rainstorm, the frogs made a lot of music. The next day they called the divine, "frog." If you enjoyed that chocolate cake you had last night, call the divine, "chocolate" for a week. Had the best sex of your life last night? If so, call the divine "sex" for a week. The divine is everything, and that means all things.*[98]

Keeney continues, "The divine is honored by the names that are associated with what you love." And, I would add, what a beautiful way for us to deepen the intimacy of our relationship with the divine—to, over time, offer our attention and care to a variety of the countless faces of the divine, and—most importantly—*to note the attention and care that come in return.*

ULTIMATELY, THERE IS nobody who is in a better position, nobody who has greater insight, knowledge, or authority, to give a name to the divine than you. Or to know the divine, to be in relationship with the divine.

Other people may have experiences and language and insights that are inspiring to you and that help to guide you in the unfolding of your own experience. But whoever you are, wherever you are, whatever you have or have not seen, felt or not felt—you are swimming in the divine; you are *of* the divine. As you turn your attention to it, as you open your mind and feelings to it, **you have the opportunity to experience the divine in a way that is as real, as authoritative, as important, as anything anybody has ever experienced or written.** Not the divine relegated to some distant past, or as exclusive to certain people or places, but the divine alive right here and now, as dynamic and present as ki has been in any moment throughout time—in, around, and through you. (*At the center of the universe dwells the Divine, and that center is really everywhere, it is within each of us …*)

I am reminded again of the words of Rabbi Zalman: "Theology is the after-thought of spiritual experience, not the other way around. We are not trying to construct some top-down authoritative system, but to nourish the seeds of our own personal spiritual experience."[99]

You—knowing the divine, naming the divine, being in direct relationship with the divine—that is where the divine beautifully, rightly, desperately belongs, the only place they/she/he/ki/it really, ultimately, can be: *with you, through you, of you.*

Who would dare claim the objective side
is more real than the subjective?[100]

—Werner Heisenberg (pioneer of quantum mechanics)

A skeleton scheme of symbols proclaims its own hollowness. It can be—nay it cries out to be—filled with something that shall transform it from skeleton into substance, from plan into execution, from symbols into an interpretation of the symbols. And if ever the physicist solves the problem of the living body, he should no longer be tempted to point to his result and say "That's you." He should say rather "That is the aggregation of symbols which stands for you in my description and explanation of those of your properties which I can observe and measure. If you claim a deeper insight into your own nature by which you can interpret these symbols—a more intimate knowledge of the reality which I can only deal with by symbolism—you can rest assured that I have no rival interpretation to propose." [101]

—Arthur Eddington (a founder of modern physics)

Tell your people they must learn to wake up their feelings.
Their heart must arise from its sleep.
It must rise and stand up.[102]

—The Elders of the Kalahari Bushmen

Live the Answers

BUT WHAT IF you're still just deluding yourself? What if it's just conceit or a childish sense of self-importance to think that you are special, a miracle, of the divine? What if the profound rightness and oneness with everything that you may feel at times is just a rush of serotonin, or just a reflection of sentimental and simple-minded ideas in our culture that have gotten into your head? (What if they are neurotic memories of the womb and the sign of a deranged mind, as Freud insisted?[103]) What if you believe you can have an intimate relationship with the universe just because you so desperately want to?

Those are good questions.

Here are some more. What if you *are*, in fact, profoundly deluded, but the delusion is imagining that you could ever be anything other than wondrous, a miracle, of the divine? What if the ridiculous conceit is to disregard your profound experiences of belonging, of love flowing through everything, of a shared heartbeat in all of life, as just a rush of serotonin, to presume your intellect knows better? What if the larger culture has indeed prejudiced your mind, and it is a reflection of its indomitable short-sightedness, violence, and reductionism that you feel compelled to deny the luminous pulse of the universe, the interconnectedness of everything? What if your desire to have an intimate relationship with the universe reflects a deep inner wisdom, a knowing? What if the yearning is a remembering, a pull homeward?

At the beginning of this chapter I noted that we can be swimming in love but not feel it. Perhaps because we are hurt, angry, distracted. Maybe we are closed to love out of fear or pain. Maybe we just don't believe it—don't believe we are worthy of being loved, don't believe love exists. Maybe we have lost "the response of the heart to what is presented to the senses." I asked, "How much is each of us deeply thirsting for love and belonging, while love and belonging are raining down on us, (are perhaps our very essence), if only we would drink?"

So many good questions.

Here's what I suggest: let them all go. Steal a rowboat and take it out on a lake on a starry night. Rise with the birds at the first hint of dawn and sit facing east for an hour. Wrap yourself in a blanket and sit outside during a snowstorm. Sit at the edge of the ocean, or on the crest of a mountain, or in a sun-drenched meadow in spring, and just be still. Or nap. Or cry. Or dance.

Forget all the questions.

Just let the answers come.

*As I was watching it [the sunrise], suddenly, in a
moment, a veil seemed to be lifted from my eyes. I
found the world wrapt in an inexpressible glory
with its waves of joy and beauty bursting and
breaking on all sides. The thick shroud of sorrow
that lay on my heart in many folds was pierced
through and through by the light of the world, which
was everywhere radiant. … There was nothing and
no one whom I did not love at that moment.*

*[The movements of the people], their forms, their
countenances seemed to be strangely wonderful to
me, as if they were all moving like waves in the
great ocean of the world. … I seemed to witness, in
the wholeness of my vision, the movements of the
body of all humanity, and to feel the beat of the
music and the rhythm of a mystic dance.*[104]

—Rabindranath Tagore

The Center of the Temple

Dear Artist of the Universe, Beloved Sculptor,
Singer, and Author of my life, …
help me believe the truth about myself,
no matter how beautiful it is. …
Amen.[105]

—Macrina WiederkehrI

I INVITE YOU to consider the fact that you are as wondrous as anything in the universe.

The poet Max Ehrmann wrote, "You are a child of the universe, no less than the trees and the stars."[106] And I would add that it is impossible for a tree or a star to be anything other than perfect. Just as it is impossible for you to be anything other than perfect.

By what measure could we ever pretend this isn't true? Or try to judge one star or tree against another? By some arbitrary human yardsticks of beauty, or brightness, or success? By some standard of the way we think they ought to be? Every star and every tree is a perfect example of one way it looks to be a tree or a star—they are not a variation on some ideal; each of them is the very standard of perfection in their own particular way.

Just as **you are not a variation on an ideal, you are the ideal, a perfect-in-every-way example of one way it looks to be human.** (*One way it looks to be love …*)

And one way it looks to be the universe. Beyond all of these invented human words and categories of star and tree and human, beyond all the boundaries we imagine between one thing and another, between "this" and "that," between us and the world:

We are all, every one of us and everything, seamless aspects of the miraculous whole of the universe.

All of us are simply (stunningly) one way the universe looks, one form this miracle has taken in its constant flow of creation and re-creation. We are not detached, somehow of a different stuff, removed observers, or residents living out our lives in this place, the universe as our backdrop—we are this place, this stuff. We are the universe, as much as anything is. You and that supernova, me and that waterfall, them and the sun and the ocean and gravity.

We are the universe, as much as anything is.

And every aspect of the universe is as miraculous as every other aspect. None of it can be separated out or cut off from the rest and deemed more or less miraculous, wondrous, extraordinary by some arbitrary standards. Imagine the hubris of us telling the universe that we have decided to separate out parts of it (a tree, a star, ourselves), and judge them as inadequate in some fundamental way, as somehow separate from and less than the miracle of everything.

We are the miracle of everything.

That miracle once took the form of the big bang and has now taken the form of you and me and everything that currently exists. The miracle and power of that first moment did not arbitrarily end a few seconds or a few years or a billion years after the big bang. We are the living embodiment of that very same miracle in a different form. We *are* the big bang—the miracle is not out there somewhere far away or in a long-ago time—that power, that miracle is us and everything that exists now.

You are not just a child of the universe, you *are* the universe, and you are as wondrous as any tree or star or other aspect of creation.

Anything that you experience as beautiful, miraculous, wondrous—that is you. The profound geography of your body, the tapestry of your thoughts and movements and feelings—all of you is of that same beauty, miracle, and wonder.

Always. You can't ever leave it behind in some place or some moment or some experience. You can't turn it off, suddenly not be of it. Whatever you are doing, wherever you go, in every moment that is you—beautiful, miraculous, and wondrous.

Whatever evokes in you a deep sense of reverence—the moon, the ocean, a temple of gold and sunlight—you are of that same sacredness, overflowing with it. There is a way that our position relative to something, or the way things can get "centered" in our attention or in our culture—for example, witnessing a ceremony in the heart of a temple, or seeing something on the news or in a movie—can create a sense of us being on the outside or on the fringe of things, not just of that specific thing, but of things in general. It can feel like what is going on "over there" or "out there" is of an entirely different quality than us and our lives—is more special, moving, profound, beautiful, important.

But in terms of your fundamental sacredness and wondrousness, you are the stars to the moon, you are the sky to the ocean, you are a waterfall to the sunlight. You are as central, as worthy, as vital as anything. You are always, always, at the center of the temple. Your life is a sacred rite unfolding in every moment. That is just what you are. You can never be less sacred than anything else.

(*Separation from creation is never really possible ... No one lives outside the walls of this sacred place, existence ... "You cannot be what I am not," the One once said to me ...*)

You are the luminous pulse of creation.

You are the miracle of everything.

Your life is a sacred rite unfolding in every moment.

You are always, *always*, at the center of the temple.

In the center of the shopping district I was suddenly overwhelmed
with the realization that I loved all those people,
that they were mine and I theirs,
that we could not be alien to one another
even though we were total strangers.
It was like waking from a dream of separateness. ...
The sense of liberation from an illusory difference
was such a relief and such a joy to me
that I almost laughed out loud. ...

Now I realize what we all are.
And if only everybody could realize this!
But it cannot be explained.
There is no way of telling people that they are
walking around shining like the sun. ...

Then it was as if I suddenly saw the secret beauty of their hearts,
the depths of their hearts where neither sin nor desire
nor self-knowledge can reach, the core of their reality,
the person that each one is in the Divine's eyes.

If only they could all see themselves as they really are.
If only we could see each other that way all the time.
There would be no more war, no more hatred,
no more cruelty, no more greed. ...
I suppose the big problem would be that
we would fall down and worship each other.[107]

—Thomas Merton

Every Ancestor in Every Breath

I am deeply grateful to Joanna Macy for her many gifts to the world,
which include the guided activity "Harvesting the Gifts of the Ancestors,"
which can be found in her book *Coming Back to Life*, coauthored with
Molly Brown. The following was entirely inspired by and adapted from
that activity.

WE ARE ABOUT to take a journey back in time. Not so far back by the scale of the universe
or this planet, but very far back in human terms. I normally offer this as a guided
meditation, where I read it aloud and include brief pauses. More than any other part of
this book, you might want to pace your reading, perhaps pausing briefly after each step, to
allow for the most meaningful experience.

We start with this moment you are in right now, wherever you are, whatever time of
day it is, whatever time of year. And now:

Imagine yourself moving back through your experiences of this day.

Now to your waking up.

Now evoke some of your experiences from the last week.

And now the last month.

Move back a season … And another … And another … Move through the entire last
year, through the last cycle of our planet around the sun, remembering some of the major
experiences and perhaps a few small, quiet moments.

Speeding up now, move back through the span of your adult years. Consider briefly
the major milestones, so many things you probably couldn't have imagined when you were
young, some that perhaps you could. You: adapting, striving, collapsing, rising. Seasons,
years, days, nights.

Move back into your early adult and teenage years—different thoughts and feelings,
a different body, a different sense of yourself, your struggles and delights. Consider some
of the places you lived and spent time, where you found refuge and challenge. Recall some
of the people.

Now move back through your childhood—you absorbing so much, delighting,
surviving, you doing your best every day depending on the circumstances and your hopes

and needs and abilities. Your body is getting smaller and smaller, until you climb up onto things in this adult-sized world, hold on to adult-sized things with both hands. Now you are so small you are carried in arms; others clean you and keep you warm and feed you.

And now return to the moment you were born, your passage into the larger world.

And keep going beyond. You are now back in your mother's womb, the warmth of your mother's body, the sounds of your mother's voice and heart vibrating through the ocean around you.

Your body is simplifying, fewer and fewer cells … until you are just one cell. And then you are two cells not yet joined.

Whatever your relationship with your biological parents, whether you even know them, their long lines of ancestors are yours as well, and you may continue to follow them. Or, if you feel a deeper connection with other parents or other people who played a more significant role in your life, their ancestors are yours also, and you can choose to follow their lines. Either way, I will refer to their parents as your grandparents—it doesn't matter whether the relationship is biological or not.

Whether you know anything about these people's lives, it doesn't matter, follow these parents or significant people back through the years of their lives. Follow them into their own young adulthood and adolescence, as they also make their way in the world as best they can, given their own circumstances and hopes and needs and abilities. This is a time before you existed in your current biology, but there was a spirit of you carried in these people—an energy, appearance, inclinations. You are rooted in their lives.

Move back with them into their childhood, the people and places, their daily lives and significant experiences. And back to the time when they were carried. And now the moment they were born. And beyond.

Into the lives of your grandparents. And now, speeding up further, notice the changing world around you. No Internet, no cell phones, soon no satellites or rockets. Into the lives of your great-grandparents, and your sixteen great-great-grandparents. No antibiotics, no television, no planes, no radio. And back another generation, and another. No cars, no electrical bulbs.

Very soon we are in the time before industrialization, with its steel and machinery and immense appetite for land and raw material. And then we leave behind European colonization and its violence and destruction and transformation of so many people and places. Many, if not all of your ancestors were deeply affected by this and involved in different ways.

And yet in the blink of an eye, none of that exists yet. Very soon countries don't even exist. And then no printing presses, no guns, no eyeglasses.

As we speed up our journey, following the lines of your ancestors, soon the Mongol Empire is behind us, and then Mohammed, and Christ, and Confucius. Until, by the Gregorian calendar, we reach the year 500 BCE (even though we actually left the Gregorian calendar behind when it was invented a thousand years later in 1584). You now have roughly 65,536 fourteenth-generation great-grandparents in the world, every one of them directly linked to you today.

Another 500 years and there are no pyramids in the Americas, 1,500 years beyond that, none in Africa, 500 more and there are no cities at all. The Buddha is now behind us. Another 700 years and there is no writing, and so everything before this is considered by later humans "prehistory." But there is so much more.

Another 200 years and there are no wheels; 5,500 years before that, there is no metalworking.

Another 1,000 years and there is no agriculture anywhere. It is also around this time that your ancestor lines have spread so wide that every human alive today shares the exact same ancestors as you. Every human alive at that time either has no descendants alive today or they are an ancestor to every human alive today.[108] To be clear, **we are only 4 percent of the way into our journey of the human story**. This means that for the remaining 96 percent of our journey, every one of our ancestors was a forager, and we all share the same ancestors.

If we were to take a mere five seconds to acknowledge each of our generations of ancestors from this point until we arrive at what people today think of as the dawn of *Homo sapiens*, it would take us *twenty-four days nonstop*. But I have us scheduled to arrive in about thirty seconds, so let's skip the formalities and pick up our pace—480 generations per second.

About 20,000 years ago the first people arrived in the Americas; 40,000 years ago the first people arrived in Australia and Europe (still engulfed in the Ice Age). It was around that time that some people started wearing shoes, meaning, as we continue back in time, every step our ancestors took, their feet were directly on the earth or plants.

Around 70,000 years ago the human population was reduced to a few thousand people. Around that same time (give or take 10,000 years), the first significant numbers of people migrated out of Africa.

We are still only 25 percent of the way into our journey. But we are scheduled to arrive right now, so in one final leap, let's span those remaining 230,000 years. We have circled back around the sun three hundred thousand times, and we have traced back some fifteen thousand generations of ancestors.

This is the point when people from our time mark the beginning of *Homo sapiens*, meaning these ancestors already had the same general anatomy and inherent capacities as us today.[109] Of course, there is no precise moment when our ancestors suddenly became humans as we know ourselves. The changes in physiology occurred over massive periods of time, and many of the things we associate with being human go back much further than 300,000 years. We've been building shelters for at least 400,000 years, and controlling fire for at least 800,000 years.[110] It has been suggested that people living 1.3 million years ago already had the language ability of a six-year-old today, meaning they could speak roughly 2,600 words and understand between 20,000 and 40,000 words.[111] We've been using tools for over 2.6 million years.[112]

Casting our gaze still further back in time, we have some 65 million years of primate ancestors before our human ones, we have 150 million years of mammal ancestors before that, and some 3.5 billion years of all of our other ancestors before the mammals.[113] At some point in the earliest periods of life there was our "last universal common ancestor," a single living creature that is an ancestor to everything alive today.

Before we begin our return journey forward in time, let us acknowledge and honor these ancestors and all that we have inherited from them, going back to the very first life, and even further back, to the very wellspring of the universe. Cosmologist Brian Swimme writes eloquently:

> *The elements were bestowed upon us by the stars, the complex compounds given to us by the young Earth, the informed sequences of the genes by the microorganisms, our limbs and organs by [more recent] life forms, and the linguistic symbols carrying our thoughts and feelings by the human venture. We could not see without the work of those who helped shape the eye; could not hear without the work of those who helped shape the ear. The universe created these gifts, lavishing them upon us; our first and deepest response is infinite gratitude.*[114]

You might take a moment to express, in quiet words or with a gesture, your gratitude to these ancestors—your earthly forebears and the living universe.

And now let's bring our attention back to the moment we left off, 300,000 years back. We are in Africa, the birthplace of humans. While our ancestors would not have thought of themselves as a new species, somehow boundaried from their grandparents and beyond, let's allow ourselves some of the thrill of what we, looking back, can nonetheless appreciate as a turning point, a vague beginning of the human story.

With that, let's begin our return journey, and as we move forward through time we're going to consider some of the gifts you have inherited from your ancestors, all of the experiences they lived, the wisdom they held, the physical and personal traits that are present in you through them.

Let's start with two gifts that you have received from every single one of your ancestors. They come from every one of your ancestors because every one of them, going all the way to the beginning of life 4.5 billion years ago, was a survivor. Spanning the vast range of lives they led, who and where they were, despite the sometimes crushing adversity and challenges they faced, every one of your ancestors survived at least long enough to pass on life to another generation. Receive this **legacy of survival** that lives within you: the skill and resilience, the knowledge and intuition, the critical support of family and community.

Similarly, let's acknowledge the incredible **wisdom and endurance of your ancestors' bodies**. Every second of every day of their lives, in countless miraculous and complex ways, their bodies sustained them in life and health. Every one of your ancestors survived the passage from the womb into the larger world, and within every generation of your ancestors there were those who also went through the fire of birthing new life. Receive from your ancestors' bodies these deep wells of wisdom and endurance, the lived experience of crossing over through water and fire.

For the vast majority of your ancestors, their lives were characterized much more by being than doing. Their lives were intimately woven with the cycles of the seasons, the stars, and the moon, the cycles of the weather and the plants and the creatures around them, the cycles of their own bodies. Receive from your ancestors **their lived experience of** *being*, **of presence, of the vastness of time and space.**

Your ancestors learned lessons from all of these relations: the plants and animals, the earth and sky. They knew a vast range of uses for different plants—medicinal, food, ceremonial. They took only what they needed and shared all that they did not. It was just the obvious and natural way of life. Receive these understandings of **living in intimate relationship with the land and life around you.** Receive from them the experience of living with very little in the way of possessions, yet very much in the way of connection, and place, and trust in receiving what they needed.

There is vast **ingenuity and creativity** in your lineage. Your ancestors learned to control fire, built shelters for themselves, invented tools and musical instruments. They birthed languages and songs and dances, stories and ceremonies and art. They adorned themselves with clothing, paints, flowers, feathers, jewelry. Feel this ingenuity and creativity alive throughout your lineage; feel it alive within you.

Your ancestors have been **voyagers**, traveling distances large and small, into the unknown, some responding to a calling and new opportunities, some forced, compelled by hunger, natural forces, violence. About 220,000 years after we began our journey forward in time from the dawn of *Homo sapiens*, people began to migrate out of Africa. Over the span of millennia some of their descendants continued that migration, some arriving as far as Australia about 40,000 years ago, others entering Europe around the same time, others reaching all the way to the Americas some 20,000 years ago.[115] Some of your ancestors were part of this long, extended journey. Others experienced sudden, massive upheavals, such as the more than ten million who were enslaved in Africa and forced to the

Americas.[116] Receive from your ancestors their **deep knowing of how to create a new life** in new places and new circumstances, how to move forward in a world full of vast possibility and pain. Receive from them an appreciation for the roots that connect us with our past, and for those put down anew.

Your ancestors have known the **depths of joy and love** in all their forms. They pass down to you these essences that you may remember them and recognize them and draw them to you.

Your ancestors have also known the depths of violence and disregard. They do not pass these on to you raw, undistilled. They pass on to you their stories of what they saw and experienced, wrapped in the greatest of tenderness and care for you and themselves. With deep respect, they tuck these stories into your heart and the marrow of your bones. **These stories are gifts.** They have the power to open a greater depth of feeling in you and a more spacious heart. They come with a greater ability to hear and see other people's pain, and to recognize people's capacity for violence, and they come with a greater ability in you to tend to both wounds and threats.

Among your ancestors are people who have done the worst of things. They passed on their pain, alienation, and afflictions in terrible ways. There are gifts here as well. To open to this part of your ancestry is to open to the possibility of greater humility and empathy, a stronger motivation to be a source of light in the world, and a deeper appreciation for the **medicines of redemption, forgiveness, and love**.

Among your ancestors have also been some of the most compassionate and caring of people. They have been there for other people, other creatures, the earth, sometimes with a quiet tenderness, sometimes with a roaring ferocity. *Your ancestors saved lives. Some of your ancestors gave their own lives.* Receive from these ancestors their **deep wells of compassion and commitment**, the pull inside you to kindness, reverence, attention, action.

And now let's pause for a moment. Let's draw close to one of your ancestors from around 10,000 years ago. Without even needing to know where or who they are or what they're doing, just pause to be with them a moment. Consider the preciousness of this person, and the preciousness of this single moment in their life—the vastness of the world and universe around them, the vastness of time before and after them, and **the vastness within this person.** Consider this person's immense complexity and humanity, the people and places in their life, the vast range of joys and pains and hopes and fears. Take in this moment and this person, knowing that your story is connected with theirs.

Now stay with this person, but pull back and feel that moment multiplied by all of the moments over the span of that person's lifetime. Feel yourself a witness to and holder of the sacredness of this person's life. And now feel that life multiplied by all of your ancestors who were alive at the same time, and then multiplied by each generation of your ancestors. Feel yourself a witness to and holder of the sacredness of each of these beings and feel all of them alive within you.

As we resume our journey forward through time, feel yourself drawing closer to your own time, until we are just a dozen or so generations out from your own, nearing the birth of people whose names you may have heard, people whose stories you may have heard parts of. And let's pause one more time. Consider for a moment if there is a gift or two in particular from your ancestors that you would like to wish for yourself, for this child yet to be born, some knowledge or traits or lived experiences of your ancestors that you want to specially distill or amplify for this being whose time is near, something you believe could be particularly helpful for this being, for you.

Call to mind what these specific gifts are. Then cast them forward in time. Feel yourself in this timeless place casting them forward.

And now, feel them arriving inside your body right now, having found you, taking root, blossoming, here for you to draw on.

And let's continue forward again, arriving into the time of your great-grandparents, and now your grandparents, and now the people who would be your parents and significant people in your life, biological and otherwise—these children who would eventually bring you into the world, the people who would care for you.

Feel all of those lines and years of ancestors going back to first life coming together in just these few people …

**And now,
feel them all come together in one person, in you …**

Move forward into the time of your life—the newest in this long family. Move forward through the years of your childhood, your adolescence, your adulthood.

Now into the past year, the past month, the last day, the dawning of this morning.

And now arrive back to yourself, to your body, to this present moment, going no further than noticing your next breath.

And now, feel your breath. And **feel every one of your ancestors taking this next breath with you.**

Feel your ancestors with you in every movement and thought.

Feel this ocean within you, which is all of your ancestors and all of their lives, sacred, profound.

Feel all of your ancestors, human and beyond, holding you in the same deep honoring and sacred witnessing you brought to them. Feel them honoring the vastness of your being and this moment, the complexity, the miracle, the beauty of you. Feel their love, their compassion, their gratitude for you.

Take a moment and, in any way that feels right to you, acknowledge your ancestors and express your gratitude for them, for all who came before you—hundreds of millennia of human ancestors, and beyond them the primates, the mammals, on and on to the single-celled, to first life, and beyond them to this planet, the stars, the universe—all of them your ancestors, all here, all part of you.

WHEN YOU ARE ready, we are going to do one more thing.

Take a moment to think about the children in your life—perhaps your own children or other children in your family, children in your neighborhood or in your friends' families.

Now consider for a moment all the children in the world today.

And now consider the cascade forward in time of their children. And their children.

Include in this circle all of the nonhuman beings that are yet to come into the world as well, and their children, generations after generations.

Now feel yourself part of the family of all these future beings' ancestors—you no longer looking back at your ancestors, feeling them within you, but you standing with them, *of* them, looking forward. **You are an ancestor to all of these future beings who will inhabit this miraculous world.**

Now feel all of the gifts alive within you that were shared with you. Without giving any of them up for yourself, without diminishing any of them, only adding to them the richness of who you are and what you have lived, I invite you to:

cast them forward
to all of the beings throughout the expanse of time
who will follow you,
along with your blessings, prayers, reverence, love.

Now sit quietly, still.
Feel your ancestors in you.
Feel yourself and these blessings alive in all the beings yet to come.

Standing at the kitchen window one day,
and looking out at where a path
wound under some maple trees, I suddenly saw
the scene with a freshness and clarity
that I'd never seen before. Simultaneously,
as though for the first time, I fully realized
I was not only on the earth but of it,
an intimate part and product of it.

It was as if a door had briefly opened.
I stood there transfixed. I remember thinking:
"Distant places on the map
such as Tibet and North Africa
are extensions of right here,
all interrelated!" [117]

—Flora Courtois

The Shared Breath of Life

PERHAPS YOU'VE HEARD the expression, "We come into the world alone and we go out of it alone." The idea is that we are ultimately, fundamentally, alone—that is the nature of us and the universe, and our job is to come to peace with it.

I remember a friend once saying this to me and I just nodded and thought about that: aloneness as an essential truth of being human, something we need to understand and embrace.

It was only later, when I found myself reflecting on her words again, that I realized how completely and absurdly I had been duped—not by her, but by our society. Her comment was consistent enough with the general worldview of the society around me that it not only didn't strike me as ludicrous or deranged, as it should have, it rang true. I was like, "Yep." I later did an Internet search and found dozens of people who have written, said, or sung some variation of this. Filmmaker Orson Welles took it to a whole other level, writing, "We're born alone, we live alone, we die alone. Only through our love and friendship can we create the illusion for the moment that we're not alone."[118] Ouch.

I imagine myself going back in time and being among some of my ancestors, speaking with some of them from millennia back, and telling them about this idea, explaining that not just one person but many people, an entire society, basically believes this to be true, experiences the world and themselves this way.

I imagine, at first, a complete disconnect, what I am telling them actually unfathomable. A chuckle, a glint of amusement in the eye, a dismissal with the hand.

"No, it's true," I persist. "Many people in that time really think and feel that way. *I* thought and felt that way about myself and the world, and often still do to some degree." (No need, for now, to go into everything that comes along with that worldview—the massive consumption, violence, and destruction.)

I imagine their continued inability to really comprehend what I'm saying, similar to when the Dalai Lama (chapter twelve) first heard about self-hatred. ("Is that some kind of nervous disorder? Are people like that very violent? How could you think of yourself that way?") I imagine them pressing me:

> *But how can we ever be apart from the earth, from the trees and sky and waters? How can we ever be apart from the sun and moon and wind and other creatures? How can we ever be separated from our own hearts and the divine within us and within everything? It just isn't possible.*

And then a deeper wondering:

How could people just close their eyes to the obvious truth around them? How could someone cut off so much of themselves, just disregard their own feelings and knowing?

I imagine their disbelief then turning to a deep sadness as what I'm saying starts to sink in, as they start to imagine what that must be like to believe that about ourselves, about the universe, "To be cut off from so much love, so much companionship, to be cut off from all of our relations, and from the heart of the universe …"

And now I imagine them turning and speaking directly to *you*, from across the millennia, tenderly taking your hand and, with tears, gazing into your eyes:

> *My poor, dear, wondrous child. What a horrible thing that has been done to you, to have been fed such cruel ideas, to be raised on them and forced to believe them, to think that that is all you are, that that is all the world is. It makes me so sad. I do not know how you bear it, being ripped away from the arms of creation and placed in such a deep, lonely place. I am sending to you immense love and strength, and my deepest wish is that you **please return home**.*

When I was the stream, when I was the
forest, when I was still the field,
when I was every hoof, foot,
fin and wing, when I
was the sky
itself,

no one ever asked me did I have a purpose, no one ever
wondered was there anything I might need,
for there was nothing
I could not
love.

It was when I left all we once were that
the agony began, the fear and questions came,
and I wept, I wept. And tears
I had never known
before.

So I returned to the river, I returned to
the mountains. I asked for their hand in marriage again,
I begged—I begged to wed every object
and creature.[119]

—Meister Eckhart

THE TRUTH, OF course, as should be at least mildly obvious to anyone who cares to reflect on it for a moment, is that it's just not possible to come into the world alone. Every one of us entered the world in the most profound and intimate of relationships with another human being. We spent months not just with that other being, but *in* and *of* that being. Everything we needed to survive and grow, everything that we were, we received through that person. The first light we saw was filtered through their body; our first tastes and smells were of that person and the food they ate, filtered through their amniotic fluid. The first sounds we heard were that person's heartbeat, the air entering and leaving their body, their stomach, their voice.[120]

And that's only the beginning. To answer those questions from our ancestral grandparents—"How can we ever be apart from the earth, from the trees and wind and other creatures, from the divine presence in our hearts and in everything?"—the obvious answer is: we can't. Ever. It's simply not possible.

The essential truth of our existence is not that we are alone, but that *we can never for a moment be even remotely alone.* We are inextricably in every moment connected to and sustained by other people, the world, the universe, our ancestors, the divine. In fact, if we want to understand the true nature of our place in the universe, we could hardly do better than the metaphor of being held in the womb. We spend our entire lives not just with the world, but *of* the world. Everything we need to survive, to grow, everything that we are, we receive through ki, is *of* ki. Any clear distinction between us and the world exists only in our minds.

Consider the air. This world holds you in a body of air that continually provides you with the essential breath of life, something you could not take for granted throughout most of the universe. Through this body of air you are connected with almost every other living being on the planet. **It is each breath entering your body, the gift of oxygen given to you by unknown plants in every part of the world, many long gone now. And it is each breath given by you, the gift of carbon dioxide to enter the bodies of unknown plants the world over, many long after you are gone.**

Consider water. Life first emerged in water, the first single-cell organisms absorbing nutrients straight from the ocean around them. You carry a "souvenir sea" within you, the liquid portion of your blood having a concentration of ions similar to seawater, with sixty thousand miles of arteries, veins and capillaries continually providing nutrients to the cells throughout your body.[121]

Indeed, more than half your body is water, and half of that is exchanged every one to two weeks.[122] **That means a quarter of what will be your body in two weeks is right now evaporating off the surface of the ocean, a lake, the soil. Part of you is floating in the sky as part of a cloud, or is raining down amidst a storm. Part of you is right now within the** body of another plant or animal, or dwelling in the cool darkness amidst rocks in the earth. And that is where a quarter of what is now your body will be in another two weeks.

Consider the 8.7 million species of plants and animals that have coevolved with us on this planet.[123] We all exist in a dazzling flow of energy and interactions that collectively sustain us. The air and water and soil all around you are alive with complex plant and animal communication through pheromones and sounds, are alive with physical linkages and the sharing of information and nutrients. More than eighty thousand plants nourish humans as food—more than one in five of all plants—as well as most animals.[124]

Consider human society. You are held by countless people around the world, through the food you eat, the clothes you wear, the home you live in, the medicines and herbs you take. They sit with you at your table, dress you and keep you warm, shelter you from the rain and cold, tend to your pain and well-being. These people are here, with you, in every moment.

And it is not just their labor and their hands that are with you, but their hearts. Certainly, their work was in part to make money to secure the things they themselves need, but for most of them there was also an element of care and pride in making these contributions. (This book is overflowing with love and well wishes for you!) There is hardly a moment in your life when you are not held to some degree in this care.

Consider the web of life within you. Only about 43 percent of your body is "human." The other 57 percent is made up of trillions of microbes living on and in your body. They depend on you, and you depend on them—they affect your immunology, your digestive and reproductive systems, your skin, even your emotional well-being.[125]

Consider also the mitochondria in your cells. They are essential to the production of energy and are part of the 43 percent of you that is "human," yet they were originally independent cells. Hundreds of millions of years ago they were absorbed into other cells in a symbiotic relationship, and were eventually bequeathed to you, *though they still have their own DNA, unique from yours.*[126]

Consider the vast range of parts to your body, very few of them open to conscious control by you, yet working in every moment to carry you forward in life. That is what all of these microbes, organelles, and organs are doing continually—they are your intimate partners, ever with you, sustaining your collective body, *giving you life, receiving life from you.*

Consider the building blocks of your body. Most of the one hundred trillion cells that make up your body are regularly replaced. Some of the neurons in your brain are with you your entire life, but the cells in your bones are replaced every ten years, those in your skin every few weeks, and those in the lining of your stomach every few days.

More than that, 98 percent of the atoms in your body are replaced every year.[127] That means **98 percent of what will be your body in one year is right now out there somewhere in the world,** much of ki still in the soil and in the bodies of plants and animals. And 98 percent of what is you right now will in one year no longer be with you, but will be out in the world again.

Through these atoms, you are connected with the first moment of creation. Every atom in your body and every atom in everything you encounter—tables, mountains, cups, shrubs, puddles, crows—every one of them has traveled from that first moment of creation across the expanses of space and time to be here now. Many of them passed through the hearts of stars where they were forged into the heavier elements necessary for life. This is what you are made of.

Going smaller still, consider quantum particles. All of the bizarre, incredible things that happen at the quantum level somehow come together to create you and a universe that it is possible for you to exist in. All of those particles that on their own have no mass but acquire mass when they are with other particles; all of them occupying an infinite number of locations simultaneously; all of them occasionally changing into other kinds of particles, and often blinking in and out of existence—**this is what you are made of, every aspect of you continually in a miraculous dance with the creative source of everything.**

CONSIDER HOW DISCONNECTED we and this society have to be, how tightly our eyes have to be closed, how strictly our experience of the world has to be gated, in order to conceive of aloneness as a core aspect of being human. Entirely the opposite of what Orson Welles wrote, belonging and connection are not temporary illusions; the great illusion/delusion is that we could ever really be alone in any fundamental way. It is only that we are embedded in the extreme materialism, reductionism, and short-sightedness of this culture that it is possible for us to miss the obvious fact of our fundamental belonging and interconnection.

(An equally ridiculous and telling expression is "There's no such thing as a free lunch." What does it say about our society that it could train us into that kind of thinking? When "free lunch" is the foundation of our existence, when "gift" and "generosity" are the foundation of the universe, when we are unavoidably, inextricably being sustained in every moment by countless other plants and beings, by this planet and the sun, by creation?)

Ecologist and activist Joanna Macy notes that the profound interconnectedness of life challenges many of the common assumptions in our society. It reveals the world "not as a battlefield or a trap" (or as a bunch of individual, separate units bumping around for a while) but as the **"wellspring of our body and mind."** To comprehend even on a basic level the vast rivers of relationship continually flowing between us and the world, is to understand the world more accurately as **"lover or larger self."**[128]

Just yesterday part of you was breathed out by flowers and grasses.

Tomorrow part of you will be breathed in by plants in crevices in the ocean.

You are the care of migrant farmworkers and weavers, of woodworkers and healers.

You contain within you 5.5 billion years of ancestors,

and every continent on this planet,
and the hearts of stars.

You are an extension of everybody and everything.

Your every thought and act is the breathing out and in of the sacred.

You exist, always, in vast, intimate, and inextricable relationship. This is the hallmark of human existence; this is you.

It was as if I had never realized before how lovely the world was. I lay down on my back in the warm, dry moss and listened to the skylark singing as it mounted up from the fields near the sea into the dark clear sky. No other music ever gave me the same pleasure as that passionately joyous singing. It was a kind of leaping, exultant ecstasy, a bright, flame-like sound, rejoicing in itself.

And then a curious experience befell me. It was as if everything that had seemed to be external and around me were suddenly within me. The whole world seemed to be within me. It was within me that the trees waved their green branches, it was within me that the skylark was singing, it was within me that the hot sun shone, and that the shade was cool. A cloud rose in the sky, and passed in a light shower that pattered on the leaves, and I felt its freshness dropping into my soul, and I felt in all my being the delicious fragrance of the earth and the grass and the plants and the rich brown soil. I could have sobbed with joy.[129]

—Forrest Reid

Love Notes

I NOTED A few sections back that:

> *Anything you experience as precious, inspiring, beautiful—all of you is of that same beauty, miracle, wonder. Whatever invokes in you a deep sense of reverence—the moon, the ocean, a temple of gold and sunlight—you are of that same sacredness, overflowing with it. You are the stars to the moon, you are the sky to the ocean, you are the waterfall to the sunlight.*

What this means is that the universe is constantly showering you with messages of how beautiful, miraculous, and wondrous you are. Because anything that moves you in these ways can be a reminder to you that that is what *you* are as well, that *you are just as beautiful*, just as miraculous, just as wondrous.

When you are struck by the beauty of a flower-filled meadow, know that *you are just as beautiful*. When you are awed by the miracle of a creek, rolling with moss and ferns, know that *you are just as miraculous*. When you are filled with wonder by a sunset, *yes, you are that wondrous.*

You and everything that inspires you are:

each a mirror of the other,
of the very same fabric, the same essence,
extensions of each other,
flowing through each other,
of the same miracle,
dancing together as one,
celebrating each other,
celebrating all that is possible, all that is *true*—the immense beauty, complexity, radiance, of both of you and of everything.

Indeed, when you enter a flower-filled meadow, it is the meeting of two gorgeous meadows—the other, and you. When you encounter a rich gathering of moss and ferns, it is the meeting of two magical bodies. When you watch a sunset, it is the meeting of two luminous wonders.

Which is another way of saying that the world is constantly bathing you in love. Because in every moment the world is offering to you so many of the experiences of love we listed earlier—tenderness, affirmation, aliveness, sanctuary, and more. It is the meadow, the moss, and the sunset reminding you that you are gorgeous, magical, and luminous, and it is so much more:

It is the soil and light winding their way to you, seeking to feed you, to give you life, to become you.

It is the rustling leaves—that is the way they say your name—saying hello as you pass.

It is the ocean, lake, and river saying to you, "You are my child; you are of me."

It is a bear, rubbing ki's back against a tree and eating all the berries ki can fit in ki's belly, affirming that your needs and the ability to meet them are both sacred gifts.

It is the dark night acknowledging the length of a day, the length of a life, and calling you to share with ki the longings and wounds of your heart.

It is a mountaintop reminding you how expansive and timeless you are—spanning horizon to horizon, companion to the wind and clouds and hawks—and how small and transient so many human concerns are.

It is the stars on a dark night, reaching into your heart with a stick to stir up awe and aliveness.

It is a snowstorm taking you by the hand and walking with you into the quietest, most peaceful place within you.

It is the petals of a flower, and their endless variety, singing to you of the exquisiteness of your sexual body, and of the life force of your sexual being.

It is a thunderstorm bending trees, chasing clouds, cracking the air, reminding you of how powerful you are, that you are thunder and rain, that you are the universe being born.

It is the changing seasons and the cycles of life and death, mourning with you your losses and the passage of time, and also affirming just how much they and your heart can hold.

It is the caress of a breeze, the sunshine, the rain—all extensions of your own sensuality, all whispering of the intimacy between you and the world.

It is squirrels and birds, in their exuberance and delight, constantly passing you love notes telling you how wonderful you are and how deserving of joy.

Love yourself as much as you can,
and all of life will mirror this love back to you.[130]

—Louise Hay

IN THEIR BOOK *Life Loves You*, Robert Holden and Louise Hay write about "mirror work." It consists of activities done while looking in a mirror, as a way of opening more to love. Hay, who had an immensely difficult early life, notes that when she first started doing mirror work, it was very uncomfortable. She judged so many things about herself, she could hardly stand to look at herself. She writes that the most difficult words for her to say were, "I love you, Louise."

But she stuck with it, and credits mirror work with changing her life. She shares about a breakthrough she had early on with this work:

> One day, I decided to try a little exercise, I looked in the mirror and said to myself, "I am beautiful and everybody loves me." Of course, I didn't believe it at first, but I was patient with myself, and it soon felt easier. Then, for the rest of the day, I said to myself wherever I went, "I am beautiful and everybody loves me." This put a smile on my face and it was amazing how people reacted to me. Everyone was so kind. That day I experienced a miracle—a miracle of self-love.[131]

The world is full of immense pain and challenge; it's often cruel. This is simply and awfully true, even when we can perhaps discern some kind of purpose that evolves from the pain and challenge. But we are also held in immense love. And miracle. And abundance. The question is, what are we open to seeing and receiving? What are we open to cultivating within ourselves, and to nourishing and affirming around us? Holden notes, "When you let life love you, you feel loveable, and when you feel loveable, you let life love you."

Holden and Hay recommend saying "I love you" to yourself in the mirror for ten days. Then you can try "Life loves me," and then "I am willing to let life love me today." You can also try the phrase Louise found so powerful, "You are beautiful and everybody loves you." Another good one, I would add, is, "Today I will let love flow through me into the world."

Be kind with yourself, don't worry if you find self-judgment showing up and taking some cheap shots at you. Just keep going. And see if you don't notice other people and the world seeming to love you a bit more each day, and you loving them a bit more right back—and you loving yourself more too.

THE WORLD IS in love with you. The world cradles you, whispers to you, holds your hand, dances with you, carries you on its shoulders, feeds and clothes and shelters you, is continually becoming you.

The world is in love with you and ki is sending you messages constantly every day, through every messenger ki can think of, trying to reach you, to get your attention. Ki is continually and urgently trying to remind you that you are a supernova, a waterfall, a snowflake, the dark night, that you are a bear, moss, music, light—that you are an exquisite, brilliant, gorgeous being.

The world loves you.

The question is whether you are open to it, to this vast love,
to the truth of how stunning you are,
how worthy of love you are.

The question is whether you are ready to start opening the love notes
instead of just letting them fall to the ground all around you.

ONE AFTERNOON AS I was absorbed in thinking about all of this, a rain of tiny yellow leaves was falling just outside the window. It filled me, this gorgeous cascade dancing in the sun, the leaves so exuberant in celebrating their beauty, so exuberant in celebrating *my* beauty, a oneness of delight between us. I cried, so moved. And then one of the leaves fell through the window and landed on my hand. There was so much love in that gesture, such a tender kiss, I wept at the beauty of us.

Maybe it was the smell of ripe tomatoes,
or the oriole singing, or that certain slant of light
on a yellow afternoon and the beans hanging thick
around me. It just came to me in a wash of
happiness that made me laugh out loud. …
I knew it with a certainty as warm and clear
as the September sunshine. The land loves us back.
She loves us with beans and tomatoes,
with roasting ears and blackberries and birdsongs.
By a shower of gifts and a heavy rain of lessons.
She provides for us and teaches us to provide for
ourselves. That's what good mothers do.[132]

—Robin Wall Kimmerer

Everything is Meant for You

ONCE, DURING A silent meditation retreat, I came inside from a very cold morning, dark with clouds. I sat down at the table with a bowl of soup. It was quiet and I was feeling very sensitive. I paused and took in the moment, the feeling of it. In an instant, I was overcome with the love in the fact of this warm nourishing soup existing here for me. Not how lucky I was to have the soup, not that the soup and I happened to be here in front of each other at that moment—I could feel that the soup was meant for me, and I was meant for it. I was filled with that truth, and I could feel an ocean of love behind that gift. I started crying, overwhelmed with the immensity of it. (And now, again, unexpectedly, as I write this I can feel it, and again I am crying.)

After a while, still held in the feeling, I started to eat, to receive this love. I looked out the window in front of me and it started to snow, gorgeous flakes drifting down. I heard in my head, "This also is for you." The words were only the most superficial representation of the immense feeling and truth that I felt behind them—so much love and wonder. I was overcome with the beauty being presented to me, and my tears flowed and my body quaked. I watched the snowflakes come down, delighted, each of them so exquisite. And I could feel each of *them* so delighted at their existence and beauty as well. And each of them so delighted at *my* delight in them, at being appreciated that way—all of us sharing the experience, all of us held in the same love, delight, miracle.

I invite you to consider that **everything you experience is *meant* for you, that this world and this life are *for* you**. That you are not incidental. That the experiences you have in the world are not you just happening to walk in on them, things that were just happening anyway and there you are coincidentally experiencing them. Each of them is *for* you.

A bird's song—the fact of you being able to hear it, your personal experience of the song, the feeling of it for you—all of these are *meant for you*. Not that the song is not also for the bird, not that it is not also for the other birds and creatures that hear it, not that it is not also for the air, the leaves, the earth that vibrates with it. Those experiences are meant for them. But *your* experience of the song is meant for you. (**Each person should vitally remember, "For my sake, the entire world was created."**) When you experience the smell of a pine tree, the power of a gust of wind, the spectacle of a sunset—each of these experiences exists for you, is meant for you.

More than that, it is not just these things—the song, the smell, the gust, the vision— each of them as passive subjects being experienced by you; it is each of them also part of the aliveness of everything, an aspect of them also in the experience of being meant for you, *and in the experience of you being meant for them*, both of you in the feeling of the shared experience. It is not you moving through some terrain, detached, with a bunch of other detached people and creatures also moving through the lifeless terrain of this world. It is you and the terrain and the world and all of life together in the experience as world and life and you and them, intimate partners, *all meant for each other.*

MAYBE WHEN YOU read that last section it felt really right to you. Maybe something inside you whispered (or cried out), "Yes!"

Or maybe you had sort of the opposite reaction. "A snowflake? Delighting in itself? Delighting in your delight? Yeah, right …"

I grant that it is entirely possible that my brain, truly unfathomable to me in its complexity, wove for me a story, a deception, about my experience that morning with those snowflakes. Our brains are notorious for filling in blanks and telling us things we want to hear. Or maybe someone read me a story about a talking snowflake when I was a child and it lodged itself in there somewhere. Maybe there was something in the soup …

I would also like to suggest (and the snowflakes are urging me on here) the possibility that there are things you do now know about snowflakes. That you do not know everything about birdsongs, the wind, or sunsets. Or about what all of this "really" is—the world, the universe, life—or about the ultimate nature of existence. That perhaps there is something about infinity that eludes you. I would like to suggest the possibility that perhaps my experience was as true as anything is ever true. That the snowflakes were, in fact, *delirious* with themselves and with me (that's the word they want me to use).

Science has vast amounts to tell us about water molecules and condensation and crystallization. And about birds and their vocalizations, trees and conifer resins, and the scattering of the sun's light in the earth's atmosphere.

And yet …

To experience a snowstorm, a bird's song, the scent of a tree, a sunset, is to live beyond the maps of these things, beyond their scientific explanations.

It is, in fact, to *release* that scientific explaining and ordering; it is to ideally release even the names and descriptions we have given them, so that we can be open to the full and direct experience of these things without being biased or distracted or boxed in.

It is to let science take us everywhere it can, and then, from atop those pillars, to leap into the vast realm beyond (and beneath and within).

Again, the words of Stephen Harrod Buhner:

> If we should recapture the response of the heart to what is presented to the senses, go below the surface of sensory inputs to what is held inside them … we would begin to experience, once more, the world as it really is: alive, aware, interactive, communicative, filled with soul, and very, very intelligent.

WHAT IF THE world loves us, is grateful for us?
What if the universe cares?
What if the waters are glad we are here?
What if the moon delights in our delight?
What if the sky feels our pain and sadness?
What if the rocks love the feel of our movement?
What if the trees celebrate us every day, breathless with our beauty?
What if we can never be alone, are always companioned?
What if we can never not be loved?

I INVITE YOU to consider, again, that perhaps the real delusion is you imagining that you could ever be anything other than wondrous, a miracle, of the divine.

Every pine needle expanded and swelled with sympathy and befriended me ...

That the ridiculous conceit is to disregard your profound experiences of belonging, of love flowing through everything, of a shared heartbeat in all of life, as just a rush of serotonin, to presume your intellect knows better.

I could hear the trees talking, the animals talking, people talking—
the sound of the universe singing to itself. Dancing ...

That the larger culture has indeed prejudiced your mind, and it is a reflection of its indomitable short-sightedness, violence, and reductionism that you feel compelled to deny the luminous pulse of the universe, the interconnectedness of everything.

It was within me that the trees waved their green branches,
that the skylark sang, that lovers and children and widows cried,
and waves crashed against cliffs, and rain evaporated into the air ...

That your desire to have an intimate relationship with the universe reflects a deep inner wisdom, a knowing.

The sweet darkness of the night, and this blanket existence around my soul,
and my heart connected to the pulse of every creature ...

And that the yearning is a remembering, a pull homeward.

The song of a bird as love.
A shower of falling leaves as kisses.
The world as our extended body, as our Beloved.
Us, them, everything, luminous, alive, love.

Psst. PS:

You can never be alone; you are always companioned.

You can never not be loved.

The trees celebrate you every day, are breathless with your beauty.

The rocks love the feel of your movement.

The sky feels your pain and sadness.

The moon delights in your delight.

The waters are glad you are here.

The universe cares about you.

The world loves you and is grateful for you!

As my glance wandered in the treetops
I became aware of undreamed beauty
in the details of the textures of leaves, stems, and branches.
Every leaf, as my attention settled on it,
seemed to glow with a greenish golden light.
Unimaginable detail of structure showed.
A nearby birdsong – the irregular arpeggio
of the siete cantos – floated down.
Exquisite and shimmering,
the song was almost visible.
Time seems suspended;
there is only now
and now is infinite.[133]

—Manuel Cordova Ríos

Your Inherent Belonging
and Sacred Place

indigenous (adj)

    1 : to be of the first people of a place

    2 : existing naturally or having always lived in a place

    3 : to be deeply, inextricably *of* a place

    From the Latin *indigena*, "**sprung from the land.**"

WE ARE EVERY one of us indigenous to this immense, gorgeous planet. We have all of us sprung forth from this land and these waters. Every nuance of our bodies is composed of the body of this planet—the earth, air, water, other plants and creatures—each transformed into our skin, blood, bones, electricity, pheromones. We are all children of this planet, children of the mountains, the rivers, the forests.

We are each of us part of an unbroken chain that connects us back to the very first forms of life on this planet, to the moment they sprang from the womb of this planet's waters. We each carry an inheritance passed down to us from those earliest ancestors; tucked away in each of our DNA are some of the patterns, the gene families, of those very first beings.

This is true not just for us but for all living beings; every form of life on Earth contains that same genetic inheritance carried forward from the first ones, every living being part of our extended family—every dragonfly, cactus, salmon—all of them connected with first life, all our relatives.[134]

For the entire history of humanity, we have existed in an intimate dance with these relatives. Who and what we are has been shaped by the 8.7 million species that share this planet with us, and those that preceded us, a brilliant coevolution, in this tiny, beautiful nook of the universe, home to all of us.

Indeed, we are children not just of this planet, but of the broader universe, of creation. Every particle in our bodies came bursting into existence in the same moment and same place as every other particle everywhere, regardless of the transformations and journeys they may have experienced since then. In a sense your body was there—every particle in your body was there, witness to and participant in the birth of all. Every particle in the most distant stars and worlds and moons is a sibling to the ones in your body, as is every particle here on this planet with you. Every particle in every canyon, breeze, mouse, starfish, volcano—each is your relation.

This is you: indigenous to this world, relation to all of life, relation to all of creation.

We are also, every one of us, descendants of indigenous people. Indeed, almost all of our ancestors were indigenous. All of our ancestors for most of the 300,000 years of human history have been indigenous. The earliest any of our ancestors took up agriculture, a major part of the shift from that way of living, was less than 12,000 years ago, and for most people an indigenous way of life continued well beyond that, for some right up until and through the present.[135]

For many of us, our connection with our indigenous roots may be deeply obscured, whether because our more recent ancestors assimilated into nonindigenous cultures, or because our indigenous cultures were violently denied to our ancestors. Author and activist Lyla June, who is of both Diné and European heritage, and who grew up with a strong connection to her Native American ancestors, writes movingly of her awakening to her obscured European indigenous roots.

> *I have been called a half-breed. I have been called a mutt. Impure. I have been told my mixed blood is my bane. That I'm cursed to have an Indian for a mother and a cowboy for a father.*
>
> *But one day, as I sat in the ceremonial house of my mother's people, a wondrous revelation landed delicately inside of my soul. It sang within me a song I can still hear today. This song was woven from the voices of my European grandmothers and grandfathers. Their songs were made of love.*
>
> *They sang to me of their life before the witch trials and before the crusades. They spoke to me of a time before serfdoms and before Roman tithes. They spoke to me of a time before the plague; before the Medici; before the guillotine; a time before their people were extinguished or enslaved. … They spoke to me of a time before the English language existed. A time most of us have forgotten.*
>
> *These grandmothers and grandfathers set the ancient medicine of Welsh bluestone upon my aching heart. Their chants danced like the flickering light of Tuscan cave-fires. Their joyous laughter echoed on and on like Baltic waves against Scandinavian shores. They blew worlds through my mind like windswept snow over Alpine mountain crests. They showed to me the vast and beautiful world of Indigenous Europe. This precious world can scarcely be found in any literature, but lives quietly within us like a dream we can't quite remember.*[136]

This is you, whoever you are, whoever your ancestors, whatever parts of the planet were home to them: you are the child of indigenous people. For countless cycles of moons and suns, for centuries upon millennia times millennia, your people have lived in intimate relationship with the land and seasons, the moon and stars, the flora and fauna, have lived on the land where their parents lived and theirs before them, have lived their ancestors' languages, stories, rituals.

To be clear, very few of us in the US are Indigenous (capital *I*), as in Native American/Indian. Only about 2 percent of us are descendants of the first people of this land. To be Indigenous to this land is to have a particular connection with place, culture, and ancestry. It comes with a particular political, economic, and historical reality, with both deep blessings and deep wounds, all very much alive and continuing to unfold. This is not about infringing on, or in any way claiming, that identity for any non-Indigenous people. This is about those other very different but also critical understandings of indigenous (lowercase *i*), which contain very important truths for all of us. It is about recognizing the beautiful and essential ways in which every one of us is indigenous to this world, to our lives, to all that is.

It is about recognizing that **you can never be separated from your inherent belonging or sacred place in the order of life and creation.**

No matter the things you've done or haven't done. No matter what your parents or grandparents have done, what your people or your society have done. No matter the things that have been done to you. No matter how wounded you or your people are.

It doesn't matter how superficial or consumptive or violent the culture is that you live in. It doesn't matter how ridiculous or ineffectual or hurtful the economic or political system is around you, or the media or the educational system, or your family.

It doesn't matter if you live in a forest or in a skyscraper, on the edge of the ocean or in a shantytown or a prison. You don't need to travel somewhere special, meet with someone special, read or watch something special, do anything special. **The truth and miracle of you are never any further than where you are, never any further than right there inside you.** You carry it with you, always and everywhere.

This truth can be obscured from us. It can be hidden or buried; it can be ignored or denied. Our culture of consumption, our economy of ownership, our language of resources and utility—these can all obscure our awareness of our fundamental sacredness and belonging. So, too, can the profound disconnect from nature that many of us experience. As well as the pace of life, the stimulation of technology, and the distraction of information. And the disregard … and the insult … and the violence directed at us personally and flowing all around us.

At the same time, many experiences and places and people can *help* us to remember who we are, to feel it, and to live into the truth of our sacredness.

Still, none of these obstacles or supports can ever change the truth.

Your belonging and sacred place in the whole of everything cannot ever be lost or diminished. You can never be an outsider to life or to the universe or to the sacred. You are as intimately indigenous to them—and they are as intimately indigenous to you—as anybody and anything that has ever existed.

The importance we place on love
has a powerful correlation with being happy.
In fact, the importance we place on love
correlates with happiness
in almost exact proportion to the degree to which
the importance we place on money
undermines our happiness.[137]

—Summary of findings by psychologists
Ed Diener and Oishi Shigehiro

Revisiting Love and Sunshine

I NOTED EARLY in this chapter:

> *Clearly we can feel love when someone tells us they love us or demonstrates their love in some way. Very clearly we can also not feel love even when we are swimming in it. Maybe we are hurt, angry, distracted. Maybe we are closed to love, out of fear or pain. Maybe we just don't believe it—don't believe we are loved, don't believe we are worthy of being loved, don't believe love exists.*
>
> *These things are true because feeling love is only partly about what is going on outside of us. It is ultimately an inner experience, and it is intimately related to our own inner processes.* **In this way love is different from, say, sunshine***, where we can only feel it when it is shining on us. It is more like gratitude, not something outside of us at all but a state of being we can potentially awaken within ourselves at any time.*

Hmm …

But then again, thinking on it a bit more, maybe love actually *is* like sunshine. Maybe it *does* need to be shining for us to feel it; it's just that it's always available for us to feel since it's *always* shining, since **we are** *always* **being loved, are swimming in love, are** *of* **love. So in allowing ourselves to feel loved we are simply opening to a love that is already always there, whether or not the source of that love is obvious.**

Maybe all the love we have ever inspired or ever been shown by other people—by anyone, ever—is still present inside us, like a well, or like drops of dew sparkling throughout us: the obvious people like family, friends, and partners, as well as the quieter ones like a neighbor, a midwife, an elementary school teacher, a kind person we passed on the street, a nurse we don't remember. Maybe all of their love is still inside us, not just something given in passing and then used up or gone, but alive and vibrant within us, always.

Maybe it's the love and dreams of our ancestors flowing through our veins, woven into our cells. Maybe their love and dreams were passed down to us along with their many other gifts, like endless points of light spread across the expanse of our beings and our lives.

Maybe it's the love of the countless people in every part of the world, people we will never know, whose lives we cannot even imagine, who at any given moment are, in their thoughts, intentions, prayers, sending love and best wishes to all beings everywhere.

Perhaps it's the love and goodwill that exists, radiating, at the core of every single person and creature, always, whether we can see it or not, whether they can see it or not.

Maybe it's the love and goodwill radiating at the core of ourselves—warm, intimate, gentle, yet powerful and unceasing, regardless of how removed from it we may sometimes feel, regardless of any barriers that may ever get thrown up around it.

Perhaps it's the vast love in the world around us: the moon cradling us, the leaves whispering to us, the flowers singing to us, the squirrels inviting us to play. Maybe it's the love in the plants and other creatures here with us, and the miraculous, continuous flow of life between us. Maybe it's the love present within our bodies: in the organs, nerves, muscles, mitochondria, DNA, and bacteria that sustain us in life and in health.

Maybe it's the miracle of creation, dancing and unfolding within us and through us. Maybe it's the miracle of the entire universe and all of time.

Perhaps it's a divine love that flows through everything, beyond what we can ever really know or comprehend.

Perhaps this divine love and all of these other loves are the same thing, all part of the same ocean. Maybe when we feel love we are just touching into this endless ocean that exists within us and around us, in everything.

Maybe that is part of our purpose, to help channel and give expression to this ocean, to be part of its flow—divine beings with a sacred purpose of love.

Maybe, ultimately, this is what we are—love.

Maybe this is all true. Maybe it's all love.

In our dreams we have seen another world,
an honest world, a world decidedly more fair
than the one in which we now live.
We saw that in this world there was no need for armies;
peace, justice and liberty were so common
that no one talked about them as far-off concepts,
but as things such as bread, birds, air, water. ...

This world was not a dream from the past,
it was not something that came to us from our ancestors.
It came from ahead, from the next step we were going to take.

And so we started to move forward to attain this dream,
make it come and sit down at our tables,
light our homes,
grow in our cornfields,
fill the hearts of our children,
wipe our sweat,
heal our history.[138]

—Marcos
(Former spokesperson for the Zapatistas,
Indigenous people of southern Mexico in rebellion)

Snowstorms, Silence,
and the Dream from Ahead

WE ARE LIVING in the most critical era in human history—and in the history of millions of other species being pushed to extinction. There is nothing in our three-hundred-thousand-year history as *Homo sapiens* that can compare with the amount of pain and suffering we are causing. World wars, plagues, famines—none of these in their vast horror can compare with what we are doing now and what we are building up to.

It has been more than a hundred pages since we sat with and considered the many people, creatures, and places that are suffering the brunt of the violence and destruction in the world today. Let us return to them now, and briefly dedicate our thoughts to them.

Let us call to mind the otters and the redwoods and the people of Bhopal, the children working the cocoa plantations of West Africa and the tin mines of Indonesia. Let us consider the billions of animals trapped in factory farms across the US, and the children, spouses, and parents who refuse to give up on corporations being held accountable under the Torture Victim Protection Act and similar laws. Let us call to mind the vast forests and mountains being sacrificed, and the waters flowing with plastic. Let us hold in our minds the millions of creatures disappearing from the world.

While it has been many pages since we explicitly acknowledged these people and places and beings, they have all been close at hand, here with us on every page. That is because their fates are entwined with our journey of reclaiming the sacred. Their well-being and very survival are woven with our own personal journeys into deeper love and joy and reverence.

As we noted in chapter twenty-four, "Exit Strategies," we cannot remedy or control effects while leaving their causes in place. We must tend to the profound wounds that are feeding this immense violence and destruction. We must tend to our hearts even as we tend to the world around us. As Greta Thunberg put it:

> *People are suffering. People are dying. Entire ecosystems are collapsing. We are in the beginning of a mass extinction, and all you can talk about is money and fairy tales of eternal economic growth. ... How dare you pretend that this can be solved with just "business as usual" and some technical solutions?*[139]

Or, as the elders of the Kalahari Bushmen put it, "Tell your people they must learn to wake up their feelings. Their heart must arise from its sleep. It must rise and stand up."[140]

Yes, it is absolutely true that, as we also noted in chapter twenty-four, it is not enough for us to tend *only* to this "inner work," we also need to actively engage in the world around us. As Martin Luther King Jr. wrote while locked in the Birmingham jail in 1963:

We must come to see that human progress never rolls in on wheels of inevitability. It comes through ... tireless efforts and persistent work.[141]

And as Frederick Douglass said in 1857:

If there is no struggle, there is no progress. ... Power concedes nothing without a demand. It never did and it never will.[142]

And as Derrick Jensen has written more recently:

I am not going to sit back and wait for every last piece of this living world to be dismembered. I'm going to fight like hell for those kin who remain—and I want everyone who cares to join me.[143]

For the children enslaved on cocoa plantations and the animals trapped in factory farms, for this gorgeous world that is our extended body, for all of us, we need to each find our way to actively support political, economic, and cultural change.

But what is also true is that our success in the world around us depends vitally on our success in the worlds *within* us as well. We need to draw closer to the sacredness of ourselves and the world, to heal these fundamental wounds that are fueling so much of this violence and destruction.

When we are disconnected from our hearts and from love and belonging—when our society insists day in and day out that we are not sacred, that we are not beautiful, wondrous, of the divine—then the violence and horror we are capable of causing in the world—the violence and horror that we *are* causing, and that we will continue to cause— is beyond words.

But this is not "just the way things are"—or just the way we are.

Everything changes when we are rooted in our sacredness.

When we can feel that we are an essential part of the universe,
that we are the breath of the divine ...

> then the toys and trinkets and numbers of this system lose their ability to buy us or distract us or sedate us.

When we can feel that we are one with the skies and rivers and trees,
that we are one with everything that is sacred ...

> then our worth, our beauty, our brilliance, are beyond proving—to ourselves or to anyone;

> then we do not need to manufacture a purpose for our lives that involves accumulating things or checking this society's boxes.

When we can feel the infinite camped in our souls,
and our hearts connected to the pulse of every creature ...

then it is easier to sit with the complexity of the world, the impermanence, the sadness and hurt;

then it is easier to keep our hearts open, to not project our inevitable pains and fears into the world, or to pass them on to others.

When we know that we are a face of creation,
and we can see the face of creation in everyone and everything around us ...

then of course we take only what we need, and of course we share all that we do not need. Because *how else would a person live?* Of course we do not trade our precious lives or cause unnecessary harm to others or to the natural world. *Why would we? So we can get more stuff? Or bigger or newer stuff?*

When we can feel the world leaning into us with overwhelming delight and admiration,
showering us with love notes ...

then striving for justice is like taking a breath. It's not about doing something because we know it's right or because we should. It's not even about doing it because we recognize that our well-being is bound up with the well-being of everyone else. When we are filled with love, then helping other people, other creatures, and the land is simply a reflexive outflowing of that love, like pulling our hand back from a fire, or responding to the cry of a child.

Robin Wall Kimmerer writes:

> *What do you suppose would happen if people believed this crazy notion that the earth loves them back? ... Knowing that you love the earth changes you, activates you to defend and protect and celebrate. But when you feel that the earth loves you in return, that feeling transforms the relationship from a one-way street into a sacred bond.*[144]

When we do not feel dependent on this system,
or on ever more money and possessions for our joy and purpose and worth;

when we recognize how little we really need, and how much we already have;

when we recognize how much we already inherently are;

when we know that we are loved;

when we know that we are love,

everything changes.

I am reminded again of the words of Henri Nouwen from chapter twelve:

The change of which I speak is the change from living life as a painful test to prove that you deserve to be loved, to living it as an unceasing "Yes" to the truth of [your] Belovedness.[145]

When we are rooted in love, then it is only obvious and natural that we strive to remake the world, that we actively strive to achieve:

carbon neutrality

        protection for ecosystems
        and restoration of wildlands

    land back and self-determination
    for Indigenous people

        reparations for African Americans

food, clean water,
and shelter for all

      not just jubilee or debt forgiveness
      for poor nations, but restitution for
      the immense violence and theft that
      has been committed against them

   health care and
   quality education for all

        protection for animals

It doesn't mean that all of this is suddenly easy. It doesn't mean that our lives are suddenly uncluttered or simple, or that the path forward is suddenly clear, or that we suddenly have extra time or energy. It doesn't mean there aren't countless other obstacles in the world around realizing these and other much-needed goals. **But with every step we take toward the sacredness of ourselves and the world, the possibilities expand, and the fundamental wounds fueling so much of the violence and destruction in the world are eased and healed.**

Let's be clear: we don't need to be some kind of fully realized, fully healed and all-the-time-in-love-with-ourselves-and-the-world beings, everything worked out and checked off. The world needs us as we are, wherever we are in our journeys. And the world also needs us to stay in that journey, to find ways to keep moving in the direction of our hearts. Besides, as Joanna Macy writes, "Action on behalf of life transforms. … It is not a question of first getting enlightened or saved and then acting. As we work to heal the earth, the earth heals us."[146]

The journey of reclaiming the sacred, this deep dive into our own hearts and the heart of the world—we do it for ourselves, that we may live in greater joy and purpose and truth. But we also do it for the larger web of life and miracle and creation, for all those whose well-being and survival are urgently, vitally, bound up in this journey.

We do it for the otters and the redwoods and the people of Bhopal.

We do it for the children working the plantations, and the animals locked in the factory farms, and the families of those who've been tortured and murdered at the behest of corporations.

We do it for the forests and mountains and waters.

We do it for the millions of species of our kin.

The journey of you embracing the sacredness of you and your life and this world is immensely self-serving, and it is also immensely selfless. It is essential. You are essential.

Our strategy should be not only to confront empire, but to lay siege to it. To deprive it of oxygen. To shame it. To mock it. With our art, our music, our literature, our stubbornness, our joy, our brilliance, our sheer relentlessness— and our ability to tell our own stories. Stories that are different from the ones we're being brainwashed to believe. The corporate revolution will collapse if we refuse to buy what they are selling—their ideas, their version of history, their wars, their weapons, their notion of inevitability.[147]

—Arundhati Roy

THIS SYSTEM IS not long for this world. It is bankrupt—morally, spiritually, ecologically, and, soon enough, economically. And one way or another it is going down. The way it is going now, it is taking a large part of the natural world and a majority of the species on earth with it. It is taking *us* down with it.

And yet ...

[deep breath]

If we pause for a moment,
if we hop off that thundering, obsessed (doomed) train,
if we take a step back—into nature, into our hearts ...

> *When we start to hear the sounds of the wind and birds,*
> *when there is space for us to feel again ...*

>> *When we notice the miracle of the world vibrating all around us*
>> *and within us ...*

There it is, right before us, an entirely different way of living and being, rooted in the wonder, reverence, and connection that have long nourished our ancestors, yet unique to our times and needs and opportunities. ("This world was not a dream from the past, it came from ahead, from the next step we were going to take."—Marcos)

This cloak of "modern society" is no older than a child at the foot of the mountain of our existence as humans. This system, and its sterile view of us and this vast, gorgeous planet, and its horrifying values ("the monstrous as success," as Kimmerer put it)—these have been very different, indeed nonexistent, in most other ages and societies, and they will be different again when we so choose.

The future is spread before us, as vast as the expanse of time behind us.

Author and activist Arundhati Roy writes:

> *Another world is not only possible, she's on her way. …*
> *On a quiet day, if I listen very carefully, I can hear her breathing.*[148]

The entire world is waiting for you expectantly, eagerly, reverently.

I DO NOT know how a person awakening to the truth of their wondrousness and belonging does or does not inherently shift something fundamental in the universe.

I *do* know that, in all of the ways we just explored, with each step we take in the direction of our hearts, the world moves in that direction as well, that our personal liberation feeds our collective liberation.

But also …

I have a profound appreciation for the worth of each of us connecting with our hearts as an end in itself. I have a deep respect for the inherent value in each of us more fully experiencing our sacredness, how beautiful and important each shift is in the universe of our beings, regardless of how it affects the world beyond.

And I know that for some of us, taking even step in the direction of loving ourselves more is the journey of a lifetime, and is an incredible mercy, for us and for the world.

And that is enough. More than enough.

You alone justify the journey to your heart, whatever that looks like—however long, or roundabout, or forward and back that journey may seem.

There is no need to search or wait, or get approval or permission, or accomplish anything. There is no more justification needed, there is no more justification possible. **You alone, right now—always—are all the justification in the universe for your own joy and love and life.**

Ultimately, this may be the most fundamental covenant that exists between you and your heart, and between you and the universe—the greatest gift you can give yourself and the world—to cultivate within yourself a knowing that:

You are the luminous pulse of creation.

You are the shared heartbeat of all of life.

There is nothing anywhere in the universe
more wondrous than you.

Give in. Give yourself the gift of your sacred being, your sacred life, this sacred world.

AS I WRITE this, there is a snowstorm outside, which has turned from a rush of tiny crystals into large flakes tumbling down as far as I can see. There is an immense peace to them, and I can feel them reaching into me. It is the promise of the love note I wrote of earlier: the snowstorm walking with me down into the quietest, most peaceful place within me.

For a moment—the one time in this book that I am asking this—please
pause reading for a moment and close your eyes and give all of your attention
to your next few breaths. Take three extended breaths in and out …

And now, I invite you to connect with this snowstorm that is before me. Feel the quiet blanket of snow descending to the earth here, right now, in front of me.

Feel ki (the snowstorm) reaching out to you, totally unconcerned with time and distance. Feel ki taking you by the hand, offering to walk with you down into the quietest, most peaceful place within you.

Feel it within you, the very same quiet and peace that is in this snowfall. Not just of the same quality, but *the exact same* quiet and peace that is here before me. That is within me. That is within every living being. Feel it within you, the very same quiet and peace that is within everything.

A JOURNALIST REPORTEDLY once asked Mother Teresa, a Catholic saint and recipient of the Nobel Peace Prize, what she says when she speaks with the divine.

She replied that she doesn't say anything, she just listens.

So the journalist asked what the divine says to her.

She replied that the divine doesn't say anything either, but also just listens.[149]

Now is the time to know
That all that you do is sacred. ...

My dear, please tell me,
Why do you still
Throw sticks at your heart
And Love?

What is it in that sweet voice inside
That incites you to fear?

Now is the time for the world to know
That every thought and action is sacred.

This is the time
For you to deeply compute the impossibility
That there is anything
But Grace.

Now is the season to know
That everything you do
Is sacred.[150]

—Hafiz

A Love Song to You

AUTHOR ARNE GARBORG once wrote, "To love a person is to learn the song that is in their heart and to sing it to them when they have forgotten."[151]

Can you hear all of these voices singing to you? Mine and so many others?

("What I would really like to do is chain you to my body, then sing for days and days."—Hafiz[152])

Can you hear them, and me, loving you, regardless of time and distance?

Can you hear all of us singing of the sacredness of You?

You are the breath of the divine.

> *You are all the beauty in the world.*

> *Every cricket, flower, plankton,*
> *and wildebeest calls you sibling.*
> *The tundra, sunlight, stillness,*
> *and song call you kin.*

> *You are your body and you are the land,*
> *and it is often impossible to know*
> *which has been injured, which is remembering,*
> *which is alive, which was dreamed, which needs care.*

You are all of your ancestors in every breath.
And you are an ancestor in the breath
of every being that will follow you.

> *You are a wonderful problem*
> *that the divine and I share:*
> *how to not faint when we see you!*

Your every act reveals the divine
and expands the divine's being.

There is a unique life force that is translated through you.
It has never existed before and will never exist again.
It can only come through you.

You are a thunderstorm raising waves and bending trees.
You are the dawn, placing a kiss of possibility
on the forehead of every creature,
on every leaf and blade of grass.

You are always, always, at the center of the temple.

You are part of an unbroken chain that connects you back
to the moment that first life sprang
from the womb of this planet's waters,
and back further, through the births and deaths of stars.
You are the birth of the universe.

The world is in love with you and is sending you
love notes constantly every day,
through every messenger ki can think of.
You are yourself a love note
from the divine to the world.

The source of everything is within you.
You are the stars,
reaching across the galaxy with their sticks to stir hearts.
You are the rustling of the leaves,
whispering the names of everything.

The wonder of you justifies existence.

You can never be an outsider to life
or to the universe
or to the sacred.

The infinite camps in your soul,

oceans are drawn from your well,

the divine lowers ki's cup into you to drink.

You alone are all the justification in the universe
for your own joy and love and life.

You are the divine and you, sitting together on a porch swing

on a dark night, silent, looking out onto the universe,

both just listening,

to whale songs and comets,

and the miracle of everything.

You are a deep remembering.
You are truth rising.

BUT WHAT IF you're still deluding yourself? **To think that all of these voices and this crazy writer are really singing to you?** To think that you are really all of this?

On the other hand, what if you really are this spectacular, this loved, this worthy of love?

What if any urge within you to deny all of this is the real insanity?

What if any urge you feel to push all of this away doesn't serve anybody—definitely not you, not your friends or loved ones, not the poor or the oppressed, not the disappearing ecosystems, not everyone around you with all of their own urges to deny their sacredness?

What if one of the most important things you can do in your entire life is to listen to these songs, to receive them deeply, and to live by them?

Again, so many good questions.

Here's what you should do: let them all go.

317

Don't even bother stealing a rowboat.

You are beyond the questions. *You are the answers.*

You are the warm blanket wrapped around these questions in the middle of a snowstorm. And you are the snowstorm kissing the questions, and consecrating them and the earth over and over again with each new flake. You are the sun rising, the miracle of "Yes" and "All of this" and "Again" laid before the questions. And you are the birds—and what else would anyone do at the dawn of a new day but sing? You are the stolen rowboat, carrying the questions out onto a lake under a starry night. And you are the lake, and you are the starry night.

You are beyond the questions.

You are the answers.

ONE DAY YOU will hear the entire universe, everything, singing to you of your wonder.

And you will find every fiber of your being singing back.

You are wondrous,

you are beautiful,

you are brilliant.

You are sacred.

Author's Note:
A Request

I spent over twelve years researching and writing this book, giving it breath and body.

I am going to do all I can to get it out in the world and in front of everyone who might appreciate it and benefit from it. At some point, though, I will have carried it as far as I can, and **the life of this book and the course of these ideas will be shaped by you and the other people who read it.**

To the degree that you found beauty in this book, or insight, or inspiration, I am asking you to **please post a review** on the website where you bought it, or on GoodReads and similar sites, and to please share your thoughts on social media.

I am proudly self-publishing this book, and those reviews and recommendations are the lifeblood of a book like this. (And I read all of the reviews!)

I also want you to know that **100% of the proceeds from this book (post-printing and distribution) are going to several nonprofits** doing important work related to the themes of this book. You can learn more about that here: www.reclaimingthesacred.net/proceeds.

This is not because I am independently wealthy (in the financial sense, that is!). In fact, I have taken on a lot of debt to bring this book into the world. It is simply clear to me that that is the spirit in which this book is meant to move in the world. I have been rewarded beyond measure in writing this book, and it feels so good to now hand it off to you, and to let the book and these ideas and any money resulting from the book flow into the world in wonder and gratitude and love.

Thank you for being part of that flow.

Last, **it would be an honor and a delight to have you join me either online or in-person** to dive more deeply into the journey of Reclaiming the Sacred. The courses and workshops I offer are highly experiential and focus on self-love and shame, vulnerability, presence and gratitude, and connecting with our emotions. Previous participants have written, "Stepping into this space every week was like stepping into a whole new world of acceptance and love and growth," "I would like to continue this class every week forever," and "I count taking this course among the best decisions of my life. Transformative is an understatement."

Please visit www.reclaimingthesacred.net/connect for more information.

Wishing you every blessing,

Jeff

Gratitude

Thank you for reading this book.
It is a blessing to me, and I hope it has been a blessing to you.

I want to also thank everyone who helped to bring this book into the world:

First of all, all those whose research, writing, and thinking are cited throughout the book, and for the vast number of people whose ideas and work and love flowed into this book in so many ways unseen and unknown to me, going back countless generations. Thank you.

And to every one of you who has taken a course or class with me, whether K-12, college, or online. You have enriched my life and deepened my vision and my experience of humanity.

Kavitha Rao, the world and my life are immensely better because of you. This book exists in part because of you. Thank you for all the ways you supported me, including being my first connection to some of the people and ideas here.

My children, Samiha and Haiden, I am so blessed to have you in my life, and I am grateful for the depth of love that I know through you. Thank you also for your helpful feedback on some of the final design. And thank you Samiha, for the back cover photo of me!

And my mother. Thank you for your love and care and devotion. Through all of the ups and downs in both our lives, you have always supported me.. (Except the time I wanted to buy that motorcycle and ride it across the country—you were right on that one!)

Lillie Pearl Allen, for your love and brilliance and dedication, and for your service in support of my own liberation and that of so many other people. You shine so brightly.

My dear friend, J Jasper, for your intelligence and insight, your camaraderie and humor, and your consistent support through the entirety of me writing this book.

My friend, , for your amazing generosity and vision in helping shepherd some of the most important and sensitive parts of this book through their final revisions.

Colette Cann, Candice Lowe-Swift, and Eva Woods-Peiró, for believing in me and opening such important doors of possibility for me to bring this work into the world.

Vassar College, for the privilege of getting to deepen this work as a scholar in residence.

Johanna Korby, for your years of work with me, helping me live in greater joy and truth.

The many amazing people who were so generous with their time and insight in giving me feedback on early drafts of the book:

Helen Whybrow, Autumn Brown, Rose Feerick, Aitabé Fornés, Dani McClain, Jessica Simkovic, Will Power, Ryan Parnes, Alex Schein, Lance Pahucki, and Sean Ritchie.

And Barbara Ditenhafer, whose early enthusiasm has been an anchor for me through many years of writing and revisions.

And the many amazing people who were so affirming and helpful in reviewing the close-to-final draft:

Warren Cornwall, Will Kemper, Tammi Price-Lall, Marta Martinez, Theresa Lyn Widman, Trip Smith, Ian Chittenden, Laura Revercomb, Colombe Nadeau-O'Shea, Nancy Case, Kathy Orchen, Arturo Pérez-Godoy, SK Kapur, Noreen Lempert, Teresa Lopez Delgado, Theo Seeds, Maria José Aldanas, Catherine Cusumano, Jenny Loeb, Michelledana Shafran, Lori Hooper, Christine Winus, Will Steiner, Martha Pita, Liz Peace, Emma Cameron, Andy Lehto, Gisela Stromeyer, Maia Martinez, Kristin Mayville, Faith Victoria Palmer.

Special thanks to Antonella DeCicci, Meg Schutte, Sarah Jaeger, and Carollynn Costella, all of whom reviewed multiple chapters!

Ruth Federman, whose depth and sacred tenor are present throughout the critical final chapter.

All of the people who gave me shelter from the storm when I needed dedicated time for reading, writing, and reflecting: Lisa Amato, Jesse Selman, and Nico, & Corina, Tammi & Steeven Price-Lall, Seth Lennon Weiner & Ngoc Nguyen, Olivia Merchant, Jade Netanya Ullman, Shelley Cahn, Zivar Amrami & Zach Friedman, Peter Forbes & Helen Whybrow, Teri Jennings, and Commander Justin of Chez Brollier.

The Wood River Community YMCA and Tira Christensen for not only giving me shelter from the storm, but giving my kids loving and enriching shelter at the same time.

The many public libraries that provided important books and materials, as well as welcoming, connected spaces to work: Howland, Grinnell, Arlington-Boardman, Boise, the Community Library (Ketchum), Warren, Joslin Memorial, Kingston, Elting Memorial, New York, Marlboro, Newburgh, and the entire amazing Mid-Hudson Library System.

Banu Akman and Cenk Bariyanik, for your unwavering enthusiasm and your critical support with the website.

Ashley Baker Reckess, for your unwavering enthusiasm—for life, and for me and my work.

Decora Sandiford and Markell Baker, for your trust, and your critical help in sustaining me and my family.

Jim Andrews, for your generous faith in my vision and writing.

Cayce Myers, for not turning me in for breaking and entering at Highlands—and for helping me with my own journey of reclaiming the sacred.

Elizabeth Romemu, one of the most insightful people I know about human psychology, and such a supportive friend at a critical time in my life.

George Smith and Sherman Teichman at Tufts University, for providing such a powerful foundation for my thinking and writing and engagement in the world.

Be Present, Inc., Kol Hai, the Isabella Freedman Jewish Retreat Center, the Rowe Camp and Conference Center, the Omega Institute, and Romemu, for providing such rich and deep experiences of growth, beauty, and spirit.

Emma Moylan, for your copy editing. (Despite her best efforts, I did insist on straying from the Chicago Manual at times, and some revisions were made after her editing, so I am fully responsible for any mistakes that made it into the final draft!)

Andreina Cardoso, for help with formatting the citations, Ahmad Talha Mahmood, for help with correspondence, Anna Repp, for the fern leaf drawing, and Hazrat Bilal, for the graphs.

There is so much helpful information about self-publishing online, but I want to offer a special thanks to Dave Chesson of Kindlepreneur for so much amazing information, and Jane Friedman for such quality guidance.

All of those who helped in ways big and small over the course of these twelve years of researching and writing this book, but who I am not remembering right now at the end of this long journey. Please know I am grateful.

Lisa Lee Andretta, for being a sounding board for some of the early chapters. But more importantly, thank you for helping me live in sacredness even as I was writing about it.

And a quiet, heartfelt appreciation for you—yes you—reading this long after I am gone from this world. I am sending so much love and best wishes to you.

I offer the whole of my heart in gratitude
for this life and this world,
for the love and beauty and miracle and depth
that is all of us and is everything.

Endnotes

ENDNOTES ARE ALSO AVAILABLE ONLINE, FOR EASIER READING AND CLICKABLE LINKS

www.reclaimingthesacred.net/endnotes

1 - Money Just Can't Buy What It Used To

[1] Lyubomirsky, S. (2008). *The how of happiness: A scientific approach to getting the life you want.* Penguin. 42; Oremus, W. (2013, July 15). A history of air conditioning. *Slate.* slate.com/articles/arts/culturebox/2011/07/a_history_of_air_conditioning.html

[2] In census data, a room by room picture of the American home. (2003, February 1). *The New York Times.* nytimes.com/2003/02/01/us/in-census-data-a-room-by-room-picture-of-the-american-home.html; TV turns on. (n.d.). *Wessel's Living History Farm.* livinghistoryfarm.org/farminginthe40s/life_27.html; US vehicle ownership growth. Federal Highway Administration. (n.d.). *Greater Auckland.* greaterauckland.org.nz/wp-content/uploads/2011/06/vehicle-ownership-rates.jpg

[3] Schumm, L. (2014, May 23). Food rationing in wartime America. *History.* history.com/news/hungry-history/food-rationing-in-wartime-america; World War II rationing. (n.d.). *United States History.* u-s-history.com/pages/h1674.html

[4] Ingraham, C. (2014, April 23). 1.6 million Americans don't have indoor plumbing. Here's where they live. *The Washington Post.* washingtonpost.com/blogs/wonkblog/wp/2014/04/23/1-6-million-americans-dont-have-indoor-plumbing-heres-where-they-live/

[5] U.S. Department of Housing and Urban Development. (2010, July 1). HUD releases 2009 American housing survey. archives.hud.gov/news/2010/pr10-138.cfm; Becker, P. C. (Ed.). (2006). *Social change in America: The historical handbook.* Bernan Press. 55; Mouawad, J., & Galbraith, K. (2009, September 19). Plugged-in age feeds a hunger for electricity. *The New York Times.* nytimes.com/2009/09/20/business/energy-environment/20efficiency.html
6 Wilson, A., & Boehland, J. (2005, July 12). Small is beautiful: US house size, resource use, and the environment. GreenBiz. greenbiz.com/news/2005/07/12/small-beautiful-us-house-size-resource-use-and-environment; Gerrity, M. (2010, June 15). New homes continue to shrink in size, says US Census Bureau Report. World Property Channel. worldpropertychannel.com/us-markets/residential-real-estate-1/real-estate-news-average-size-of-new-homes-new-home-sales-us-census-bureau-new-home-sales-report-nahb-david-crowe-new-home-construction-2689.php
7 U.S. Department of Housing and Urban Development. HUD releases; DOT releases new NHTS showing vehicles in households outnumber drivers. (2003, May 26). *Bureau of Transportation Statistics.* bts.gov/newsroom/2003-statistical-releases; Hu, P. S., & Reuscher, T. R. (2004). Summary of travel trends: 2001 national household travel survey. *Federal Highway Administration.* 11. nhts.ornl.gov/2001/pub/STT.pdf; Owen, D. (2010). The efficiency dilemma. *The New Yorker.* 78. https://www.davidowen.net/files/the-efficiency-dilemma.pdf

[8] Cell phone ownership hits 91% of adults. (2013, June 6). *Pew Research Center.* pewresearch.org/fact-tank/2013/06/06/cell-phone-ownership-hits-91-of-adults/; Smith, A. (2015, April 1). U. S. Smartphone use in 2015. *Pew Research Center.* pewinternet.org/2015/04/01/us-smartphone-use-in-2015/; Channick, R. (2014, July 8). 40% of homes now without a landline. *Chicago Tribune.* chicagotribune.com/business/chi-landlines-survey-20140708-story.html

[9] United States personal income per capita, 1929–2008: Inflation adjusted (2008$). (n.d.) *Demographia.* demographia.com/db-pc1929.pdf

[10] Median income 1947 to 1997 has increased 58%: Measuring 50 years of economic change. *US* (1998). *Bureau of the Census, Current Population Reports P60–203.* 7. census.gov/prod/3/98pubs/p60-203.pdf; Median income 1997 to 2011 has remained roughly the same: Historical income tables: People. Table p15, all races. (n.d.). *The United States Census Bureau.* census.gov/data/tables/time-series/demo/income-poverty/historical-income-people.html

[11] Gabriel, T. (2014, April 20). 50 years into the war on poverty, hardship hits back. *The New York Times.* nytimes.com/2014/04/21/us/50-years-into-the-war-on-poverty-hardship-hits-back.html; Geography of Poverty. (2014, Feb 28). USDA Economic Research Service. ers.usda.gov/topics/rural-economy-population/rural-poverty-well-being/geography-of-poverty.aspx; Sedlak, A. J., Mettenburg, J., Basena, M., Petta, I., McPherson, K., Greene, A., & Li, S. (2010). Fourth national incidence study of child abuse and neglect (NIS-4): Report to Congress. *US Department of Health and Human Services, Administration for Children and Families.* 11–12. childhelp.org/wp-content/uploads/2015/07/Sedlak-A.-J.-et-al.-2010-Fourth-National-Incidence-Study-of-Child-Abuse-and-Neglect-NIS%E2%80%934.pdf; Grohol, J. M. (2011, November 2). The vicious cycle of poverty and mental health. *PsychCentral.* psychcentral.com/blog/the-vicious-cycle-of-poverty-and-mental-health#1; Entin, E. (2011, October 26). Poverty and mental health: Can the 2-way connection be broken? *The Atlantic.* theatlantic.com/health/archive/2011/10/poverty-and-mental-health-can-the-two-way-connection-be-broken/247275; Durury, N. (2009). Keeping gender on the agenda: Gender based violence, poverty and development. *The Irish Joint Consortium on Gender-Based Violence.* realizingrights.org/pdf/Keeping_Gender_on_the_Agenda.pdf; Rector, R., & Sheffield, R. (2011). Air conditioning, Cable TV, and an Xbox: What is poverty in the United States today? *The Heritage Foundation.* thf_media.s3.amazonaws.com/2011/pdf/bg2575.pdf

[12] Rector & Sheffield. Air conditioning, Cable TV, and an Xbox.

[13] Lyubomirsky. *The how of happiness*. 42.

[14] Layard, R., Mayraz, G., & Nickell, S. (2010). Does relative income matter? Are the critics right? *International Differences in Well-Being*, *28*, 141–42. doi.org/10.1093/acprof:oso/9780199732739.003.0006;   Myers, D. G. (2000). The funds, friends, and faith of happy people. *American Psychologist*, *55*(1), 56–67. doi.org/10.1037/0003-066x.55.1.56

[15] Diener, E., & Seligman, M. E. (2004). Beyond money: Toward an economy of well-being. *Psychological Science in the Public Interest*, *5*(1), 3.

[16] Connery, H.S. (Ed.). (2011). *Alcohol use and abuse.* Harvard Health Publications.

[17] Ehrenreich, B. (2009). *Bright-sided: How positive thinking is undermined America*. Metropolitan Books. 3;   Cohen, E. (n.d.). CDC: Antidepressants most prescribed drugs in US *CNN Health*. cnn.com/2007/HEALTH/07/09/antidepressants/index.html

[18] Katz, J. (2017, June 5). Drug deaths in American are rising faster than ever. *The New York Times*. nytimes.com/interactive/2017/06/05/upshot/opioid-epidemic-drug-overdose-deaths-are-rising-faster-than-ever.html

[19] Brooks, D. (2021, July 29). What's ripping american families apart? *The New York Times*. nytimes.com/2021/07/29/opinion/estranged-american-families.html

[20] Crime > violent crime > murder rate: Countries compared. (n.d.). *NationMaster*. nationmaster.com/country-info/stats/Crime/Violent-crime/Murder-rate;   World Health Organization. (2004, Dec.). *Death and DALY estimates, Deaths 2002, GI-W157, GI-W158*. who.int/entity/healthinfo/statistics/bodgbddeathdalyestimates.xls;   Brooks. What's ripping american families apart?;   Bose, J., Hedden, S. L., Lipari, R. N., & Park-Lee, E. (2018). Key substance use and mental health indicators in the United States: Results from the 2017 national survey on drug use and health. *Substance Abuse and Mental Health Services Administration (SAMHSA)*. 44. samhsa.gov/data/sites/default/files/cbhsq-reports/NSDUHFFR2017/NSDUHFFR2017.pdf

[21] Diener & Seligman. Beyond money. 3. That statistic is based on the 1980s, but the situation has only worsened since then: Gray, P. (2010, January 26). The decline of play and rise in children's mental disorders. *Psychology Today*. psychologytoday.com/us/blog/freedom-learn/201001/the-decline-play-and-rise-in-childrens-mental-disorders;   Mann, D. (2013, December 4). More than 6 percent of US teens take psychiatric meds: Survey. *WebMD*. webmd.com/mental-health/news/20131204/more-than-6-percent-of-us-teens-take-psychiatric-meds-survey

[22] Easterlin, R. A. (1995). Will raising the incomes of all increase the happiness of all? *Journal of Economic Behavior & Organization*, *27*(1), 38–40. doi.org/10.1016/0167-2681(95)00003-b

[23] Nigeria tops happiness survey. (2003, October 2). *BBC News*. news.bbc.co.uk/2/hi/africa/3157570.stm

[24] Lane, R. E. (2005). *Loss of happiness in market democracies*. Yale University Press. 5.

[25] Inglehart, R. (2000). Globalization and postmodern values. *The Washington Quarterly*, *23*(1), 218. doi.org/10.1162/016366000560665

[26] Lyubomirsky, S., & Boehm, J. K. (2010). Human motives, happiness, and the puzzle of parenthood. *Perspectives on Psychological Science*, *5*(3), 338. doi.org/10.1177/1745691610369473;   Biswas-Diener, R., & Diener, E. (2006). The subjective well-being of the homeless, and lessons for happiness. *Social Indicators Research*, *76*(2), 185–205. doi.org/10.1007/s11205-005-8671-9

[27] See Chapter 3.

[28] Diener, E., Horwitz, J., & Emmons, R. A. (1985). Happiness of the very wealthy. *Social Indicators Research*, *16*(3), 263–274. doi.org/10.1007/bf00415126

[29] Bundervoet, T. (2013, March 28). Poor but happy? *The World Bank*. blogs.worldbank.org/africacan/poor-but-happy

[30] Note, these studies do not themselves make the distinction I make about being relevant only to people living above the abundance point. I make that distinction because neither of these studies was designed to address the impact of increased spending on basic needs for people living in extreme poverty. DeLeire, T., & Kalil, A. (2010). Does consumption buy happiness? Evidence from the United States. *International Review of Economics*, *57*(2), 163–176. doi.org/10.1007/s12232-010-0093-6;   Dunn, E. W., Aknin, L. B., & Norton, M. I. (2008). Spending money on others promotes happiness. *Science*, *319*(5870), 1687–1688. doi.org/10.1126/science.1150952

[31] Inglehart, R., & Klingemann, H. D. (2000). Genes, culture, democracy, and happiness. *Culture and Subjective Well-Being*, 165;   Lykken, D., & Tellegen, A. (1996). Happiness is a stochastic phenomenon. *Psychological Science*, *7*(3), 186–189. doi.org/10.1111/j.1467-9280.1996.tb00355.x

[32] Lyubomirsky, S., King, L., & Diener, E. (2005). The benefits of frequent positive affect: Does happiness lead to success? *Psychological Bulletin*, *131*(6), 803–855. doi.org/10.1037/0033-2909.131.6.803;   De Neve, J.-E., & Oswald, A. J. (2012). Estimating the influence of life satisfaction and positive affect on later income using sibling fixed-effects. *SSRN Electronic Journal*. doi.org/10.2139/ssrn.2184708

[33] Huddleston, C. (2019, June 14). Why happy people earn more money. yahoo.com/now/why-happy-people-earn-more-100000135.html;   Lyubomirsky, King, & Diener. The benefits of frequent positive affect. 803–855;   Diener, E., & Oishi, S. (2000). Money and happiness: Income and subjective well-being across nations. *Culture and Subjective Well-Being*, 194.

[34] Lane. *Loss of happiness*. 25.

[35] Babington, D. (2007, June 15). Americans less happy today than 30 years ago: study. *Reuters*. reuters.com/article/idUSL1550309820070615

[36] Lane. *Loss of happiness*. 137.

[37] See the Calcutta study in Chapter 6 for one example.

2 - Twins, Lotteries, and Human Resilience... Set Point

[1] Diener, E., Oishi, S., & Lucas, R. E. (2003). Personality, culture, and subjective well-being: Emotional and cognitive evaluations of life. *Annual Review of Psychology*, *54*(1), 404–5. doi.org/10.1146/annurev.psych.54.101601.145056;   The history of positive psychology. Companion website – PSYCH 243: Introduction to well-being and positive psychology. (2009). *Pennsylvania State University*. courses.worldcampus.psu.edu/welcome/psych243/samplecontent/positive_05.html;   Ehrenreich, B. (2009). *Bright-sided: how positive thinking is undermined America*. Metropolitan Books. 147.

[2] Ben-Shahar, T. (2008). *Happier*. McGraw-Hill. 169.

[3] Cardoso, S. H. (2006, December 15). Hardwired for happiness. *Dana Foundation*. dana.org/Cerebrum/2006/Hardwired_for_Happiness/

[4] McMahon, D. M. (2007). *Happiness: A history*. Grove Press. 1.

[5] Lykken, D., & Tellegen, A. (1996). Happiness is a stochastic phenomenon. *Psychological Science*, *7*(3), 186–189.doi.org/10.1111/j.1467-9280.1996.tb00355.x;   Lykken, D. (2000). *Happiness: The nature and nurture of joy and contentment*. St. Martin's Griffin. 59.

[6] Lyubomirsky, S. (2011). *Hedonic adaptation to positive and negative experiences*. Oxford University Press. 205;   Lyubomirsky, S., Sheldon, K. M., & Schkade, D. (2005). Pursuing happiness: The architecture of sustainable change. *Review of General Psychology*, *9*(2), 113. doi.org/10.1037/1089-2680.9.2.111;   Lykken & Tellegen. Happiness is. 186–

189;   Inglehart, R., & Klingemann, H. D. (2000). Genes, culture, democracy, and happiness. *Culture and Subjective Well-Being*, 165–183.

[7] Lyubomirsky. Hedonic adaptation. 205;   Diener, E., Lucas, R. E., & Scollon, C. N. (2006). Beyond the hedonic treadmill: Revising the adaptation theory of well-being. *American Psychologist*, 61(4), 305–314. doi.org/10.1037/0003-066x.61.4.305

[8] Brickman, P., Coates, D., & Janoff-Bulman, R. (1978). Lottery winners and accident victims: Is happiness relative? *Journal of Personality and Social Psychology*, 36(8), 917–927. doi.org/10.1037/0022-3514.36.8.917

[9] Lanchester, J. (2006, February 19). Pursuing happiness. *The New Yorker*. newyorker.com/archive/2006/02/27/060227crbo_books?currentPage=all

[10] Gilbert, D. (2004, February). *The surprising science of happiness* [Video]. TED. ted.com/talks/dan_gilbert_asks_why_are_we_happy/transcript?language=en#t-1153000

[11] Lyubomirsky. Hedonic adaptation. 205;   Chancellor, J., & Lyubomirsky, S. (2011). Happiness and thrift: When (spending) less is (hedonically) more. *Journal of Consumer Psychology*, 21(2), 132. doi.org/10.1016/j.jcps.2011.02.004

[12] Meyers, D. G. (1993). *The Pursuit of happiness*. Harper-Collins. 48.

[13] Diener, E., & Diener, C. (1996). Most people are happy. *Psychological Science*, 7(3), 181–185. doi.org/10.1111/j.1467-9280.1996.tb00354.x;   Lane, R. E. (2005). *The loss of happiness in market democracies*. Yale University Press. 39.

[14] Not a "point" but a "range": Watson, D. (2001). Positive Affectivity. In C. R. Snyder, C.R. & Lopez, H.J. (Eds.) *Handbook of Positive Psychology*. (2001). United Kingdom: Oxford University Press. 116;   Lanchester. Pursuing happiness;   Haidt, J. (2006). *The happiness hypothesis: Finding modern truth in ancient wisdom*. Basic Books. 90; Seligman, M. E. (2002). *Authentic happiness: Using the new positive psychology to realize your potential for lasting fulfillment*. Simon and Schuster. 47;   Ben-Shahar. *Happier*. 137.   Not set but changeable: Diener, Lucas, & Scollon. Beyond the hedonic treadmill. 308–310;   Diener, E. (2008). Myths in the science of happiness, and directions for future research. *The science of subjective well-being*, 510;   Lykken. *Happiness*. 60;   Begley, S. (2008). *Train your mind, change your brain: How a new science reveals our extraordinary potential to transform ourselves*. Random House Digital, Inc. 230.   Multiple set points: Diener, Lucas, & Scollon. Beyond the hedonic treadmill. 307–308.

3 - How Much Is Enough? ... The Abundance Point

[1] Graham, C. (2008). Happiness and health: Lessons and questions for public policy. *Health Affairs*, 27(1), 76. doi.org/10.1377/hlthaff.27.1.72;   Easterlin, R. A. (2003). Explaining happiness. *Proceedings of the National Academy of Sciences*, 100(19), 11180–11181. doi.org/10.1073/pnas.1633144100

[2] Graham. Happiness and health. 76;   Easterlin. Explaining happiness. 11180–11181;   Easterlin, R. A. (2005). A puzzle for adaptive theory. *Journal of Economic Behavior & Organization*, 56(4), 516–517. doi.org/10.1016/j.jebo.2004.03.003

[3] Easterlin. A puzzle for adaptive theory. 516–517;   Rainwater, L., & Smeeding, T. M. (2005). *Poor kids in a rich country: America's children in comparative perspective*. Russell Sage. 155–158.

[4] Schor, J. B. (1999). *The overspent American: Why we want what we don't need*. Harper Perennial. 14. I adjusted for inflation using the US Bureau of Labor Statistics CPI Inflation Calculator, data.bls.gov/cgi-bin/cpicalc.pl

[5] Schor. *The overspent American*. 7. 1995 dollars converted to 2021 using the US Bureau of Labor Statistics CPI Inflation Calculator.

[6] Annual social and economic supplement. (2009.). *US Census Bureau*. census.gov/data/datasets/time-series/demo/cps/cps-asec.2009.html;   Milanović, B. (2012). *The haves and the have-nots: A brief and idiosyncratic history of global inequality*. Basic Books. 168–169.

[7] Lewis, M. (2014, November 12). Extreme wealth is bad for everyone—especially the wealthy. *The New Republic*. newrepublic.com/article/120092/billionaires-book-review-money-cant-buy-happiness

[8] Yeager, J. (2010). *The cheapskate next door: The surprising secrets of Americans living happily below their means*. Three Rivers Press. 31.

[9] Taylor, P. (2010). The fading glory of the television and telephone. *Pew Research Center*. pewresearch.org/wp-content/uploads/sites/3/2011/01/Final-TV-and-Telephone.pdf

[10] Things we can't live without: The list has grown in the past decade. (2006, December 14). *Pew Research Center*. pewsocialtrends.org/2006/12/14/luxury-or-necessity/

[11] Davies, R., & Smith, W. (1998). The basic necessities survey: The experience of ActionAid Vietnam. *Actionaid Giving People Choices*. 5-6. mande.co.uk/wp-content/uploads/1998/BasicNecessitiesSurveyAAV1998.pdf

[12] Gilbert, D. (2006). *Stumbling on happiness*. Vintage Books.

[13] Gilbert. *Stumbling on happiness*. 225.

[14] Gilbert. 224–228.

[15] Diener, E., & Seligman, M. E. (2004). Beyond money: Toward an economy of well-being. *Psychological Science in the Public Interest*, 5(1), 5;   Inglehart, R. (2000). Globalization and postmodern values. *The Washington Quarterly*, 23(1), 217–218. doi.org/10.1162/016366000560665;   Kesebir, P., & Diener, E. (2008). In pursuit of happiness: Empirical answers to philosophical questions. *Perspectives on Psychological Science*, 3(2), 122. doi.org/10.1111/j.1745-6916.2008.00069.x;   Frey, B. S., & Stutzer, A. (2002). What can economists learn from happiness research? *Journal of Economic Literature*, 40(2), 416. doi.org/10.1257/jel.40.2.402

[16] A few notes on the inexactness of that number, beyond what is discussed in the text: 1) A number of the articles cited in the previous endnote are from the late 90s and early 2000s. $10,000 at that time translated to 2021 dollars would be more in the range of $14–15,000. Martin Seligman suggested in 2002 that the number was closer to $8,000, which would be $11,300 in 2021 (Seligman, M. E. (2002). *Authentic happiness: Using the new positive psychology to realize your potential for lasting fulfillment*. Simon and Schuster. 53). 2) At the same time, some of those researchers were using a higher bar than what we are considering here. Inglehart for example, cited $10,000 as the point not just where the "impact of rising income" drops off significantly, but where it "stops." 3) A graph based on 1960 levels of happiness suggests that the key amount is more like $7,900 in 2021 dollars (Di Tella, R., & MacCulloch, R. (2010). Happiness Adaptation to Income beyond 'Basic Needs'. In Diener, E., Helliwell, J., & Kahneman D. (Eds.), *International Differences in Well-Being*. 233. Oxford University Press.). 4) These studies haven't been nearly so detailed as to look at individual income and happiness worldwide. They looked at metrics like "average per capita income" and average level of happiness for each country. They also often looked at GNP per capita, which isn't quite the same thing as average per capita income though they are often used interchangeably. Likewise, there are issues with looking at average national income instead of median national income because the average can be significantly skewed for a country with large income gaps between rich and poor, like the US. 5) The studies in poorer countries "undersample the illiterate portion of the public and oversample the urban areas and the more educated strata." They attempted to weight the samples accordingly, but the more educated and more urbanized groups in low-income countries "tend to have orientations relatively similar to those found in the publics of industrial societies." (Inglehart, R., et al., (2000). World values surveys and European values surveys 1981–1984, 1990–1993 and 1995–1997. *Inter-university Consortium for Political and Social Research*. web.stanford.edu/group/ssds/dewidocs/icpsr2790_superseded/cb2790.pdf)

[17] Diener, E., Sandvik, E., Seidlitz, L., & Diener, M. (1993). The relationship between income and subjective well-being: Relative or absolute? *Social Indicators Research*, 28(3), 205. My numbers are based on Fig.2, estimating the levels of income at which thee were half point increases in reported levels of happiness, starting with the first identified point, which I estimated at $3,000, then $4,000, $9,000, and $22,000. Those figures were per household, which was an average of 2.6 people per household (Chao, E. L., & Utgoff, K. P. (2006). 100 Years of US consumer spending. *US Department of Labor.* bls.gov/opub/uscs/1984–85.pdf), so per person those figures become: $1,154, $1,538, $3,462, and $8,462. Those were in 1984 dollars, which I converted to 2021 dollars (1/1984 to 1/2021) using the US Bureau of Labor Statistics CPI Inflation Calculator: $2,962, $3,948. $8,887, and $21,722. Rounding brings us back to the same numbers we started with.

[18] Inglehart. Globalization and postmodern values. 218.

[19] Reproduced and adapted from Inglehart, R., & Klingemann, H. D. (2000). Genes, culture, democracy, and happiness. *Culture and Subjective Well-Being*, 168. Another helpful chart showing a similar curve can be found in Diener & Seligman. Beyond money. 5.

[20] As we will explore in a later chapter, happiness (or "subjective well-being") is generally considered to have three components, positive affect and negative affect, basically good and bad feelings, and life satisfaction. Betsey Stevenson and Justin Wolfers of the Wharton School in an important and thorough 2008 paper, confirmed that the life satisfaction component of happiness continues to increase beyond the $75,000 per household figure, and based on that have argued that, "If there is a satiation point, we are yet to reach it." While this is technically correct, they confirmed that the already slight impact continues to be less and less with higher income. The increase in this, just one of three components of happiness, is so marginal at that point that it led to Quinn's comment about it no longer being very relevant. Kahneman, D., & Deaton, A. (2010). High income improves evaluation of life but not emotional well-being. *Proceedings of the National Academy of Sciences*, 107(38), 16489–16492. doi.org/10.1073/pnas.1011492107; Stevenson, B., & Wolfers, J. (2013). Subjective well-being and income: Is there any evidence of satiation? *American Economic Review*, 103(3), 598–604; Thompson, J. H., & Quinn, A. (2014, July 25). Money buys happiness for some people some of the time. *The Pursuit of Happiness.* thepursuitofhappiness.com/insight/money-buys-happiness-people-time. The latest in the debate about a "satiation point" are Easterlin, R. A., McVey, L. A., Switek, M., Sawangfa, O., & Zweig, J. S. (2010). The happiness-income paradox revisited. *Proceedings of the National Academy of Sciences of the United States of America*, 107(52), 22463–22468. doi.org/10.1073/pnas.1015962107; Stevenson, B., & Wolfers J. (2013). Subjective well-being and income: Is there any evidence of satiation? *NBER Working Paper Series.* doi.org/10.3386/w18992; Ed Diener confirms Stevenson and Wolfer's findings in Diener, E., Tay, L., & Oishi, S. (2013). Rising income and the subjective well-being of nations. *Journal of Personality and Social Psychology*, 104(2), 275. doi.org/10.1037/a0030487 Average 2010 household size: Lofquist, D., et al. (2012). Households and families: 2010 2010 census briefs. *US Census Bureau.* census.gov/prod/cen2010/briefs/c2010br-14.pdf

[21] Federal poverty level (FPL). (n.d.). *HealthCare.gov.* healthcare.gov/glossary/federal-poverty-level-FPL; Federal poverty level = "the minimum income required for a household to buy such things as food, clothing and shelter": Warne, L., & Ostria, M. (2013, November 7). How differences in the cost of living affect low-income families. *National Center for Policy Analysis.* ncpathinktank.org/pub/ib133; 2015 dollars converted to 2021 using the US Bureau of Labor Statistics CPI Inflation Calculator.

[22] Aknin, L. B., Norton, M. I., & Dunn, E. W. (2009). From wealth to well-being? Money matters, but less than people think. *The Journal of Positive Psychology*, 4(6), 525–526.

[23] $78.3 billion in total personal income, divided by 132.3 million people in 1940 = $592/person, converted to 2021 dollars using the US Bureau of Labor Statistics CPI Inflation Calculator. Income: Series A 134–144. (1945). *Historical Statistics of the United States, 1789–1945: A Supplement to the Statistical Abstract of the United States*, Bureau of the Census. 13. www.census.gov/prod2/statcomp/documents/HistoricalStatisticsoftheUnitedStates1789-1945.pdf; Population: 1940-2010 how has America changed? (2010). *US Census Bureau.* 1. census.gov/content/dam/Census/library/visualizations/2012/comm/1940_census_change.pdf

[24] Almost half the world lives on less than $2.50/day x 365 days = $912.50/year. 80% of the world lives on less than $10/day = 3,650/year. Shah, A. (n.d.). Poverty facts and statistics. *Global Issues.* globalissues.org/article/26/poverty-facts-and-stats

[25] Diener, E., & Biswas-Diener, R. (2005). Psychological empowerment and subjective well-being. *PsycEXTRA Dataset*, 132. doi.org/10.1037/e597202012-007; Biswas-Diener, R. (2008). Material wealth and subjective well-being. *The Science of Subjective Well-Being*, 310–15.

[26] Diener, E., & Diener, C. (1996). Most people are happy. *Psychological Science*, 7(3), 181. doi.org/10.1111/j.1467-9280.1996.tb00354.x; Inglehart. Globalization and postmodern values. 217; Inglehart & Klingemann. Genes, culture, democracy, and happiness. 168.

[27] Alter, A. (2014, January 24). Do the poor have more meaningful lives? *The New Yorker.* newyorker.com/business/currency/do-the-poor-have-more-meaningful-lives

[28] Biswas-Diener, R., & Diener, E. (2009). Making the best of a bad situation: Satisfaction in the slums of Calcutta. *Social Indicators Research Series, 38*, 339–341. doi.org/10.1007/978-90-481-2352-0_13 PPP for India generally involves multiplying by 3 (Milanović. The haves and the have-nots. 168.)

[29] Bundervoet, T. (2013, March 28). Poor but happy? *The World Bank.* blogs.worldbank.org/africacan/poor-but-happy

[30] Biswas-Diener, & Diener. Making the best. *38*, 349.

[31] Easterlin. A puzzle for adaptive theory. 516.

[32] Milanović. *The haves and the have-nots.*168–169; Cohen, P. (2015, January 19). Oxfam study finds richest 1% is likely to control half of global wealth by 2016. *The New York Times.* nytimes.com/2015/01/19/business/richest-1-percent-likely-to-control-half-of-global-wealth-by-2016-study-finds.html

[33] Table A-2, Denavas-Walt, C., & Proctor, B. D. (2014, September 16). Income and poverty in the United States: 2013. *Census Bureau.* census.gov/library/publications/2014/demo/p60-249.html

[34] Kahneman, & Deaton. High income. 16489–16492. Table A-2, Denavas-Walt, & Proctor. Income and poverty.

4 - The *Hap* in Our Happiness: ... Circumstances

[1] Melville, H. (2021). *Moby-Dick or the whale.* Penguin Group. 253.

[2] McMahon, D. M. (2007). *Happiness: A history.* Grove Press. 3–7.

[3] Swahili: -a bahati – relates back to luck or fortune. (Madan, A. C. (2006). English-Swahili and Swahili-English Dictionary. Germany: Asian Educational Services. 256.) Chinese: xing fú – relates back to lucky, fortunate (Su, Q. G. (2019, February 20). *Daily mandarin lesson: "happy" in Chinese.* ThoughtCo. thoughtco.com/kuai-le-happy-2278649) Farsi: khosh bakht – relates back to good fortune or luck (Shojaaee, K. [Farsi Wizard.] *Farsi / Persian Lesson: Expressing Happiness and Sadness.* [Video]. YouTube. youtube.com/watch?v=qBY6OLsXP1M&t=85s) Japanese: shiawase – relates back to good fortune (Abe, N. (2019, January 31). *What does Shiawase mean in Japanese?* ThoughtCo. thoughtco.com/shiawase-meaning-and-characters-2028801) Arabic: saeida, sa'eed, saad, sa'd – relates back to lucky, fortunate, blessed (*Saad (name).* (n.d.). Wikimedia

Foundation. en.wikipedia.org/wiki/Saad_(name)) Harper, D. (n.d.). *Happy*. Etymonline: Online Etymology Dictionary. www.etymonline.com/search?q=happy

[4] Lyubomirsky includes in her definition of circumstances many things that we *do* have some degree of control over, like whether we are married, the kind of work we do, and our income. She writes that all of these circumstances combined account for 8 to 15 percent of our happiness. Given the significance of some of the circumstances that *are* to some degree within our control, it seems fair to assume that the impact of circumstances *outside* our control is roughly half of her estimate, so in the range of 4–8 percent. Lyubomirsky, S., Sheldon, K. M., & Schkade, D. (2005). Pursuing happiness: The architecture of sustainable change. *Review of General Psychology*, 9(2), 117–21. doi.org/10.1037/1089-2680.9.2.111

[5] Wallis, C. (2009, July 8). The science of happiness turns 10. what has it taught? *Time*. content.time.com/time/health/article/0,8599,1908173,00.html

[6] Diener, E. (2008). Myths in the science of happiness, and directions for future research. *The science of subjective well-being*, 493–514; Diener, M. L., & Diener McGavran, M. B. (2008). What makes people happy?: A developmental approach to the literature on family relationships and well-being. *The Science Of Subjective Well-Being*. 347–375; Tolman, R. M., & Rosen, D. (2001). Domestic violence in the lives of women receiving welfare: Mental health, substance dependence, and economic well-being. *Violence Against Women*, 7(2), 141–158; Poutiainen, M., & Holma, J. (2013). Subjectively evaluated effects of domestic violence on well-being in clinical populations. *International Scholarly Research Notices*, 1–8. doi.org/10.1155/2013/347235; Bellis, M. A., Hughes, K., Jones, A., Perkins, C., & McHale, P. (2013). Childhood happiness and violence: A retrospective study of their impacts on adult well-being. *BMJ Open*, 3(9).

[7] Moray, P. (2023, May 26.) Personal communication. May 26, 2022.

[8] Seery, M. D., Holman, E. A., & Silver, R. C. (2010). Whatever does not kill us: Cumulative lifetime adversity, vulnerability, and resilience. *Journal of Personality and Social Psychology*, 99(6), 1025–1041. doi.org/10.1037/a0021344

[9] The following are just a few examples of this. Diener, E., & Biswas-Diener, R. (2002). Will money increase subjective well-being? *Social Indicators Research*, 57(2), 121. doi.org/10.1023/A:1014411319119; Williams, M. T. (2015, September 6). The link between racism and PTSD. *Psychology Today*. psychologytoday.com/us/blog/culturally-speaking/201509/the-link-between-racism-and-ptsd ; Wortham, J. (2015, June 24). Racism's psychological toll. *The New York Times Magazine*. nytimes.com/2015/06/24/magazine/racisms-psychological-toll.html ; The Sentencing Project. (2018, April 19). Report to the United Nations on Racial Disparities in the U.S. Criminal Justice System. sentencingproject.org/publications/un-report-on-racial-disparities ; Foster, D. (n.d.). *How Long Will I Live - Gender, Race, and Education*. Blueprint Income. blueprintincome.com/tools/life-expectancy-calculator-how-long-will-i-live/info/gender-other-determinants-of-longevity ; Human Rights Campaign Foundation. (n.d.). *Sexual Assault and the LGBTQ Community*. hrc.org/resources/sexual-assault-and-the-lgbt-community ; Machles, M., Cochran, C., Hill, A.M. & Brewer, S. (2019, Oct 18). 1 in 3 American Indian and Alaska Native women will be raped, but survivors rarely find justice on tribal lands. *USA Today*; US Dept. of Health and Human Services, Office of Minority Health. (n.d.). *Infant Mortality and American Indians/Alaska Natives*. minorityhealth.hhs.gov/omh/browse.aspx?lvl=4&lvlID=38#:~:

[10] Henrich, J., Heine, S., & Norenzayan, A. (2010). The weirdest people in the world? *The Behavioral and Brain Sciences*. 33(2-3), 61–83.

[11] Fredrickson, B. L., Cohn, M. A., Coffey, K. A., Pek, J., & Finkel, S. M. (2008). Open hearts build lives: Positive emotions, induced through loving-kindness meditation, build consequential personal resources. *Journal of Personality and Social Psychology*, 95(5), 1060. doi.org/10.1037/a0013262

[12] Diener. Myths. 494–96.

[13] Evidence for our ability to intentionally increase our happiness will be presented throughout the following chapters, but this article provides a superb overview: Lyubomirsky, S. (2011). *Hedonic adaptation to positive and negative experiences*. Oxford University Press. 200–224; This article presents a good summary of some of the evidence: Lyubomirsky, Sheldon, & Schkade. Pursuing happiness. 124–126; The Seligman statistic is from BBC. (2006). Episode 2: Rewiring Our Brains, *The Happiness Formula*. 1:13. The even more significant results are documented in Holden, R. (2010). *Be happy!* Hay House, Inc. xi-xvi.

[14] Harper, D. (n.d.). *Happy*.

5 - Spouses and Kids and Sex, Oh My! ... Kith and Kin

[1] Ladinsky, D. (2002). *Love poems from God: Twelve sacred voices from the East and West*. Penguin. 175.

[2] Lyubomirsky, S., & Boehm, J. K. (2010). Human motives, happiness, and the puzzle of parenthood. *Perspectives on Psychological Science*, 5(3), 328. doi.org/10.1177/1745691610369473

[3] Biswas-Diener, R., & Diener, E. (2006). The subjective well-being of the homeless, and lessons for happiness. *Social Indicators Research*, 76(2), 185–205. doi.org/10.1007/s11205-005-8671-9

[4] Diener, E. Frequently answered questions: Q: What is your advice to those who want to be happy? *Psychology Department, University of Illinois at Urbana-Champaign*. psych.uiuc.edu/~ediener/faq.html

[5] Diener, E., & Seligman, M. E. P. (2002). Very happy people. *Psychological Science*, 13(1), 83. doi.org/10.1111/1467-9280.00415

Some scientists have asserted that our brains are hardwired for us to be social creatures, and some believe we evolved our large brains *exactly because* of our need to be social and to be successful in navigating group dynamics. (Goleman, D. (2006). *Social intelligence: The new science of human relationships*. Bantam Books; Dalai Lama XIV, & Cutler, H. C. (1998). *The art of happiness: A handbook for living*. Riverhead Books. 285, 289.)

Brené Brown writes that "We are biologically, cognitively, physically, and spiritually wired to love, to be loved, and to belong. When those needs are not met, we don't function as we were meant to. We break. We fall apart. We numb. We ache. We hurt others. We get sick." (Brown, B. (2010). *The gifts of imperfection: Let go of who you think you're supposed to be and embrace who you are*. Simon and Schuster. 26).

[6] Putnam, R. D. (2000). *Bowling alone: The collapse and revival of American community*. Simon & Schuster. 332; Dreifus, C. (2008, April 22). The smiling professor. *The New York Times* nytimes.com/2008/04/22/science/22conv.html; Diener, & Seligman. Very happy people. 83. Prominent social psychologist, Roy Baumeister, has suggested that having intimate bonds with other people is perhaps a necessary condition for happiness, and Ed Diener and Martin Seligman agree that it's at least necessary for someone to be "very happy." (Baumeister, R. F. (1991). *Meanings of life*. Guilford Press. 213; Diener, & Seligman. Very happy people. 83. People also generally experience their most positive moods of each day when they are socializing with friends or family (Dunn, E., & Norton, M. (2014). *Happy money: The science of happier spending*. Simon and Schuster. 67).

[7] Srivastava, S., Angelo, K. M., & Vallereux, S. R. (2008). Extraversion and positive affect: A day reconstruction study of person–environment transactions. *Journal of Research in Personality*, 42(6), 1613–1618. doi.org/10.1016/j.jrp.2008.05.002

[8] Holt-Lunstad, J., Smith, T. B., & Layton, J. B. (2010). Social relationships and mortality risk: A meta-analytic review. *PLoS Medicine*, 7(7): e1000316. doi.org/10.1371/journal.pmed.1000316

[9] Helliwell, J. F., & Huang, H. (2013). Comparing the happiness effects of real and on-line friends. *PloS one*, 8(9), e72754. doi.org/10.1371/journal.pone.0072754

[10] Murphy, M. (n.d.). Effects of solitary confinement on the well-being of prison inmates. *Applied Psychology OPUS*. wp.nyu.edu/steinhardt-appsych_opus/effects-of-solitary-confinement-on-the-well-being-of-prison-inmates; Méndez, J. (2011, October 18). Solitary confinement should be banned in most cases, UN expert says. *United Nations*. news.un.org/en/story/2011/10/392012-solitary-confinement-should-be-banned-most-cases-un-expert-says; Perkins, B. (2015, April 30). Torture or safety mechanism? Lawmakers looking closely at the use of solitary confinement in state prisons. *The New York State Senate*. nysenate.gov/newsroom/in-the-news/bill-perkins/torture-or-safety-mechanism-lawmakers-looking-closely-use-solitary

[11] Goode, E. (2015, August 3). Solitary confinement: Punished for life. *The New York Times*. nytimes.com/2015/08/04/health/solitary-confinement-mental-illness.html

[12] Dunn. & Norton. *Happy money.* 6–7.

[13] Brooks, D. (2016, August 9). The great affluence fallacy. *The New York Times*. nytimes.com/2016/08/09/opinion/the-great-affluence-fallacy.html

[14] Jayson, S. (2012, October 17). Young parents, older adults change face of cohabitation. *USA Today*. usatoday.com/story/news/nation/2012/10/17/cohabitation-divorced-families-parents/1623117; Facts for features: Unmarried and single Americans week Sept. 15-21, 2013 Census Bureau, U. S. C. (2013, July 30). *The United States Census Bureau.* census.gov/newsroom/facts-for-features/2013/cb13-ff21.html

[15] Luscombe, B. (2018, November 26). The divorce rate is dropping. That may not actually be good news. *Time.* time.com/5434949/divorce-rate-children-marriage-benefits; Cherlin, A. J. (1992). *Marriage, divorce, remarriage.* Harvard University Press. 28.

[16] Putnam. *Bowling alone.* 333.

[17] Lyubomirsky, S. (2011). *Hedonic adaptation to positive and negative experiences.* Oxford University Press. 216.

[18] Blanchflower, D.G. & Oswald, A. J. (2004). Well-being over time in Britain and the USA. *Journal of Public Economics*, 88(7–8), 1373–1381. doi.org/10.1016/S0047-2727(02)00168-8

[19] Diener, E. (2008). Myths in the science of happiness, and directions for future research. *The science of subjective well-being*, 493–514.

[20] Stutzer, A., & Frey, B. S. (2006). Does marriage make people happy, or do happy people get married? *The Journal of Socio-Economics*, 35(2), 326–347. doi.org/10.1016/j.socec.2005.11.043; Lyubomirsky, & Boehm. (2010). Human motives. 329; Diener, E., & Biswas-Diener, R. (2011). *Happiness: Unlocking the mysteries of psychological wealth*. John Wiley & Sons. 56–57.

[21] Diener, & Biswas-Diener. *Happiness.* 56–57; Diener. Myths. 493–514; Lyubomirsky.& Boehm. (2010). Human motives. 329.

[22] Stein, J. (2005, January 9). Marriage: Is there a hitch? *Time.* content.time.com/time/magazine/article/0,9171,1015873-3,00.html

[23] Blanchflower, & Oswald. Well-being over time. 1373–1381; Sostek, A. (2010, May 23). Which comes first: sex or happiness? *Pittsburgh Post-Gazette.* post-gazette.com/pg/10143/1060183-51.stm

[24] Mooney, C. (2014, February 26). You share 98.7 percent of your DNA with this sex-obsessed ape. *Mother Jones.* motherjones.com/politics/2014/02/evolution-creationism-bonobos-neanderthals-denisovans-chromosome-two

[25] Waal, F., & Lanting, F. (1998). *Bonobo: The forgotten ape.* University of California Press. 101–113; Handwerk, B. (2008, February 12). Gorillas photographed mating face-to-face-a first. *National Geographic.* news.nationalgeographic.com/news/2008/02/080212-gorilla-sex_2.html

[26] O'Brien, M., & Kellan, A. (2011, March 8). Peaceful bonobos may have something to teach humans. *Phys.org.* phys.org/news/2011-03-peaceful-bonobos-humans.html; Waal, F. B. (1995). Bonobo sex and society. *Scientific American*, 272(3), 82–88. doi.org/10.1038/scientificamerican0395-82; Waal, & Lanting. *Bonobo.* 101–13.

[27] O'Brien, & Kellan. Peaceful bonobos.

[28] Raffaele, P. (2006, November). The smart and swinging bonobo. *Smithsonian Magazine.* smithsonianmag.com/science-nature/the-smart-and-swinging-bonobo-134784867/

[29] Blair, E. (2022, June 28). The book about sex that everybody should read. *The New York Times Magazine.* nytimes.com/2022/06/28/magazine/sex-ed-books-teens-parents.html

[30] Wiens, S. (2013, May 3). To parents of small children: Let me be the one who says it out loud. *The Huffington Post.* huffingtonpost.com/steve-wiens/let-me-be-the-one-who-says-it-out-loud_b_3209305.html

[31] Newport, F. (2003, August 19). Desire to have children alive and well in America. *Gallup.* news.gallup.com/poll/9091/desire-children-alive-well-america.aspx; Lyubomirsky, S., & Boehm, J. K. (2010). Human motives, happiness, and the puzzle of parenthood. *Perspectives on Psychological Science*, 5(3), 330. doi.org/10.1177/1745691610369473

[32] Mesure, S. (2011, October 23). Children, a bundle of joy? *Independent.* independent.co.uk/life-style/health-and-families/features/children-a-bundle-of-joy-2017935.html; Senior, J. (2010, July 2). All Joy and no fun. *New York Magazine.* nymag.com/news/features/67024; Wallis, C. (2009, July 8). The science of happiness turns 10. what has it taught? *Time.* content.time.com/time/health/article/0,8599,1908173,00.html

[33] Kahneman, D., & Krueger, A. B. (2006). Developments in the measurement of subjective well-being. *Journal of Economic Perspectives*, 20(1), 13. doi.org/10.1257/089533006776526030

[34] Carey, B. (2010, May 22). Families' every fuss, archived and analyzed. *The New York Times.* nytimes.com/2010/05/23/science/23family.html?pagewanted=1&_r=1&sq=elinor percent20ochs&st=cse&scp=1

[35] Diener, & Biswas-Diener. *Happiness.* 62; Senior. All Joy and no fun; Baumeister, R. F., & Vohs, K. D. (2002). The pursuit of meaningfulness in life. *Handbook of positive psychology, 1*, 612.

[36] Lyubomirsky.& Boehm. (2010). Human motives. 330; Senior. All Joy and no fun; Hosley, D. (n.d.). When do most couples start having sex again after their baby is born? *BabyCenter.* babycenter.com/404_when-do-most-couples-start-having-sex-again-after-their-baby_11805.bc. That was based on a 2003 study: Sex Survey. (2003). *Baby Center.* babycenter.com/sexsurvey-2003

[37] Senior. All Joy and no fun.

[38] Senior; Lyubomirsky.& Boehm. (2010). Human motives. 330; Are we happy yet? (2006, February 13). *Pew Research Center.* pewresearch.org/social-trends/2006/02/13/are-we-happy-yet/

[39] Diener, & Biswas-Diener. *Happiness.* 62.

[40] Diener, & Biswas-Diener. 61–63.

[41] Senior. All Joy and no fun.

[42] Lane, R. E. (2005). *The loss of happiness in market democracies.* Yale University Press. 77.

[43] Lane. *The loss of happiness.* 77.

[44] Lewis, D., Al-Shawaf, L., Russell, E., & Buss, D. (2015) Friends and happiness: An evolutionary perspective on friendship. *Friendship and Happiness.* doi.org/10.1007/978-94-017-9603-3_3

45 Lewis, Al-Shawaf, Russell, & Buss. Friends and happiness;   Lane. *The loss of happiness*. 80–87.

46 Helliwell, & Huang. Comparing the happiness effects. e72754;   Diener, & Biswas-Diener. *Happiness*. 51.

47 Aked, J., Marks, N., Cordon, C., & Thompson, S. (2008). Five ways to wellbeing. *Center for Wellbeing (The New Economics Foundation)*. 5, 6. neweconomics.org/uploads/files/five-ways-to-wellbeing-1.pdf;   Helliwell, & Huang. Comparing the happiness effects. e72754

48 Helliwell, & Huang. Comparing the happiness effects. e72754;   Diener, & Biswas-Diener. *Happiness*. 51.

49 Markman, A. (2013, January 16). When is it good to have a few close friends? *Psychology Today*.psychologytoday.com/intl/blog/ulterior-motives/201301/when-is-it-good-have-few-close-friends

50 Lyubomirsky. Hedonic adaptation. 216;   Lyubomirsky, S., Sheldon, K. M., & Schkade, D. (2005). Pursuing happiness: The architecture of sustainable change. *Review of General Psychology*, 9(2), 124–126. doi.org/10.1037/1089-2680.9.2.111

51 Mineo, L. (2017, April 11). Good genes are nice, but joy is better. *The Harvard Gazette*. news.harvard.edu/gazette/story/2017/04/over-nearly-80-years-harvard-study-has-been-showing-how-to-live-a-healthy-and-happy-life/

52 DeAngelis, B. Note: there is a slightly different version of this quote on DeAngelis' website, but the one presented here is far more common, though I could not find it in that exact form in her transcripts. barbaradeangelis.com/advice_LandI.asp#4

6 - More Sleep, Less Cow: Physical Health ...

1 Graham, C. (2008). Happiness and health: Lessons and questions for public policy. *Health Affairs*, 27(1), 78-80. doi.org/10.1377/hlthaff.27.1.72

2 Graham. Happiness and health. 78–80.

3 Riis, J., Loewenstein, G., Baron, J., Jepson, C., Fagerlin, A., & Ubel, P. A. (2005). Ignorance of hedonic adaptation to hemodialysis: A study using ecological momentary assessment. *Journal of Experimental Psychology: General*, 134(1), 3–9. doi.org/10.1037/0096-3445.134.1.3

4 Graham. Happiness and health. 73;   Dreifus, C. (2008, April 22). The smiling professor. *The New York Times* nytimes.com/2008/04/22/science/22conv.html

5 Goldman, L. (2011, April/May). Like mother, like daughter? *Natural Health*. 68.

6 Zhivotovskaya, E.(2008, August 21). Healthy minds reside in healthy bodies. *Positive Psychology* News. positivepsychologynews.com/news/emiliya-zhivotovskaya/20080821972;   Dunn, E., & Norton, M. (2014). *Happy money: The science of happier spending*. Simon and Schuster. 62.

7 Babyak, M., Blumenthal, J. A., Herman, S., Khatri, P., Doraiswamy, M., Moore, K, Craighead, W. E., Baldewicz, T., & Krishnan, K. R. (2000). Exercise treatment for major depression: maintenance of therapeutic benefit at 10 months. *Psychosomatic Medicine*, 62(5), 633–638. doi.org/10.1097/00006842-200009000-00006 ;   Reynolds, G. (2011, August 31). Prescribing exercise to treat depression. *The New York Times*. well.blogs.nytimes.com/2011/08/31/prescribing-exercise-to-treat-depression;   Bloke334. (2009, November 23). Robert Holden 8 week happiness documentary part 2 [Video]. YouTube. youtube.com/watch?v=nQwxMtyfomM

8 The New York Times. (n.d.). 7 habits for a healthy heart. nytimes.com/guides/well/how-to-prevent-heart-disease;   How much physical activity do adults need? (n.d.). *Centers for Disease Control and Prevention*. cdc.gov/physicalactivity/everyone/guidelines/adults.html

9 Healthy eating plate vs. USDA's MyPlate. (2019, September 24). *Harvard T. H. Chan School of Public Health*. hsph.harvard.edu/nutritionsource/healthy-eating-plate-vs-usda-myplate

10 Mujcic, R., & J Oswald, A. (2016). Evolution of well-being and happiness after increases in consumption of fruit and vegetables. *American Journal Of Public Health*, 106(8), 1504–1510. doi.org/10.2105/AJPH.2016.303260

11 Snider, S. (2016, December 30). How to save money by going vegetarian. *USNews*. money.usnews.com/money/personal-finance/articles/2016-12-30/how-to-save-money-by-going-vegetarian

12 The Protein Myth. (n.d.). *Physicians Committee for Responsible Medicine*. pcrm.org/health/diets/vsk/vegetarian-starter-kit-protein

13 Lowering cholesterol with a plant-based diet. (n.d.). *Physicians Committee for Responsible. Medicine*. pcrm.org/good-nutrition/nutrition-information/lowering-cholesterol-with-a-plant-based-diet;   Healthy eating plate vs. USDA's MyPlate;   Calcium. (n.d.). *Harvard T. H. Chan School of Public Health*. hsph.harvard.edu/nutritionsource/what-should-you-eat/calcium-and-milk

14 Munro, D. (2013, September 17). Chipotle and Credit Suisse fire shots across the bow of US food and agricultural industries. *Forbes*. forbes.com/sites/danmunro/2013/09/17/chipotle-and-credit-suisse-fire-shots-across-the-bow-of-u-s-food-and-agricultural-industries/

15 Jacka, F. N., Kremer, P. J., Leslie, E. R., Berk, M., Patton, G. C., Toumbourou, J. W., & Williams, J. W. (2010). Associations between diet quality and depressed mood in adolescents: Results from the Australian Healthy Neighbourhoods Study. *Australian & New Zealand Journal of Psychiatry*, 44(5), 435–442. doi.org/10.3109/00048670903571598 ;   Sánchez, A., Verberne, L., De Irala, J., Ruíz, M., Toledo, E., Serra, L., & Martínez, M. A. (2011). Dietary fat intake and the risk of depression: The sun project. *PLoS ONE*, 6(1). doi.org/10.1371/journal.pone.0016268

16 Kahneman, D., Krueger, A. B., Schkade, D. A., Schwarz, N., & Stone, A. A. (2004). A survey method for characterizing daily life experience: The day reconstruction method. *Science*, 306(5702), 1776–1780. doi.org/10.1126/science.1103572. Related studies are cited in Lyubomirsky, S., & Boehm, J. K. (2010). Human motives, happiness, and the puzzle of parenthood. *Perspectives on Psychological Science*, 5(3), 328. doi.org/10.1177/1745691610369473

17 Salvas, M. (2009, June 24). Does sleep really matter? *Positive Psychology News Daily*. positivepsychologynews.com/news/marie-josee-salvas/200906242571

18 Sapolsky, R. M. (2004). *Why zebras don't get ulcers: The acclaimed guide to stress, stress-related diseases, and coping*. Holt paperbacks. 235

19 Wolchover, N. (2011, February 16). Busting the 8-hour sleep myth: Why you should wake up in the night. *LiveScience*. livescience.com/12891-natural-sleep.html;   Hegarty, S. (2012, February 22). The myth of the eight-hour sleep. *BBC News*. bbc.com/news/magazine-16964783

20 How much sleep do we really need? (n.d.). *Sleep Foundation*. sleepfoundation.org/article/how-sleep-works/how-much-sleep-do-we-really-need;   Jones, M. (2011, April 15). How little sleep can you get away with? *The New York Times Magazine*. nytimes.com/2011/04/17/magazine/mag-17Sleep-t.html

21 Jones. (2011, April 15). How little sleep.

22 Kagan, W. (2011, October 26). In search of sleep. *Chronogram Media*. chronogram.com/hudsonvalley/in-search-of-sleep/Content?oid=2175545;   Salvas. (2009, June 24). Does sleep really matter?;   Brody, J. E. (2013, June 17). Cheating ourselves of sleep. *The New York Times*. well.blogs.nytimes.com/2013/06/17/cheating-ourselves-of-sleep/

23 Hegarty. (2012, February 22). The myth of the eight-hour sleep.

[24] Mwinnyaa, G., Porch, T., Bowie, J., & Thorpe, R. J., Jr (2018). The association between happiness and self-rated physical health of African American Men: A population-based cross-sectional study. American Journal Of Men's Health, 12(5), 1615–1620. doi.org/10.1177/1557988318780844; Oaklander, M. (n.d.). Do happy people really live longer? Time. time.com/collection/guide-to-happiness/4217052/do-happy-people-really-live-longer; LaMotte, S. (2019. October 5). Being happier will help you live longer, so learn how to be happier. CNN. cnn.com/2019/09/30/health/happiness-live-longer-wellness/index.html; Diener, E., & Chan, M. Y. (2011). Happy people live longer: Subjective well-being contributes to health and longevity. Applied Psychology: Health and Well-Being, 3(1), 32. doi.org/10.1111/j.1758-0854.2010.01045.x ; Lyubomirsky, S., King, L., & Diener, E. (2005). The benefits of frequent positive affect: Does happiness lead to success? Psychological Bulletin, 131(6), 825. doi.org/10.1037/0033-2909.131.6.803
[25] Mineo, L. (2017, April 11). Good genes are nice, but joy is better. The Harvard Gazette. news.harvard.edu/gazette/story/2017/04/over-nearly-80-years-harvard-study-has-been-showing-how-to-live-a-healthy-and-happy-life/

7 - A "Fool's Life" or a "Labor of Love"? Work ...

[1] Merton, T. (2009). Conjectures of a guilty bystander. Image. 81
[2] Reynolds, G. (2013, September 20). Ask well: Sleep or exercise? The New York Times. well.blogs.nytimes.com/2013/09/20/ask-well-sleep-or-exercise/
[3] Ben-Shahar, T. (2008). Happier. McGraw-Hill. 154; Dunn, E., & Norton, M. (2014). Happy money: The science of happier spending. Simon and Schuster. 56.
[4] Thoreau, H. D., & Carradine, J. (1964). Walden. Houghton Mifflin. 4
[5] Kubey, R., & Csikszentmihalyi, M. (2013). Television and the quality of life: How viewing shapes everyday experience. Routledge. 24.
[6] United States average weekly hours. (n.d.). Trading Economics. tradingeconomics.com/united-states/average-weekly-hours; Rosnick, D. & Weisbrot, M. (2006). Are shorter work hours good for the environment? Center for Economic and Policy Research. cepr.net/documents/publications/energy_2006_12.pdf
[7] Saad, L. (2014, August 29). The "40-Hour" workweek is actually longer -- by seven hours. Gallup. news.gallup.com/poll/175286/hour-workweek-actually-longer-seven-hours.aspx
[8] Warr, P. (1982). Psychological aspects of employment and unemployment. Psychological Medicine, 12(1), 7. doi.org/10.1017/s0033291700043221 ; Aked, J., Marks, N., Cordon, C., & Thompson, S. (2008). Five ways to wellbeing. Center for Wellbeing (The New Economics Foundation). 11. neweconomics.org/uploads/files/five-ways-to-wellbeing-1.pdf
[9] Diener, E., & Biswas-Diener, R. (2011). Happiness: Unlocking the mysteries of psychological wealth. John Wiley & Sons. 72; Ben-Shahar. Happier. 102; Adams, S. (2013, October 10). Unhappy employees outnumber happy ones by two to one worldwide. Forbes. forbes.com/sites/susanadams/2013/10/10/unhappy-employees-outnumber-happy-ones-by-two-to-one-worldwide; Brooks, K. (2012, June 29). Job, career, calling: Key to happiness and meaning at work? Psychology Today. psychologytoday.com/blog/career-transitions/201206/job-career-calling-key-happiness-and-meaning-work
[10] Americans' job satisfaction falls to record low. (2010, January 6). USA Today. usatoday.com/money/workplace/2010-01-05-job-satisfaction-use_N.htm
[11] Preidt, R. (2010, October). Recession may be changing Americans 'attitudes toward work. Bloomberg Business Week. businessweek.com/lifestyle/content/healthday/644692.html; Rosenfeld, S. (2013, June 18). 70 percent of Americans 'emotionally disconnected' at work: shocking poll reveals workforce Zombieland. AlterNet. alternet.org/2013/06/70-percent-americans-emotionally-disconnected-work/
[12] Kasser, T., Cohn, S., Kanner, A. D., & Ryan, R. M. (2007). Some costs of American corporate capitalism: A psychological exploration of value and Goal Conflicts. Psychological Inquiry, 18(1), 15. doi.org/10.1080/10478400701386579 ; Kasser, T. & Brown, K.W. On time, happiness, and ecological footprints. In Take Back Your Time, (John de Graaf, Ed.) Berrett-Koehler, 2000, pp. 109–10.
[13] Clark, A. E. (2012). Happiness, habits and high rank: Comparisons in economic and social life. SSRN, 12. doi.org/10.2139/ssrn.2131154; Graham, C. (2008). Happiness and health: Lessons and questions for public policy. Health Affairs, 27(1), 77. doi.org/10.1377/hlthaff.27.1.72
[14] Baumeister, R. F., & Vohs, K. D. (2002). The pursuit of meaningfulness in life. Handbook of positive psychology, 1, 617; Clark. (2012). Happiness, habits and high rank. 12. Further support comes the following two sources: Lucas, R. E., Clark, A. E., Georgellis, Y., & Diener, E. (2004). Unemployment alters the set point for life satisfaction. Psychological Science, 15(1), 8–13. doi.org/10.1111/j.0963-7214.2004.01501002.x; Frey, B. S. (2010). Happiness: A revolution in economics. MIT Press; One day of paid work a week is all we need to get mental health benefits of employment. Cambridge University. cam.ac.uk/employmentdosage
[15] Clark, A. E., & Oswald, A. J. (2002). A simple statistical method for measuring how life events affect happiness. International Journal of Epidemiology, 31(6), 1141–1143. doi.org/10.1093/ije/31.6.1139
[16] Charles, K. K. (2002). Is retirement depressing?: Labor force inactivity and psychological well-being in later life. National Bureau of Economic Research Working Paper Series. doi.org/10.3386/w9033 ; Beck, S. H. (1982). Adjustment to and satisfaction with retirement. Journal of Gerontology, 37(5), 616–624. doi.org/10.1093/geronj/37.5.616
[17] Dunn, & Norton. Happy money. 64.
[18] Census Bureau estimates show average one-way travel time to work rises to all-time high. (2021, March 18). US Census Bureau. census.gov/newsroom/press-releases/2021/one-way-travel-time-to-work-rises.html; Burd, C. , Burrows,, M., & McKenzie, B. (2021, March 18). travel time to work in the United States: 2019, American community survey reports, ACS-47. US Census Bureau. census.gov/library/publications/2021/acs/acs-47.html; Dunn, & Norton. Happy money. 63–65, 68–69.
[19] Griffiths, S. (2021, March 21). Study finds the key to happiness: working one day a week. The Times. thetimes.co.uk/article/study-finds-the-key-to-happiness-working-one-day-a-week-0vmp8l3zb
[20] Norton, A. (2019, June 19). How much work makes us happy? Not much, study shows. WebMD. webmd.com/mental-health/news/20190619/how-much-work-makes-us-happynot-much-study-shows
[21] White, C. (n.d.). The idols of environmentalism. Orion. orionmagazine.org/article/the-idols-of-environmentalis/
[22] Retailleau, U., Doumenc, I., & Hamilton, G. (n.d.) Workers of the World Relax: The Jevons Paradox. vimeo.com/2450335
[23] Surowiecki, J. (2005, November 20). No work and no play. The New Yorker. newyorker.com/archive/2005/11/28/051128ta_talk_surowiecki
[24] Rauch, E. (n.d.). Productivity and the workweek. groups.csail.mit.edu/mac/users/rauch/worktime/
[25] Williams, J. C., & Boushey, H. (2010). The three faces of work-family conflict: The poor, the professionals, and the missing middle. SSRR, ii, iii. doi.org/10.2139/ssrn.2126314

[26] Noah, T. (2010, September 3). The United States of inequality. *Slate*. slate.com/articles/news_and_politics/the_great_divergence/features/2010/the_united_states_of_inequality/introducing_the_great_divergence.html

[27] 1,652 in 1975: Hours and weeks worked by wage fifth, all workers, 1975-2013. (n.d.) *Economic Policy Institute*. epi.org/files/2015/hours-and-weeks-worked-data-from-1975-2013.xlsx ; 1,767 in 2020: Average annual hours actually worked per worker. (n.d.). *OECD*. stats.oecd.org/index.aspx?DataSetCode=ANHRS ; Similar numbers: Mishel, L. (2013, January 30). Vast majority of wage earners are working harder, and for not much more. *Economic Policy Institute*. epi.org/publication/ib348-trends-us-work-hours-wages-1979-2007/

[28] Miller, G.E. (2022, January 30). The U.S. is the most overworked developed nation in the world. *20 Something Finance*. 20somethingfinance.com/american-hours-worked-productivity-vacation; Levitz, E. (2016, October 18). Europeans Work 19 Percent Fewer Hours Than Americans Do. The Cut. thecut.com/2016/10/europeans-work-19-percent-fewer-hours-than-americans-do.html

[29] De Graaf, J. (2003). *Take back your time: Fighting overwork and time poverty in America*. Berrett-Koehler Publishers. ix.

[30] McKibben, B. (2007). *Deep economy: The wealth of communities and the durable future*. Macmillan Publishers. 115–16.

[31] McKibben, B. 115–16.

[32] Rauch, E. Productivity and the workweek.

[33] Brody, L. (2008, October). Which way to happy? Two authors weigh in. *Oprah.com*. oprah.com/spirit/Which-Way-to-Happy-Two-Experts-Weigh-In

[34] Schor, J. B. (1999). *The overspent American: Why we want what we don't need*. Harper Perennial. 113.

[35] West, B. (20109, Feb 13). This Harvard Study Says the Happiest People Have More Time and Less Money. *Forbes*. forbes.com/sites/briannawiest/2019/02/13/this-harvard-study-says-the-happiest-people-have-more-time-and-less-money

[36] Robinson, J. (2011, May 31). Overworked and underplayed: The incredible shrinking vacation. *The Huffington Post*.huffingtonpost.com/joe-robinson/vacation-time_b_868655.html

8 - The Hard Test of Our Wisdom: TV & Internet

[1] Arnheim, R. (2006). Film as Art. United Kingdom: University of California Press. 195.

[2] US Bureau of Labor Statistics. (2019). Average hours per day spent in selected leisure and sports activities by age. *US Bureau of Labor Statistics*.bls.gov/charts/american-time-use/activity-leisure.htm; US Bureau of Labor Statistics. (2021, July 22). American time Use Survey Summary. *US Bureau of Labor Statistics*. Table 3.bls.gov/news.release/atus.nr0.htm; Koblin, J. (2016, June 30). How much do we love tv? Let us count the ways. *The New York Times*. nytimes.com/2016/07/01/business/media/nielsen-survey-media-viewing.html; Hubbard, K. (2021 July 22). Outside of Sleeping, Americans Spend Most of Their Time Watching Television. *US News*. usnews.com/news/best-states/articles/2021-07-22/americans-spent-more-time-watching-television-during-covid-19-than-working

[3] On average, we actually watched more TV than worked during the coronavirus pandemic. Table 1. Time spent in primary activities and percent of the civilian population engaging in each activity, averages per day by sex, 2021 annual averages. US Bureau of Labor Statistics. bls.gov/news.release/atus.t01.htm

[4] More than Half the Homes in U.S. Have Three or More TVs. (2009, July 20). Nielsen. nielsen.com/us/en/insights/article/2009/more-than-half-the-homes-in-us-have-three-or-more-tvs/

[5] Bronson, P. (2006, October 23). How we spend our leisure time. *Time*.content.time.com/time/nation/article/0,8599,1549394,00.html; Robinson, J. & Godbey, G. (2000*). Time for life: The surprising ways Americans use their time*. Pennsylvania State University Press. 238–39, 247.

[6] Kahneman's Day Reconstruction Method study (Kahneman, D. (Dec, 2004). A Survey Method for Characterizing Daily Life Experience: The Day Reconstruction Method, *Science*. 1778) surprised people by showing that when people are asked to rate how happy they were the day before when doing different activities (apparently a fairly reliable measure), people indicated they had been significantly happier while watching TV than had been previously reported. For reasons revealed later in the chapter, I am citing the results of a study that looks at overall impact on happiness, not "in-the-moment" happiness: Frey, B. S. (2010). *Happiness: A revolution in economics*. MIT press. 283–313.

[7] Frey, B. S. *Happiness: A revolution*. 287, 295.

[8] Csikszentmihalyi, M. (1999). If we are so rich, why aren't we happy? *American Psychologist*, *54*(10), 825.doi.org/10.1037/0003-066x.54.10.821

[9] Seligman, M. E. (2003). *Authentic happiness*. Nicholas Brealey Publishing. 176; Kubey, R., & Csikszentmihalyi, M. (2013). *Television and the quality of life: How viewing shapes everyday experience*. Routledge. 172.

[10] Frey, B. S. *Happiness: A revolution*. 305.

[11] Frey, B. S. 300, 302, 305.

[12] Herr, N. (n.d.). Television and Health. *California State University, Northridge (CSUN)*. csun.edu/science/health/docs/tv&health.html

[13] Frey, B. S. *Happiness: A revolution in economics*. 300, 302, 305.

[14] Schor, J. B. (1999). *The overspent American: Why we want what we don't need*. Harper Perennial. 80.

[15] One study indicates that it is the *type* of exposure, not the amount, that is correlated with negative body image, with soap operas, movies, and music videos having the greatest impact. Tiggemann, M., & Pickering, A. S. (1996). Role of television in adolescent women's body dissatisfaction and drive for thinness. *International Journal of Eating Disorders*, *20*(2), 199–203. doi.org/10.1002/(sici)1098-108x(199609)20:2<199::aid-eat11>3.0.co;2-z ; However, other studies and articles document the impact of quantity, regardless of the type of exposure. Eating disorders: Body image and advertising. (2000, April 25). *HealthyPlace*. healthyplace.com/eating-disorders/articles/eating-disorders-body-image-and-advertising; Kassow, D. (2004, August 10). TV and body Image. *The New York Times*.nytimes.com/2004/08/10/science/l-tv-and-body-image-583448.html

[16] Fox, K. (1997). Mirror, mirror. *Social Issues Research Centre*. sirc.org/publik/mirror.html

[17] Hall, R. H. (2016). Internet use and happiness. *Lecture Notes in Computer Science*, 37–45. doi.org/10.1007/978-3-319-39396-4_4 ; Trilling, D. (2016, November 17). The internet makes people happier: New research. *The Journalist's Resource*. journalistsresource.org/studies/society/internet/internet-happiness-elderly-senior-citizens

[18] Trilling. The internet; Richtel, M. (2012, Jan 25). Does Technology Affect Happiness? *New York Times*. bits.blogs.nytimes.com/2012/01/25/does-technology-affect-happiness/

[19] Information: Trilling. The internet; Hall. Internet use and happiness; Hall, R. *Lecture*. 37–45; Gaming: Vitelli, R. (2014, February 10). Are there benefits in playing video games? *Psychology Today*. psychologytoday.com/us/blog/media-spotlight/201402/are-there-benefits-in-playing-video-games

[20] Hanh, T. N. (1992). *Peace is every step the path of mindfulness in everyday life*. Bantam. 13–14.

[1] Feuerstein, R. (Ed.) (2008). *Yoga Gems: A Treasury of Practical and Spiritual Wisdom from Ancient and Modern Masters*. United Kingdom: Random House Publishing Group.

9 – Free Your Mind

[1] Doty, J. (2016, February 11). The magic shop of the brain. *On Being*. onbeing.org/programs/james-doty-the-magic-shop-of-the-brain-nov2018/
[2] Holden, R. (2010). *Be happy!* Hay House, Inc. xi–xvii.
[3] Holden. *Be happy!* 86–87.
[4] Baumeister, R. F. (1991). *Meanings of life*. Guilford Press. 212.
[5] Acosta, H. N., & Marcenaro, O. D. (2020). The relationship between subjective well-being and self-reported health: *Applied Research in Quality of Life*, 1–21. doi.org/10.1007/s11482-020-09852-z
[6] Baumeister. *Meanings of life*. 211; Summarized very similarly in Diener, E., Lucas, R. E., & Oishi, S. (2002). Subjective well-being: The science of happiness and life satisfaction. *The Oxford Handbook of Positive Psychology*. 2, 68.
[6] Lyubomirsky, S. (2001). Why are some people happier than others? The role of cognitive and motivational processes in well-being. *American Psychologist*, 56(3), 244. doi.org/10.1037/0003-066x.56.3.239
[7] Lyubomirsky. Why are some people happier. 244.
[8] Begley, S. (2008). *Train your mind, change your brain: How a new science reveals our extraordinary potential to transform ourselves*. Random House Digital, Inc. 241.
[9] Much of this research is cited in the following chapters. Regarding the link between "our habitual outlook on life" and our ability to stay positive in the face of stressors, there is a summary of the "findings across studies" in: Ong, A. D., Bergeman, C. S., Bisconti, T. L., & Wallace, K. A. (2006). Psychological resilience, positive emotions, and successful adaptation to stress in later life. *Journal of Personality and Social Psychology*, 91(4), 730–749. doi.org/10.1037/0022-3514.91.4.730 ; Brief readings/citation on the question of our ability to shift our perception include: Lykken, D. (2000). *Happiness: The nature and nurture of joy and contentment*. St. Martin's Griffin (Lykken writes that the entire book is aimed at recanting his earlier claim that "because happiness has strong genetic roots, 'trying to be happier is like trying to be taller.'" p. 60 is a good summary of his position); Diener, E., Lucas, R. E., & Oishi, S. (2002). Subjective well-being. 2, 69; Ben-Shahar, T. (2008). *Happier*. McGraw-Hill, 102, 107–8; Dalai Lama XIV, Cutler, H. C. (2008). *Art of happiness in a troubled world*. Doubleday Books. 241–44.
[10] Begley. *Train your mind, change your brain*. 230–41.
[11] Begley. 230.

10 - Nurturing a Positive Relationship ... Emotions

[1] Chödrön, P. (2006). *Practicing Peace in Times of War*. United States: Shambhala. 54
[2] LeDoux J. E. (2012). Evolution of human emotion: A view through fear. *Progress in Brain Research, 195,* 431–442.doi.org/10.1016/B978-0-444-53860-4.00021-0
[3] Al-Shawaf, L., Conroy-Beam, D., Asao, K., & Buss, D. M. (2015). Human Emotions: An Evolutionary Psychological Perspective. Emotion Review, 8(2), 173–86. doi.org/10.1177/1754073914565518 ; Novacek, M. (n.d.). The rise of mammals. *PBS*. pbs.org/wgbh/evolution/library/03/1/l_031_01.html; Sci News. (2021, March 1). 5-Million-Year-Old Primate Fossils Uncovered in Montana. sci-news.com/paleontology/purgatorius-fossils-montana-09402.html; Rafferty, J. P. (n.d.). Just how old is homo sapiens? *Encyclopedia Britannica*. britannica.com/story/just-how-old-is-homo-sapiens; Gibbons, A. (2010, March). The human family's earliest ancestors. *Smithsonian Magazine*. smithsonianmag.com/science-nature/the-human-familys-earliest-ancestors-7372974
[4] Izard C. E. (2009). Emotion theory and research: highlights, unanswered questions, and emerging issues. *Annual review of psychology*, 60, 1–25. doi.org/10.1146/annurev.psych.60.110707.163539 ; Peil-Kauffman, K., (2020). The purpose of emotion An overlooked self-regulatory sense *Research OUTREACH*, 115. researchoutreach.org/articles/emotion-overlooked-self-regulatory-sense
[5] Tamir, M., Schwartz, S. H., Oishi, S., & Kim, M. Y. (2017). The secret to happiness: Feeling good or feeling right? *Journal of Experimental Psychology: General*, 146(10), 1448–59. doi.org/10.1037/xge0000303 ; Ford, B. Q., Lam, P., John, O. P., & Mauss, I. B. (2018). The psychological health benefits of accepting negative emotions and thoughts: Laboratory, diary, and longitudinal evidence. *Journal of personality and social psychology*, 115(6), 1075–1092. doi.org/10.1037/pspp0000157 ; Kaufman, S. B. (2020, April 20). Post-traumatic growth: Finding meaning and creativity in adversity. *Scientific American*.blogs.scientificamerican.com/beautiful-minds/post-traumatic-growth-finding-meaning-and-creativity-in-adversity; Bonanno, G. A. (2008). Loss, trauma, and human resilience: Have we underestimated the human capacity to thrive after extremely aversive events? *Psychological Trauma: Theory, Research, Practice, and Policy*, S(1), 102. doi.org/10.1037/1942-9681.s.1.101 ; Ong, A. D., Bergeman, C. S., Bisconti, T. L., & Wallace, K. A. (2006). Psychological resilience, positive emotions, and successful adaptation to stress in later life. *Journal of Personality and Social Psychology*, 91(4), 743. doi.org/10.1037/0022-3514.91.4.730
[6] Horwitz, A. V., & Wakefield, J. C. (2007). *The loss of sadness: How psychiatry transformed normal sorrow into depressive disorder*. Oxford University Press. vii, 225; Diener, E. (2008). Myths in the science of happiness and directions for future research. In M. Eid & R. J. Larsen (Eds.), *The science of subjective well-being* (pp. 493–514). Guilford Press. 507; Gruber, J., Mauss, I. B., & Tamir, M. (2011). A dark side of happiness? How, when, and why happiness is not always good. *Perspectives on Psychological Science*, 6(3), 226–227. doi.org/10.1177/1745691611406927 ; Ben-Shahar, T. (2008). *Happier*. McGraw-Hill. 91
[7] Bonanno, G. A., Papa, A., Lalande, K., Westphal, M., & Coifman, K. (2004). The importance of being flexible: The ability to both enhance and suppress emotional expression predicts long-term adjustment. *Psychological Science*, 15(7), 482–487. doi.org/10.1111/j.0956-7976.2004.00705.x
[8] Dunn, B. D., Billotti, D., Murphy, V., & Dalgleish, T. (2009). The consequences of effortful emotion regulation when processing distressing material: A comparison of suppression and acceptance. *Behaviour research and therapy*, 47(9), 761–73. doi.org/10.1016/j.brat.2009.05.007
[9] Dalai Lama XIV, Cutler, H. C. (2008). *Art of happiness in a troubled world*. Doubleday Books. 223; Sirgy, M.J. (2012, May 4). Effects of Social Comparisons on Subjective QOL. The Psychology of Quality of Life. Vol 50 of the Social Indicators Research Series. 223–33.
[10] Brown, B. (2010, June). *The power of vulnerability* [Video]. TED Talk. ted.com/talks/brene_brown_the_power_of_vulnerability
[11] Chödrön, P. (2019). Living beautifully: An inspirational journal. Shambala. 28.
[12] Dunn, B. D., Billotti, D., Murphy, V., & Dalgleish, T. The consequences. 761–773.
[13] Lucy E Cousins, L.E. (2018, Feb) Are there downsides to always trying to be positive? hcf.com.au/health-agenda/body-mind/mental-health/downsides-to-always-being-positive
[14] Somé. S. (n.d.). Embracing Grief. sobonfu.com/articles/writings-by-sobonfu-2/embracing-grief/

[15] Cousins, L.E. Are there downsides.

[16] Hạnh Nhất, Chödzin Sherab, & McLeod, M. (2012). *You are here: Discovering the magic of the present moment.* Shambhala Library. 2.

[17] Rumi. (2004). *Selected Poems.* Penguin Books. 109.

[18] Kaufman, S. B. Post-traumatic growth.

[19] Chödrön, P. (2010). *The wisdom of no escape: And the path of loving-kindness.* Shambhala Publications. 1.

[20] Seery, M. D., Holman, E. A., & Silver, R. C. (2010). Whatever does not kill us: Cumulative lifetime adversity, vulnerability, and resilience. *Journal of Personality and Social Psychology*, 99(6), 1025–1041. doi.org/10.1037/a0021344 ; Kaufman, S. B. Post-traumatic growth.

[21] Dalai Lama XIV, & Cutler, H. C. Art of happiness. 220–22.

[22] Seery, Holman, & Silver. Whatever does not kill us: Cumulative lifetime adversity, vulnerability, and resilience. 1025–1041.

[23] Fredrickson, B. L., Tugade, M. M., Waugh, C. E., & Larkin, G. R. (2003). What good are positive emotions in crises? A prospective study of resilience and emotions following the terrorist attacks on the United States on September 11th, 2001. *Journal of Personality and Social Psychology, 84*(2), 365–376. doi.org/10.1037//0022-3514.84.2.365 ; Tennen, H., & Affleck, G. (2002). Benefit-finding and benefit-reminding. In C. R. Snyder & S. J. Lopez (Eds.), *Handbook of positive psychology* (pp. 584–597). Oxford University Press; Troy, A. S., Shallcross, A. J., Brunner, A., Friedman, R., & Jones, M. C. (2018). Cognitive reappraisal and acceptance: Effects on emotion, physiology, and perceived cognitive costs. *Emotion, 18*(1), 58–74. doi.org/10.1037/emo0000371

[24] Eppard, M. (2020, Nov 13). Personal communication.

[25] Post, S. G., & Neimark, J. (2006). *Why good things happen to good people.* Broadway Books. 17.

[26] Mangelsdorf, J., Eid, M., & Luhmann, M. (2019). Does growth require suffering? A systematic review and meta-analysis on genuine posttraumatic and postecstatic growth. *Psychoolgicall Bulletin, 145*(3), 302–338. doi: 10.1037/bul0000173

[27] Collier, L. (2016, November). Growth after trauma. *American Psychological Association.*apa.org/monitor/2016/11/growth-trauma

[28] Collier. Growth after trauma.

[29] Held, B. (2007, October 22). Combating the tyranny of the positive attitude. *This I Believe.*thisibelieve.org/essay/34187/

11 - Living in Mystery, Magic, and Miracle

[1] Lewis, C. S. (2014). God in the Dock. United States: Eerdmans Publishing Company. 13.

[2] White, E. B. (2018). Charlotte's Web. United States: Penworthy Company, LLC. 108, 109.

[3] Wilber, K. (2001). *Quantum questions: Mystical writings of the world's great physicists.* Shambhala Publications. 3–9.

[4] Wilber. *Quantum questions.* 135.

[5] Wilber. 8.

[6] Schweitzer, A., Powers, J. (1923). *Christianity and the Religions of the World,.* George H. Doran Company. 80. google.com/books/edition/Christianity_and_the_Religions_of_the_Wo/CTsPAQAAIAAJ?hl=en&gbpv=0

[7] Swimme, B. (1984). *The universe is a green dragon: A cosmic creation story.* Bear & Company. 37–38.

[8] The quarks in neutrons and protons are held together by gluons, which themselves are inherently massless, yet they possess energy, which contributes around 99 percent of the neutrons' and protons' mass. Ball, P. (2008). Nuclear masses calculated from scratch. *Nature.* doi.org/10.1038/news.2008.1246 ; The masses of the individual elementary particles are shown in this nice graphic: Elementary particles included in the Standard Model. (2019, September 17). *Wikipedia.* en.wikipedia.org/wiki/Elementary_particle#/media/File:Standard_Model_of_Elementary_Particles.svg

[9] Folger, T. (2005, June 4). If an electron can be in two places at once, why can't you? *Discover Magazine.* discovermagazine.com/2005/jun/cover; Jeffrey C. (2015, June 04). Experiments suggests that reality doesn't exist until it is measured. *New Atlas.* gizmag.com/quantum-theory-reality-anu/37866

[10] Kane, G. (2006, October 9). Are virtual particles really constantly popping in and out of existence? Or are they merely a mathematical bookkeeping device for quantum mechanics? *Scientific American.* scientificamerican.com/article/are-virtual-particles-rea

[11] I have not found the source of the original quote, though it shows up in various forms all over the web, perhaps in part due to the fact that he said it in Danish, so these are all translations. Seth, L. (2012, December 14). Ask a quantum mechanic. *NPR.* npr.org/2012/12/14/167255707/ask-a-quantum-mechanic

[12] Markoff, J. (2015, October 21). Sorry, Einstein. Quantum study suggests 'spooky action' is real. *The New York Times.* nytimes.com/2015/10/22/science/quantum-theory-experiment-said-to-prove-spooky-interactions.html

[13] Battersby, S. (2008, November 20). It's confirmed: Matter is merely vacuum fluctuations. *New Scientist.* newscientist.com/article/dn16095-its-confirmed-matter-is-merely-vacuum-fluctuations. And Kane. Are virtual particles really; The mass of gluons: Ball. Nuclear masses calculated from scratch.

[14] Howell, E. (2022, Jan 10). What is the Big Bang Theory? space.com/25126-big-bang-theory.html

[15] Chödrön, P. (2010). *The Wisdom of No Escape: And The Path of Loving-Kindness.* United Kingdom: Shambhala. 36.

[16] Lyubomirsky, S. (2011). *Hedonic adaptation to positive and negative experiences.* Oxford University Press. 215; Wallis, C. (2009, July 8). The science of happiness turns 10. what has it taught? *Time.* content.time.com/time/health/article/0,8599,1908173,00.html; Post, S. G., & Neimark, J. (2006). Why good things happen to good people. Broadway Books. 34–35.

[17] Lyubomirsky, S., Sheldon, K. M., & Schkade, D. (2005). Pursuing happiness: The architecture of sustainable change. *Review of General Psychology, 9*(2), 199–220.

[18] Hạnh Nhất, Chödzin Sherab, & McLeod, M. (2012). *You are here: Discovering the magic of the present moment.* Shambhala Library. 5.

[19] André,C. (2019, January 15). Proper Breathing Brings Better Health. *Scientific American.* scientificamerican.com/article/proper-breathing-brings-better-health; Cuda, G. (2010, Dec 6). Just Breathe: Body Has A Built-In Stress Reliever NPR. npr.org/2010/12/06/131734718/just-breathe-body-has-a-built-in-stress-reliever

[20] Aked, J., Marks, N., Cordon, C., & Thompson, S. (2008, October 22). Five ways to wellbeing. *New Economics Foundation.*neweconomics.org/2008/10/five-ways-to-wellbeing

[21] Begley, S. (2008). *Train your mind, change your brain: How a new science reveals our extraordinary potential to transform ourselves.* Random House Digital, Inc. 233–238.

[22] Swimme, B. T. (1999). *Hidden heart of the cosmos: Humanity and the new story.* Orbis Books. 39.

[23] Davis, P. (2020, Feb 1). How Big Is the Solar System? solarsystem.nasa.gov/news/1164/how-big-is-the-solar-system/

[24] King, B. (2014, September 17). 9,096-Is that all? *Sky & Telescope.* skyandtelescope.com/astronomy-resources/how-many-stars-night-sky-09172014; Cassan, A., Kubas, D., Beaulieu, J.-P., Dominik, M., Horne, K., Greenhill, J., Wambsganss, J., Menzies, J., Williams, A., Jørgensen, U. G., Udalski, A., Bennett, D. P., Albrow, M. D., Batista, V.,

Brillant, S., Caldwell, J. A., Cole, A., Coutures, C., Cook, K. H., … Wyrzykowski, Ł. (2012). One or more bound planets per Milky Way star from microlensing observations. *Nature*, *481*(7380), 167–169.doi.org/10.1038/nature10684
[25] Atkinson, N. (2009, July 7). How often are new stars born in the Milky Way? *Universe Today*. universetoday.com/34380/how-often-are-new-stars-born-in-the-milky-way; Dunbar, B. (2006, January 5). Milky Way churns out seven new stars per year, scientists say. *NASA*. nasa.gov/centers/goddard/news/topstory/2006/milkyway_seven.html; Cassan, … Wyrzykowski. One or more. 167–69; Rabie, P. (2020, October 30). Scientists pinpoint how many planets in the Milky Way could host life. *Inverse*. inverse.com/science/how-many-planets-host-life
[26] Eicher, D. J. (2019, July 1). How many galaxies are in our group? *Astronomy*. astronomy.com/magazine/greatest-mysteries/2019/07/49-how-many-galaxies-are-in-our-group
[27] What is the local group? (2018, December 6). *EarthSky*. earthsky.org/astronomy-essentials/galaxy-universe-location
[28] The universe within 14 billion light years the visible universe. (n.d.). *An Atlas of Universe*. atlasoftheuniverse.com/universe.html
[29] Garber, M. (2013, November 19). How many stars are there in the sky? *The Atlantic*. theatlantic.com/technology/archive/2013/11/how-many-stars-are-there-in-the-sky/281641/
[30] Universe has 2 trillion galaxies, astronomers say. (2016, October 13). *The Guardian*. theguardian.com/science/2016/oct/13/hubble-telescope-universe-galaxies-astronomy
[31] Northcutt, P. & Kennard, d. (Directors). (2011). *Journey of the Universe* (Film). Northcutt Productions.
[32] Greene, B. (2000). The Elegant Universe: Superstrings, Hidden Dimensions, and the Quest for the Ultimate Theory. United Kingdom: Vintage Books. 13.
[33] Nhat Hanh, T. (2019, December 10). Thich Nhat Hanh on the practice of mindfulness. *Lion's Roar*. lionsroar.com/mindful-living-thich-nhat-hanh-on-the-practice-of-mindfulness-march-2010/
[34] Dive Training. (2000, Sept 4). From Ripples to Rogues: How Waves Are Formed. dtmag.com/thelibrary/ripples-rogues-waves-formed; National Weather Service. (n.d.). Wind, Swell and Rogue Waves. weather.gov/jetstream/waves; Heller, E. J. (2013). *Why You Hear What You Hear: An Experiential Approach to Sound, Music, and Psychoacoustics*. United Kingdom: Princeton University Press. 15.
[35] Michaelson, J. (2013). *Evolving Dharma: Meditation, Buddhism, and the Next Generation of Enlightenment*. United States: North Atlantic Books. xvi, 7–19; André. Proper Breathing.
[36] Lisa Marie Basile, L.M. (2020, May 18). Is Cortisol Good or Bad? endocrineweb.com/cortisol-good-bad; André. Proper Breathing; Aked, Marks, Cordon, & Thompson. Five ways to wellbeing.
[37] Joseph Goldstein, J. (2019, March 23). These Are Not "Your" Thoughts. *Tricycle*. tricycle.org/trikedaily/joseph-goldstein-mindfulness-consciousness; Rinpoche, D.P. (2020, Nov 9). Break the Chains of Thought. Lion's Roar. lionsroar.com/break-the-chains-of-thought
[38] Chödrön, P. (2013, summer). Meditating with Emotions. *Tricycle*. tricycle.org/magazine/meditating-emotions
[39] Michaelson. *Evolving Dharma*. 24.
[40] Halliwell, E. (2011, May 10). Meditation is an emotional rollercoaster. *The Guardian*. theguardian.com/commentisfree/belief/2011/may/10/meditation-journey-relaxation
[41] Ladinsky, D. (2002). *Love poems from God: Twelve sacred voices from the East and West*. Penguin. 249.
[42] Angelou, M. (2014, May 23). twitter.com/DrMayaAngelou
[43] Miller, G. (2010). *Spent: Sex, evolution, and consumer behavior*. Penguin Books.
[44] Clegg, B. (2012). *The universe inside you: The extreme science of the human body from quantum theory to the mysteries of the brain*. Icon Books Ltd. 1–2.
[45] Clegg. *The universe inside you*. 2.
[46] Schirber, M. (2009, April 16). The chemistry of life: The human body. *Live Science*. livescience.com/3505-chemistry-life-human-body.html; Clegg. 40; Origin of the elements. (2000, August 9). *Guide to the Nuclear Wall Chart*. www2.lbl.gov/abc/wallchart/chapters/10/0.html
[47] All about stars. (n.d.). *Scholastic*. scholastic.com/teachers/articles/teaching-content/all-about-stars/
[48] How elements are formed. (2009, October 22). *Science Learning Hub*. sciencelearn.org.nz/resources/1727-how-elements-are-formed
[49] Clegg. *The universe inside you*. 40; Schirber. The chemistry of life.
[50] Gee, H. (1999, November 22). Size and the single sex cell. *Nature*. nature.com/articles/news991125-4
[51] Eveleth, R. (2013, October 24). There are 37.2 trillion cells in your body. Smithsonian Magazine. smithsonianmag.com/smart-news/there-are-372-trillion-cells-in-your-body-4941473/
[52] The cells in your body. (n.d.). *Science NetLinks*. sciencenetlinks.com/student-teacher-sheets/cells-your-body/
[53] Choi, C. Q. (2010, February 3). Brute force: Humans can sure take a punch. *Live Science*. livescience.com/6040-brute-force-humans-punch.html; How hard are human teeth and enamel? (2017, August 10). *Western Pennsylvania Ohio Valley Oral Maxillofacial Surgery*. westernpaoms.com/hard-human-teeth/
[54] Your amazing brain. (2021, February 10). *National Geographic Kids*. kids.nationalgeographic.com/science/article/your-amazing-brain
[55] Radford, T. (2004, May 27). Secrets of human hair unlocked at Natural History Museum in London. *The Guardian*. theguardian.com/uk/2004/may/27/sciencenews.research
[56] Morrison, J. (2014, March20). Human nose can detect 1 trillion odours. *Nature*. nature.com/news/human-nose-can-detect-1-trillion-odours-1.14904
[57] Roland, J. (2019, May 23). How far can the human eye see? *Healthline*. healthline.com/health/how-far-can-the-human-eye-see; Starwatch: The furthest thing you can see with the naked eye. (2019, August 25). *The Guardian*. theguardian.com/science/2019/aug/25/starwatch-the-furthest-thing-you-can-see-with-the-naked-eye
[58] Fischetti, M. (2021, April 1). Our bodies replace billions of cells every day. *Scientific American*. scientificamerican.com/article/our-bodies-replace-billions-of-cells-every-day/
[59] Narasimhan, K. (2004). Scaling up neuroscience. *Nature Neuroscience*, *7*(5), 425. doi.org/10.1038/nn0504-425
[60] Computation power: Human brain vs supercomputer (2019, April 10). *Foglets*. foglets.com/supercomputer-vs-human-brain/
[61] Shankland, S. (2020, March 4). El Capitan supercomputer to blow past rivals, with 2 quintillion calculations per second. *CNET*. cnet.com/news/el-capitan-supercomputer-blow-past-rivals-with-2-quintillion-calculations-per-second/
[62] Computation power.
[63] Shankland. El Capitan supercomputer.
[64] The thermodynamics of brains and computers. (n.d.). *Duke*. webhome.phy.duke.edu/~hsg/363/table-images/brain-vs-computer.html
[65] Computation power.
[66] Coffey, D. (2021, July 19). How does DNA know which job to do in each cell? livescience.com/how-dna-turns-on-off.html
[67] Bouchard, R. P. (2019, January 25). Is DNA like a blueprint, a computer program, or a list of ingredients? *Medium*. medium.com/the-philipendium/is-dna-like-a-blueprint-a-computer-program-or-a-list-of-ingredients-1484b34a9121
[68] Epigenetics: Fundamentals. (2018, March 15). *What is Epigenetics?/Fundamentals*.

whatisepigenetics.com/fundamentals/; Wein, H. (2010, September 27). Stress hormone causes epigenetic changes. *National Institutes of Health*. nih.gov/news-events/nih-research-matters/stress-hormone-causes-epigenetic-changes; What is Epigenetics?/Epigenetics. (2020, August 3). *Centers for Disease Control and Prevention*. cdc.gov/genomics/disease/epigenetics.htm; Krol, K. M., Moulder, R. G., Lillard, T. S., Grossmann, T.H., & Connelly, J. J. (2019). Epigenetic dynamics in infancy and the impact of maternal engagement. *Science Advances, 5*(10). 10.1126/sciadv.aay0680; Holloway, T., & González-Maeso, J. (2015). Epigenetic mechanisms of serotonin signaling. *ACS Chemical Neuroscience, 6*(7), 1099–1109. doi.org/10.1021/acschemneuro.5b00033
[69] Clegg. *The universe inside you.* 223–24.
[70] Clegg. 223; Machemer, T. (2021, January 13). Many identical twins actually have slightly different DNA. *Smithsonian Magazine*. smithsonianmag.com/smart-news/identical-twins-can-have-slightly-different-dna-180976736/
[71] Bouchard. Is DNA like a blueprint.
[72] Pistoi, S. (2020, February 6). DNA is not a blueprint. *Scientific American*. blogs.scientificamerican.com/observations/dna-is-not-a-blueprint/
[73] Miller, H. (1953). *Plexus*. Grove Press. 53.
[74] Post & Neimark. *Why good things happen.* 34.
[75] Lyubomirsky. Hedonic adaptation to positive and negative experiences. 215; Folkman, S. (2011). *The Oxford Handbook of stress, health, and coping.* Oxford University Press; Wallis. The science of happiness turns 10. What has it taught?; Post & Neimark. *Why good things happen.* 34-35.
[76] Holden, R. (2010). *Be happy!* Hay House, Inc. 116–121.
[77] Lyubomirsky, S., Dickerhoof, R., Boehm, J. K., & Sheldon, K. M. (2011). Becoming happier takes both a will and a proper way: An experimental longitudinal intervention to boost well-being. *Emotion, 11*(2), 393.doi.org/10.1037/a0022575; Wallis. The science of happiness turns 10. what has it taught?
[78] Lyubomirsky, Sheldon, & Schkade. Pursuing happiness. 199–220.
[79] Holden. *Be happy!* 116–121.
[80] Holden. 115–116.
[81] Nhat Hanh. Thich Nhat Hanh on the practice.

12 - The Heart of Happiness: Loving Ourselves

[1] Holden, R. (2010). *Be happy!* Hay House, Inc. 77.
[2] Ben-Shahar, T. (2008). *Happier*. McGraw-Hill. 141, 144.
[3] Baumeister, R. F. (1991). *Meanings of life*. Guilford Press. 216. It is noteworthy that this seems to be less true in cultures that are less individualistic (Diener, E., Lucas, R. E., & Oishi, S. (2002). Subjective well-being: The science of happiness and life satisfaction. *The Oxford Handbook of Positive Psychology*. 2, 68).
[4] Baumeister, R. F., Campbell, J. D., Krueger, J. I., & Vohs, K. D. (2003). Does high self-esteem cause better performance, interpersonal success, happiness, or healthier lifestyles? *Psychological Science in the Public Interest, 4*(1), 28. doi.org/10.1111/1529-1006.01431 . That article and the following have great overviews of the many ways self-esteem as it has been traditionally understood is *not* linked with any other major positive outcomes besides happiness. Bronson, P. (2007, February 9). How not to talk to your kids. *New York Magazine*. nymag.com/news/features/27840 ; "Self-love," as I am using the term in this book, is very different from self-esteem in general, but seems comparable to "true self-esteem," "a self-determined and autonomous way of evaluating oneself that is not dependent on particular outcomes or social approval," or "optimal self-esteem," "which is founded on stable and noncontingent self-evaluations." Neff, K. D., & Vonk, R. (2009). Self-compassion versus global self-esteem: Two different ways of relating to oneself. *Journal of Personality, 77*(1), 25.doi.org/10.1111/j.1467-6494.2008.00537.x ; Self-compassion is best understood in the context of this book as a critical component of self-love. Self-compassion is defined "in terms of three main components: self-kindness, a sense of common humanity, and mindfulness when considering personal weakness or hardships." (Neff & Vonk.)
[5] Likewise, in reviewing a range of prominent perspectives on happiness, Carol Ryff found that "self-acceptance or self-esteem" was the "most recurrent criterion for positive well-being." Lyubomirsky, S., Tkach, C., & DiMatteo, M. R. (2005). What are the differences between happiness and self-esteem? *Social Indicators Research, 78*(3), 364. doi.org/10.1007/s11205-005-0213-y ; Lyubomirsky, S., King, L., & Diener, E. (2005). The benefits of frequent positive affect: Does happiness lead to success? *Psychological Bulletin, 131*(6), 825. doi.org/10.1037/0033-2909.131.6.803
[6] The Oprah Winfrey show finale. (2011, May 25). *Oprah.com*.oprah.com/oprahshow/The-Oprah-Winfrey-Show-Finale_1/6
[7] Brown, B. (2010). *The gifts of imperfection: Let go of who you think you're supposed to be and embrace who you are.* Simon and Schuster. 38–39; Seltzer, L. (2008, September 10). The path to unconditional self-acceptance. *Psychology Today*. psychologytoday.com/blog/evolution-the-self/200809/the-path-unconditional-self-acceptance
[8] Brown. *The gifts of imperfection.* 39; Hartling, L. M., Rosen, W., Walker, M., & Jordan, J. V. (2000). *Shame and humiliation: From isolation to relational transformation.* Wellesley College. 2; Graham, J. (n.d.). Bulletin #4422, Violence part 2: Shame and humiliation. *University of Maine Cooperative Extension.* umaine.edu/publications/4422e; VanScoy, H. (2016, May 17). Shame: The quintessential emotion. *PsychCentral.* psychcentral.com/lib/shame-the-quintessential-emotion#1
[9] Brown, B. (2010, June). *The power of vulnerability* [Video]. TED Talk. ted.com/talks/brene_brown_the_power_of_vulnerability
[10] Graham. Violence part 2 (Shame and Pride/Shame). Brené Brown writes that "While it feels like shame hides in our darkest corners, it actually tends to lurk in all of the familiar places, including appearance and body image, motherhood, family, parenting, money and work, mental and physical health, addiction, sex, aging and religion." (Brown, B. (2008). *I thought it was just me (but it isn't): Telling the truth about perfectionism, inadequacy, and power.* Gotham Books. xii.) Robert Holden writes that "Our addiction to this belief, that we are not good enough, feeds all other addictions, including over-dependency in relationships, extreme competitiveness, punishing social comparison, envy and jealousy, pained perfectionism, constant self-judgment, lack of assertiveness, and a desperate never-ending pursuit of happiness. This addiction also feeds other more physical addictions that provide temporary comfort and escape, such as overeating, alcoholism, compulsive sex, and illegal drugs." (Holden, R. (2011). *Happiness now!: Timeless wisdom for feeling good fast.* Hay House, Inc. 63.) Donald Nathanson writes that the effects of shame go far beyond our individual lives. "Shame—our reaction to it and our avoidance of it"—is "a primary force in social and political evolution." (Nathanson, D. L. (1994). *Shame and pride: Affect, sex, and the birth of the self.* WW Norton & Company. 15–16).
[11] Hartling, Rosen, Walker, & Jordan. *Shame and humiliation.* 2; Harper, D. (n.d.). *Shame.* Etymonline: Online Etymology Dictionary. etymonline.com/word/shame
[12] Brown, B. (2006). Shame resilience theory: A grounded theory study on women and shame. *Families in Society, 87*(1), 46.doi.org/10.1606/1044-3894.3483

[13] Dearing, R. L., Stuewig, J., & Tangney, J. P. (2005). On the importance of distinguishing shame from guilt: Relations to problematic alcohol and drug use. *Addictive Behaviors*, *30*(7), 1393, 1400–1401. doi.org/10.1016/j.addbeh.2005.02.002 ; Brown. *The gifts of imperfection*. 41–42; Robinson, B. E. (2014). *Chained to the desk*. New York University Press. 48; Benson, A. L. (2008). *To buy or not to buy: Why we overshop and how to stop*. Shambhala Publications. 13; Goodtherapy. (2019, July 29). Inadequacy. goodtherapy.org/learn-about-therapy/issues/inadequacy; Brown. *The gifts of imperfection*. 41–42.

[14] Hartling, Rosen, Walker, & Jordan. *Shame and humiliation*. 4; Holden. *Happiness now!* 62; Gilbert, P. (2000). The relationship of shame, social anxiety and depression: The role of the evaluation of social rank. *Clinical Psychology & Psychotherapy*, *7*(3), 176. doi.org/10.1002/1099-0879(200007)7:3<174::aid-cpp236>3.0.co;2-u

[15] Goodtherapy. Inadequacy.

[16] Holden. *Happiness now!* 62.

[17] Brown, B. (2012). *Daring greatly: How the courage to be vulnerable transforms the way we live, love, parent, and lead*. Penguin. 21–23. Kreger, R. (2012. Jan 5). Shame Is at the Root of Narcissistic, Borderline Disorder. *Psychology Today*. psychologytoday.com/blog/stop-walking-eggshells/201201/shame-is-the-root-narcissistic-borderline-disorder; Grossman, H. (n.d.). Compass of Shame. 2. southdown.on.ca/publications/articles/Compass-of-Shame.pdf

[18] Firestone, L. (2011, September 26). What drives jealousy? *Psychology Today*. psychologytoday.com/blog/compassion-matters/201109/what-drives-jealousy

[19] Graham. Violence part 2 (Shame and Violence/Violence as the Absence of Love).

[20] Hartling, Rosen, Walker, & Jordan. *Shame and humiliation*. 4; Graham. Violence part 2 (Shame and Violence/Violence as the Absence of Love); Kreger. Shame Is at the Root; Brown. *The gifts of imperfection*. 41–42.

[21] Graham. Violence part 2 (Introduction; and Shame and Pride).

[22] Hartling, Rosen, Walker, & Jordan. *Shame and humiliation*. 4.

[23] Abramson, A. (2020, July 23). The science of shame. *Elemental*. elemental.medium.com/the-science-of-shame-e1cb32f6f2a

[24] Holden. *Happiness now!* 62.

[25] Bragg, H. L. (2003). *Child protection in families experiencing domestic violence*. US Department of Health and Human Services, Administration for Children and Families, Administration on Children, Youth, and Families, Children's Bureau, Office on Child Abuse and Neglect. 25–26; Domestic Abuse Project. (n.d.). Compelling Reasons Women Stay. domesticabuseproject.com/get-educated/compelling-reasons-women-stay/

[26] Brown. Shame resilience theory. 50.

[27] Brown. *Daring greatly*. 85, 86, 101, 102.

[28] Giordano, S. (2018). Understanding the emotion of shame in transgender individuals—some insight from Kafka. *Life Sciences, Society and Policy*, *14*(23). doi.org/10.1186/s40504-018-0085-y ; Scandurra C., Mezza, F., Maldonato, N. M., Bottone, M., Bochicchio, V., Valerio, P., Vitelli, R. (2019, June 25). Health of non-binary and genderqueer people: A systematic review. *Front. Psychol*. frontiersin.org/articles/10.3389/fpsyg.2019.01453/full

[29] Adams, L. M., & Miller, A. B. (2022). Mechanisms of Mental-Health Disparities Among Minoritized Groups: How Well Are the Top Journals in Clinical Psychology Representing This Work?. *Clinical psychological science : a journal of the Association for Psychological Science*, *10*(3), 387–416. doi.org/10.1177/21677026211026979

[30] Holden. *Be happy!* 185.

[31] VanScoy. Shame: The quintessential emotion.

[32] Kaufman, G. (2004). *The psychology of shame: Theory and treatment of shame-based syndromes*. Springer Publishing Company. 5, 16, 21. Graham. Violence part 2 (Shame and Violence/Violence as the Absence of Love; and Shame as Affect and Social Shame/Social Shame).

[33] Breines, J. (2016, Jan 30). Why You Have to Love Yourself First. Psychology Today. psychologytoday.com/us/blog/in-love-and-war/201601/why-you-have-love-yourself-first

[34] Waheed, N. (2013). *Salt*. Self-published. 221.

[35] Salzberg, S. (2015, Nov 9). The Self-Hatred Within Us. onbeing.org/blog/the-self-hatred-within-us/

[36] hooks, b. (2018). *All about love: New visions*. HarperCollins Publishers. 57, 66.

[37] Neff, K. (2011, May 29). The motivational power of self-compassion. *The Huffington Post*. huffingtonpost.com/kristin-neff/self-compassion_b_865912.html

[38] Brown. *Daring greatly*. 73; Graham. Violence part 2; Lutwak, N., Panish, J. B., Ferrari, J. R., & Razzino, B. E. (2001). Shame and guilt and their relationship to positive expectations and anger expressiveness. *Adolescence*, *36*(144), 641–53.

[39] Neff. The motivational power; Neff, K. (2011, April 25). Self-compassion: Treating yourself as you'd treat a good friend. *The Huffington Post*.huffingtonpost.com/kristin-neff/the-golden-rule-in-revers_b_850465.html; Leary, M. R., Tate, E. B., Adams, C. E., Batts Allen, A., & Hancock, J. (2007). Self-compassion and reactions to unpleasant self-relevant events: The implications of treating oneself kindly. *Journal of Personality and Social Psychology*, *92*(5), 902. doi.org/10.1037/0022-3514.92.5.887 ; Parker-Pope, T. (2011, Feb 28). Go Easy on Yourself, a New Wave of Research Urges. *New York Times*. well.blogs.nytimes.com/2011/02/28/go-easy-on-yourself-a-new-wave-of-research-urges; Brown. *The gifts of imperfection*. 41–42.

[40] Rogers, C. R., & Kramer, P. D. (1995). *On becoming a person: A therapist's view on psychotherapy*. Houghton Mifflin. 17

[41] Brown. *The gifts of imperfection*. 41.

[42] hooks. *All about love*. 57.

[43] Brown. *The gifts of imperfection*. xi.

[44] Brown. The power of vulnerability.

[45] I have sought, without success, to determine if there is a term for this or if there is research documenting it, but I include it because it has been so clearly identified by people I have worked with.

[46] Corbett, H. (2012, September 12). Understanding male vulnerability. *Redbook*. redbookmag.com/health-wellness/advice/brene-brown-shame-vulnerability

[47] Brown. *The gifts of imperfection*. xi.

[48] Kolin, A. M. (1999). *Rumi: Whispers of the beloved*. Harpercollins Publishers. 84

[49] Goldstein. E. (2010, June). You Don't Deserve to be Happy? An Interview with David Simon, M.D. blogs.psychcentral.com/mindfulness/2010/06/you-dont-deserve-to-be-happy-an-interview-with-david-simon-m-d/

[50] Seltzer. The path; Neff, K. D. (2011). Self-compassion, self-esteem, and well-being. *Social and Personality Psychology Compass*, *5*(1), 9.doi.org/10.1111/j.1751-9004.2010.00330.x ; Kaufman. *The psychology of shame*. 34–46; VanScoy. Shame: The quintessential emotion; Firestone. What drives jealousy?

[51] Who wrote "risk"? Is the mystery solved? (2017, May 9). *The Official Anais Nin Blog*. anaisninblog.skybluepress.com/2013/03/who-wrote-risk-is-the-mystery-solved

[52] Brown. The power of vulnerability.

[53] Woods, H., & Proeve, M. (2014). Relationships of mindfulness, self-compassion, and meditation experience with shame-proneness. *Journal of Cognitive Psychotherapy*, *28*(1), 20–33. doi.org/10.1891/0889-8391.28.1.20 ; Westerman, G., McCann, E., & Sparkes, E. (2020). Evaluating the effectiveness of mindfulness and compassion-based

programs on shame and associated psychological distress with potential issues of salience for adult survivors of childhood sexual abuse: A systematic review. *Mindfulness, 11*(8), 1827–1847. doi.org/10.1007/s12671-020-01340-7 ; Millière, R., Carhart-Harris, R. L., Roseman, L., Trautwein, F., & Berkovich-Ohana, A. (2018). Psychedelics, meditation, and self-consciousness. *Frontiers in Psychology, 9*, 1475. doi.org/10.3389/fpsyg.2018.01475 ; Khanna, S., & Greyson, B. (2015). Near-death experiences and posttraumatic growth. *Journal of Nervous & Mental Disease, 203*(10), 749–50. doi.org/10.1097/nmd.0000000000000362 ; Schouborg, G. (2011). Transcending the shamed self. *Journal of Consciousness Exploration & Research, 2*(9), 1438–1462; Allen, S. (2018). The science of awe. *Greater Good Science Center at UC Berkeley.* 35–36. ggsc.berkeley.edu/images/uploads/GGSC-JTF_White_Paper-Awe_FINAL.pdf; Lee, R. G., & Wheeler, G. (2013). *The voice of shame: Silence and connection in psychotherapy.* Taylor & Francis. 48–50.

[54] Dolezal, L., & Lyons, B. (2017). Health-related shame: An affective determinant of health? *Medical Humanities, 43*(4), 257–263. doi.org/10.1136/medhum-2017-011186 ; Ferguson, T. J. (2005). Mapping shame and its functions in relationships. *Child Maltreatment, 10*(4), 381. doi.org/10.1177/1077559505281430

[55] Nhat Hanh, T. (2010). *Reconciliation: Healing the inner child.* Random House Inc. 1, 2.

[56] Brown. Shame resilience theory. 50.

[57] Brown. *The gifts of imperfection.* 10, 40.

[58] One frame for understanding this difference between paths of avoidance and approach: Grasso, D. J., Cohen, L. H., Moser, J. S., Hajcak, G., Foa, E. B., & Simons, R. F. (2012). Seeing the silver lining: Potential benefits of trauma exposure in college students. *Anxiety, Stress & Coping, 25*(2), 117–136.doi.org/10.1080/10615806.2011.561922

[59] Brené Brown has proposed, based on her research, that "shame resilience … is the sum of four things: (a) the ability to recognize and accept personal vulnerability/recognizing shame and understanding its triggers, (b) the level of critical awareness regarding social/cultural expectations and the shame web; (c) the ability to form mutually empathic relationships that facilitate reaching out to others; and (d) the ability to 'speak shame' or possess the language and emotional competence to discuss and deconstruct shame." Brown. Shame resilience theory. 47-49; Brown. *Daring greatly.* 75.

[60] Holden. *Happiness now! 55.*

[61] Brown. *The gifts of imperfection.* 9, 10, 25, 38, 40; Brown. Shame resilience theory. 47.

[62] One of Brené Brown's four components of "shame resilience" is "the ability to form mutually empathic relationships that facilitate reaching out to others." Brown. Shame resilience theory. 47-49.

[63] Brown. Shame resilience theory. 48.

[64] For some people therapy carries a stigma that actually can leave us feeling worse for having seen one. (Brown. Shame resilience theory. 51.). Therapists may also not be sensitive to dynamics related to race, sexual orientation, gender identity, etc. (hooks, b. (2003). *Rock My Soul.* Simon and Schuster. 205; hooks, b. (2001). *Salvation: Black People and Love.* William Morrow. 90–91; Meyer, O. L., & Zane, N. (2013). The influence of race and ethnicity in clients' experiences of mental health treatment. *Journal of Community Psychology, 41*(7), 884–901. doi.org/10.1002/jcop.21580). On the effectiveness of therapy: Norcross, J. C., & Wampold, B. E. (2011). Evidence-based therapy relationships: Research conclusions and clinical practices. *Psychotherapy, 48*(1), 98–102. doi.org/10.1037/a0022161 ; Boyce, C. J., & Wood, A. M. (2010). Money or mental health: The cost of alleviating psychological distress with monetary compensation versus psychological therapy. *Health Economics, Policy and Law, 5*(4), 509–16. doi.org/10.1017/s1744133109990326

[65] Werber, C. (2018, October 15). Psychologists who studied shame around the world say it's an essential part of being human. *Quartz.*qz.com/1420754/these-psychologists-studied-shame-around-the-world-and-now-think-its-an-essential-part-of-human-evolution; Tomasello M. (2014). The ultra-social animal. *European journal of social psychology, 44*(3), 187–94. doi.org/10.1002/ejsp.2015

[66] Kassam, K. S., & Mendes, W. B. (2013). The effects of measuring emotion: physiological reactions to emotional situations depend on whether someone is asking. *PloS one, 8*(7), e64959. doi.org/10.1371/journal.pone.0064959

[67] hooks. *All about love.* 54.

[68] Brown has noted, "Reactions are much stronger and responses less effective when the shame emerges around an area the person has not acknowledged or somewhat opened to." Brown, B. (2006). Brown. Shame resilience theory. 48.

[69] Welwood, J. (1997). *Love and awakening: Discovering the sacred path of intimate relationship.* HarperCollins Publishers. 142.

[70] Nouwen, H. J. M. (1992). *Life of the beloved.* United States: Crossroad. 106.

[71] Salzberg, S. (2011, Jan 6). Buddha Nature. rebelbuddhabook.com/2011/01/buddha-nature/

[72] Walcott, D. (1986). Collected Poems, 1948–1984. United Kingdom: Farrar, Straus and Giroux. 328.

13 - The Scourge of Rampant Happiness

[1] Wilson, E. G. (2008). Against Happiness: In Praise of Melancholy. Farrar, Straus and Giroux. 5, 8, 9.

[2] Holden, R. (2010). *Be happy!* Hay House, Inc. 106; Holden, R. (2011). *Happiness now!: Timeless wisdom for feeling good fast.* Hay House, Inc. 36.

[3] Nietzsche, F. (2001). *The Gay Science.* Cambridge University Press (Williams, Bernard, Ed.) 191-92; Nietzsche, F. (1974). *The Gay Science.* Vintage Books (Kaufmann, Walter, trans.). §338.

[4] McMahon, D. M. (2007). *Happiness: A history.* Grove Press. 437.

[5] Chödrön, P. (1996). *Awakening loving-kindness.* Shambhala Publications. 2.

[6] Chödrön. *Awakening loving-kindness.* 63.

[7] Note: I wasn't able to find the original source of this popular quote, but have chosen to include it because it does seem a faithful summary of his philosophy, especially as articulated in the Preface and Chapter 2 of *Civilization and Ethics*, 1929. See also: Schweitzer, A. (1967). *Civilization and ethics.* 246; Schweitzer, A. (1998). Schweitzer, A. (2009). *Out of My Life and Thought: An Autobiography.* Johns Hopkins University Press. 157.

[8] Emerson, R. W. (2014, November 29). The purpose of life is not to be happy but to matter. *Quote Investigator.* quoteinvestigator.com/category/ralph-waldo-emerson/

[9] Holden. *Be happy!* 110.

[10] Wilson. *Against Happiness.* 5, 8, 21–22.

[11] Dalai Lama XIV. (n,d). A biased mind cannot grasp reality. *The 14th Dalai Lama.* dalailama.com/messages/religious-harmony

[12] Dalai Lama XIV. (n.d.). Compassion and the individual. *The 14th Dalai Lama.* dalailama.com/messages/compassion

[13] Dalai Lama XIV, Cutler, H. C. (2008). *Art of happiness in a troubled world.* Doubleday Books. 269, 273; Ben-Shahar, T. (2008). *Happier.* McGraw-Hill. 127; Dunn, E., & Norton, M. (2014). *Happy money: The science of happier spending.* Simon and Schuster. 123.

[14] Happy parents make happy kids. (2004, October 21). *HowStuffWorks.* health.howstuffworks.com/pregnancy-and-parenting/happy-parents-make-happy-kids.htm

[15] Neihart, M. (1998). Creativity, the arts, and madness. *Roeper Review, 21*(1), 47–50. doi.org/10.1080/02783199809553930 ; Begley, S. (2008). *Happiness: enough already.* Newsweek.

newsweek.com/happiness-enough-already-93989; Warr, P. (2007). *Work, happiness, and unhappiness*. Laurence Erlbaum Associates, Inc. 423.

[16] Minkel, J. R. (2006, December 18). Happiness: Good for creativity, bad for single-minded focus. *Scientific American*. scientificamerican.com/article.cfm?id=happiness-good-for-creati; Lyubomirsky, S., King, L., & Diener, E. (2005). The benefits of frequent positive affect: Does happiness lead to success? *Psychological Bulletin, 131*(6), 830–831. doi.org/10.1037/0033-2909.131.6.803 ; Warr. *Work, happiness, and unhappiness*. 423.

[17] Lyubomirsky, King, & Diener. The benefits of frequent positive affect. 822–23, 832.

[18] Oishi, S., Diener, E., & Lucas, R. E. (2007). The optimum level of well-being: Can people be too happy? *Perspectives on Psychological Science, 2*(4), 346–360. doi.org/10.1111/j.1745-6916.2007.00048.x ; Kashdan, T., & Biswas-Diener, R. (2013, July 2). What happy people do differently. *Psychology Today*.psychologytoday.com/articles/201306/what-happy-people-do-differently

[19] More specifically, the literature indicates that people who are "depressed" are more likely to be hostile and abusive, not "unhappy" people. But the literature also indicates that "unhappy" people are more likely to abuse alcohol and drugs, and that people who abuse drugs and alcohol are more likely to abuse others. Myers, D. G., & Diener, E. (1995). Who is happy?. *Psychological Science, 6*(1), 11; Lyubomirsky, King, & Diener. The benefits of frequent positive affect. 824; Watson, S. (n.d.). Alcohol and depression. *WebMD*. webmd.com/depression/alcohol-and-depresssion; Alcohol and domestic violence. (n.d.). *Stop Violence Against Women*. www1.umn.edu/humanrts/svaw/domestic/link/alcohol.htm; Adults Surviving Child Abuse. (n.d.). Impact of Child Abuse. asca.org.au/about/resources/impact-of-child-abuse.aspx; An estimated 40 percent of people who sexually abuse children were sexually abused themselves. (Vanderbilt, H. (1992). *Incest: A chilling report*. Lear Pub. 18) Roughly 50 percent of men who physically abuse their spouse or partner were themselves physically abused as children. (Straus, V.J. (n.d.). There is Life After Abuse/Why Batterers Do What They Do. thereislifeafterabuse.com/Page8.html) And mothers who are abused by their spouse or partner are more than twice as likely to abuse their children. (Straus, M. A., Gelles, R. J., & Asplund, L. M. (1990). Physical violence in American families: Risk factors and adaptations to violence in 8,145 families. *Canadian Journal of Sociology, 6*(3), 253.doi.org/10.2307/3340687).

[20] Wilson. *Against Happiness*. 104, 31–32, 35–36, 52–54, 85, 89, 91–95, 101–4, 123–28.

[21] Holden. *Be happy!* 96–103.

[22] Holden. 106.

[23] Walker, A. (1994). *In search of our mothers' gardens*. Duke University Press. 36.

[24] Holden. *Be happy!* 111, 196.

[25] Holden. 111.

[26] Waheed, N. (2013). *Salt*. Self-published. 217.

[27] Holden. *Be happy!* 108.

[28] Holden. 196.

[29] Kashdan, T. (2010, September 30). The problem with happiness. *The Huffington*. huffingtonpost.com/todd-kashdan/whats-wrong-with-happines_b_740518.html; Gruber, J., Mauss, I. B., & Tamir, M. (2011). A dark side of Happiness? How, when, and why happiness is not always good. *Perspectives on Psychological Science, 6*(3), 226–27. doi.org/10.1177/1745691611406927 ; Lyubomirsky, S., Sheldon, K. M., & Schkade, D. (2005). Pursuing happiness: The architecture of sustainable change. *Review of General Psychology, 9*(2), 113.

[30] Gruber, J., Mauss, I. B., & Tamir, M. A dark side. 226–27; Diener, E., & Diener, C. (1996). Most people are happy. *Psychological Science, 7*(3), 181–185. jstor.org/stable/40062938; Biswas-Diener, R., Vittersø, J. & Diener, E. (2005). Most people are pretty happy, but there is cultural variation: The Inughuit, The Amish, and The Maasai. *Journal of Happiness Studies, 6*(3), 205. doi.org/10.1007/s10902-005-5683-8 ; Diener, E., & Seligman, M. E. P. (2002). Very happy people. *Psychological Science, 13*(1), 84. doi.org/10.1111/1467-9280.00415

[31] Gruber, J., Mauss, I. B., & Tamir, M. A dark side. 222–33; Lyubomirsky, Sheldon, & Schkade. Pursuing happiness. 113; Schooler, J., Ariely, D., & Loewenstein, G. (2003). The pursuit and assessment of happiness can be self-defeating. *The Psychology of Economic Decision*. 19-20, 55–6.

[32] Holden. *Happiness now!* 133.

[33] Schooler, Ariely, & Loewenstein. The pursuit. 66.

[34] Gruber, J., Mauss, I. B., & Tamir, M. A dark side. 227.

14 - Dancing with the Universe [Purpose]

[1] Smith, Z. (2016, October 29). Zadie Smith: Dance lessons for writers. *The Guardian*. theguardian.com/books/2016/oct/29/zadie-smith-what-beyonce-taught-me.

[2] A common definition of subjective well-being goes something like this: "experiencing high levels of pleasant emotions and moods, low levels of negative emotions and moods, and high life satisfaction." Diener, E., Oishi, S., & Lucas, R. E. (2009). Subjective well-being: The science of happiness and life satisfaction. *The Oxford Handbook of Positive Psychology*. 63. doi.org/10.1093/oxfordhb/9780195187243.013.0017

[3] Having a sense of meaning isn't a *guarantee* that we'll be happy, but Roy Baumeister has argued that it is a *prerequisite* for happiness. (Baumeister, R. F. (1991). *Meanings of life*. Guilford Press .214) Meaning has been described as a "defining feature of positive mental health." (Ryff, C. D., & Singer, B. (1998). The role of purpose in life and personal growth in positive human health. In P. T. P. Wong & P. S. Fry (Eds.), *The Human Quest For Meaning: A Handbook of Psychological Research And Clinical Applications*. Lawrence Erlbaum Associates Publishers. 216; Emmons, R. A. (2003). Personal goals, life meaning, and virtue: Wellsprings of a positive life. *Flourishing: Positive Psychology and the Life Well-Lived.*, 106. doi.org/10.1037/10594-005) "A strong sense of meaning is associated with life satisfaction and happiness, and a lack of meaning is predictive of depression and disengagement." (Emmons. Personal goals. 107–8.)

[4] Ben-Shahar, T. (2008). *Happier*. McGraw-Hill. 170. The connection between life satisfaction and conscious judgment based on our sense of what it means to live a good or successful life is explored in Pavot, W., & Diener, E. (1993). Review of the satisfaction with life scale. *Psychological Assessment, 5*(2), 164. doi.org/10.1037/1040-3590.5.2.164

[5] Baumeister. *Meanings of life*. 215.

[6] Brooks, C. A. (2008, May). How to buy happiness. *The Atlantic*. theatlantic.com/family/archive/2021/04/money-income-buy-happiness/618601/

[7] Lane, R. E. (2005). *The loss of happiness in market democracies*. Yale University Press. 73.

[8] Baumeister. *Meanings of life*. 233–34.

[9] Culey, S. A. (2019). Transition Point: From Steam to the Singularity. United Kingdom: Matador. 664.

[10] Adler, N. & the Psychosocial Working Group. (1997, November). Purpose of Life. *MacArthur Research Network on SES & Health*. macses.ucsf.edu/research/psychosocial/purpose.php

[11] Smith, E. E. (2013, January 9). There's more to life than being happy. *The Atlantic*. theatlantic.com/health/archive/2013/01/theres-more-to-life-than-being-happy/266805/

[12] Ben-Shahar. *Happier.* 71;   Steger, M. F., & Kashdan, T. B. (2006). Stability and specificity of meaning in life and life satisfaction over one year. *Journal of Happiness Studies, 8*(2), 163. doi.org/10.1007/s10902-006-9011-8

[13] Ali L. (2008). Having kids makes you happy. *Newsweek, 152* (1), 62–63;   Senior, J. (2014). *All joy and no fun: The paradox of modern parenthood.* Hachette UK;   Baumeister, R. F., & Vohs, K. D. (2002). The pursuit of meaningfulness in life. *Handbook of positive psychology, 1*, 612;   Baumeister. *Meanings of life.*

[14] Meyers, D. G., & Diener, E. (1995). Who is happy? *Psychological Science, 6*(1), 16. doi.org/10.1111/j.1467-9280.1995.tb00298.x ;   Diener, E., & Biswas-Diener, R. (2011). *Happiness: Unlocking the mysteries of psychological wealth.* John Wiley & Sons. 114–124.

[15] Denning, S. (2011, August 11). Think your job is bad? Try one of these! *Forbes.* forbes.com/sites/stevedenning/2011/08/11/think-your-job-is-bad-try-one-of-these;   Denning, Steve, (2011, September 12). The ten happiest jobs. *Forbes.* forbes.com/sites/stevedenning/2011/09/12/the-ten-happiest-jobs; Evolving for success [part one]. (2001, February 26). *HBS Working Knowledge.* hbswk.hbs.edu/item/2026.html; Evolving for success [part two]. (2001, March 11). *HBS Working Knowledge.* hbswk.hbs.edu/item/2097.html;   Kanter, R. M. (2013, April 10). The happiest people pursue the most difficult problems. *Harvard Business Review.* blogs.hbr.org/2013/04/to-find-happiness-at-work-tap;   Joslyn, H. (2002, October 10). Nonprofit employees are more satisfied than other workers with their jobs, says new Brookings survey – but problems loom. *The Chronicle of Philanthropy.* philanthropy.com/article/nonprofit-employees-are-more-satisfied-than-other-workers-with-their-jobs-says-new-brookings-survey-a-but-problems-loom;   Smith, J. (2012, February 9). The happiest and unhappiest industries to work in. *Forbes.* forbes.com/sites/jacquelynsmith/2012/02/09/the-happiest-and-unhappiest-industries-to-work-in;   Oswald A. (2003). If you want an enjoyable job, what should you do? *Sunday Times.* www2.warwick.ac.uk/fac/soc/economics/staff/faculty/oswald/jobsatisfaction2003.pdf;   Millennial mindset: Deloitte survey finds workers who frequently volunteer are happier with career progression. (2011, June 3). *CSRWire.* csrwire.com/press_releases/32364-millennial-mindset-deloitte-survey-finds-workers-who-frequently-volunteer-are-happier-with-career-progression;   Virtue rewarded: Helping others at work makes people happier. (2013, July 29). *University of Madison-Wisconsin News.* news.wisc.edu/virtue-rewarded-helping-others-at-work-makes-people-happier; Dunn, E., & Norton, M. (2014). *Happy money: The science of happier spending.* Simon and Schuster. 122.

[16] Jacobs, T. (2014, April 3). Meaningful work boosts happiness, even for lawyers. *Pacific Standard.* psmag.com/navigation/politics-and-law/meaningful-work-boosts-happiness-even-lawyers-78120

[17] Grimm, R., Spring, K., & Dietz, N. (2007). The health benefits of volunteering: A review of recent research. *Corporation for National and Community Service.* 1–17. americorps.gov/sites/default/files/evidenceexchange/FR_2007_TheHealthBenefitsofVolunteering_1.pdf;   Post & Neimark. *Why good things happen.* 49, 54, 68–72;   One study by Sonja Lyubomirsky sought to very clearly tease out the question of causation versus correlation when it comes to helping people. (Does helping people make us happier or do happier people tend to help people more?). She had people commit five acts of kindness a week. She found that if the acts were small and spread out over the week they just blended in with the rest of the people's lives and had no impact. But for people whose acts were larger or done in a single day, there was a significant increase in happiness. She also found that over time it was important to vary the acts so they didn't come to feel like a burden. (Lyubomirsky, S. (2008). *The how of happiness: A scientific approach to getting the life you want.* Penguin. 127–129;)   A study by Martin Seligman randomly assigned students to either do something pleasurable like see a movie or eat some ice cream, or to do some kind of volunteer work, like helping at a soup kitchen. The volunteers experienced more happiness afterward. (Honigsbaum, M. (2004, April 4). On the happy trail. *The Guardian.* theguardian.com/society/2004/apr/04/mentalhealth.observermagazine).

[18] Post & Neimark. *Why good things happen.* 53.

[19] Post & Neimark. 8;   Davis, J.L. (2005). The science of good deeds. WebMD. webmd.com/balance/features/science-good-deeds

[20] Dunn, E. W., Aknin, L. B., & Norton, M. I. (2008). Spending money on others promotes happiness. *Science, 319*(5870), 1687–1688. doi.org/10.1126/science.1150952 ;   Dunn, & Norton. *Happy money.* 109–10, 136–37.

[21] Dunn, Aknin, & Norton. Spending money on others. 1687–1688;   Dunn, & Norton. *Happy money.* 110.

[22] Dunn, & Norton. *Happy money.* 113;   Dunn, Aknin, & Norton. Spending money on others. 1687–1688;   (plus Supporting Online Materials to that study, p. 1. sciencemag.org/cgi/content/full/319/5870/1687/DC1)

[23] Dunn, & Norton. *Happy money.* 113;   Aknin, L.B., (2013) Prosocial spending and well-being: Cross-cultural evidence for a psychological universal. *Journal of Personality and Social Psychology*, 2013, *104*(4), 638;   Dunn, Aknin, & Norton. Spending money on others. 1687–1688;   (plus Supporting Online Materials, p. 1.)

[24] Post, S. G., & Neimark, J. (2006). Why good things happen to good people. Broadway Books. 1;   Lyubomirsky, S., Sheldon, K. M., & Schkade, D. (2005). Pursuing happiness: The architecture of sustainable change. *Review of General Psychology, 9*(2), 125.

[25] Dunn, & Norton. *Happy money.* 124–25.

[26] Chance, Z., & Norton, M. (n.d.). I give, therefore I have: giving and subjective wealth. 10. citeseerx.ist.psu.edu/viewdoc/download;jsessionid=DE3E6BB57EDB19FCB2DBD3C723319EAC?doi=10.1.1.348.5049&rep=rep1&type=pdf

[27] Dunn, & Norton. *Happy money.* 60–61.

[28] Chance & Norton. I give, therefore. 4–6

[29] Martin, J. (2014, December 18). Finding god in all things. *On Being.* onbeing.org/programs/james-martin-finding-god-in-all-things-2

[30] Sheldon, K. M., Ryan, R. M., Deci, E. L., & Kasser, T. (2004). The independent effects of goal contents and motives on well-being: It's both what you pursue and why you pursue it. *Personality and Social Psychology Bulletin, 30*(4), 475–76, 484–85. doi.org/10.1177/0146167203261883 ;   Chancellor, J., & Lyubomirsky, S. (2011). Happiness and thrift: When (spending) less is (hedonically) more. *Journal of Consumer Psychology, 21*(2), 135–36. doi.org/10.1016/j.jcps.2011.02.004 ;   The following article examines the impact of whether one lives in an individualistic culture like the US, and one that is more collectivistic, in which case meeting cultural norms might have a greater influence on happiness: Oishi, S. (2000). Goals as cornerstones of subjective well-being: Linking individuals and cultures. *Culture and Subjective Well-Being.* 104–107. The following suggests that the relationship between intrinsic and extrinsic goals may be largely independent of culture: Emmons. Personal goals. 114.

[31] Pepper, J. (1992). *How to be happy.* Gateway Books. 160.

Happiness Coda

[1] Steiner, S. (2012, February 1). Top five regrets of the dying. *The Guardian.* theguardian.com/lifeandstyle/2012/feb/01/top-five-regrets-of-the-dying

[2] Holden, R. (2010). *Be happy!* Hay House, Inc. 135–36.

[3] Holden. *Be happy!* 77.

15 - Money Strikes Back ...

[1] Melville, H. (2013). Moby-Dick: Or, the Whale. United States: Penguin Books. 7.

[2] Lane, R. E. (2005). *The loss of happiness in market democracies.* Yale University Press. 74.

[3] Diener, E., & Biswas-Diener, R. (2002). Will money increase subjective well-being? *Social Indicators Research, 57*(2), 121. doi.org/10.1023/A:1014411319119

[4] Launius, C. (2012, July 9). Is marriage becoming a marker of class? *Working-Class Perspectives.* workingclassstudies.wordpress.com/2012/07/09/is-marriage-becoming-a-marker-of-class; Krugman, P. (2012, February 9). Money and morals. *The New York Times.* nytimes.com/2012/02/10/opinion/krugman-money-and-morals.html; Renzetti, C. M. (2009). *Economic stress and domestic violence.* awnet.org/sites/default/files/materials/files/2016-09/AR_EconomicStress.pdf

[5] Diener, & Biswas-Diener. Will money increase subjective well-being? 121; Ladd, H. F. (2012). Education and poverty: Confronting the evidence. *Journal of Policy Analysis and Management, 31*(2), 203–27. doi.org/10.1002/pam.21615 ; Barriers to Higher Education, White House Task Force on Middle Class Families. (2009, Sept). whitehouse.gov/assets/documents/MCTF_staff_report_barriers_to_college_FINAL.pdf

[6] Zumbrun, J. (2014, October 7). SAT scores and income inequality: How wealthier kids rank higher. *Wall Street Journal.* blogs.wsj.com/economics/2014/10/07/sat-scores-and-income-inequality-how-wealthier-kids-rank-higher

[7] A study looking at the Senate in the late 1980s and early 1990s found that constituents in the upper 1/3 of income distribution received 50 percent more weight than those in the middle third in the voting decisions of their senators, and views of the constituents in the bottom third received *no weight at all.* Bartels, L. M. (2018). *Unequal democracy: The political economy of the new gilded age.* Russell Sage Foundation. 254; Bartels, L. M. (2002). *Economic inequality and political representation.* Princeton University. russellsage.org/sites/all/files/u4/Bartels percent20EIPR.pdf

[8] Gilens, M., & Page, B. (2014). Testing Theories of American Politics: Elites, Interest Groups, and Average Citizens. *Perspectives on Politics, 12*(3), 564–581. doi:10.1017/S1537592714001595

[9] According to the Center on Budget and Policy Priorities, entitlement benefits in 2010 came to $1.8 trillion. The top 20 percent receives 10 percent, or $180 billion. The middle 60 percent receives 58 percent, over $1 trillion. The bottom 20 percent receives 32 percent, or $576 billion. According to the Urban Institute-Brookings Institution Tax Policy Center, tax expenditure benefits (deductions and exemptions) in 2011 totaled $1.1 trillion. The top 20 percent receives 66 percent, or $726 billion. (The top 1 percent alone receives 23.9 percent, or $263 billion.) The middle 60 percent receives 31 percent, or $341 billion. The bottom 20 percent received 2.8 percent, or $31 billion. In total, entitlement and expenditure benefits come to $2.9 trillion. The top 20 percent receives $906 billion, or 31 percent of the total benefits. The middle 60 percent receives $1.4 trillion, or 48 percent of the total benefits. The bottom 20 percent receives $607 billion, or 21 percent of the total benefits. In other words, the top 20 percent receives 50 percent more than the bottom 20 percent, and almost 100 percent more per person than the average person in the middle 60 percent. The top 1 percent receives on average nearly 900 percent more per person than the average person in the bottom 20 percent, and nearly 1200 percent more per person than the average person in the middle 60 percent. (This assumes that the top 1 percent receives a proportional share, 5 percent, of what the top 20 percent receives in entitlement benefits.) Sherman, A. Greenstein, R. & Ruffing, K. (2012, February 11). Contrary to "entitlement society" rhetoric, over nine-tenths of entitlement benefits go to elderly, disabled, or working households. *Center on Budget and Policy Priorities.* cbpp.org/cms/index.cfm?fa=view&id=3677#_ftnref5 ; Goozner, M. (2012, September 18). "Makers" take lion's share of tax breaks. *The Fiscal Times.* thefiscaltimes.com/Articles/2012/09/18/Makers-Take-Lions-Share-of-Tax-Breaks.aspx#page1

[10] Sivanathan, N., & Pettit, N. C. (2010). Protecting the self through consumption: Status goods as affirmational commodities. *Journal of Experimental Social Psychology, 46*(3), 564–70. doi.org/10.1016/j.jesp.2010.01.006 ; Koles, B., Wells, V., & Tadajewski, M. (2017). Compensatory consumption and consumer compromises: A state-of-the-art review. *Journal of Marketing Management, 34*(1), 96–133. doi.org/10.1080/0267257x.2017.1373693

[11] Twitchell, J. B. (1999). *Lead us into temptation: The triumph of American materialism.* Columbia University Press. 244.

[12] Diener, & Biswas-Diener. Will money increase subjective well-being? 121

[13] Kristof, K. M. (2005, January 14). Study: Money can't buy happiness, security either. *Los Angeles Times.* latimes.com/archives/la-xpm-2005-jan-14-fi-richpoll14-story.html

[14] Walton, A. G. (2015, January 26). Why the super-successful get depressed. *Forbes.* forbes.com/sites/alicegwalton/2015/01/26/why-the-super-successful-get-depressed

[15] Diener, & Biswas-Diener. Will money increase subjective well-being? 157–158.

[16] Lieber, R. (2015, January 9). Growing up on easy street has its own dangers. *The New York Times.* nytimes.com/2015/01/10/business/growing-up-on-easy-street-has-its-own-dangers.html

[17] Diener, & Biswas-Diener. Will money increase subjective well-being? 124.

[18] Inglehart, R., & Klingemann, H. D. (2000). Genes, culture, democracy, and happiness. *Culture and subjective well-being,* 165; Lykken, D., & Tellegen, A. (1996). Happiness is a stochastic phenomenon. *Psychological Science, 7*(3), 187. doi.org/10.1111/j.1467-9280.1996.tb00355.x

[19] Twist, L. (2017). *The Soul of Money: Transforming Your Relationship with Money and Life.* United States: W. W. Norton. 138.

[20] Sheldon, K. M., Ryan, R. M., Deci, E. L., & Kasser, T. (2004). The independent effects of goal contents and motives on well-being: It's both what you pursue and why you pursue it. *Personality and Social Psychology Bulletin, 30*(4), 475, 485. doi.org/10.1177/0146167203261883

[21] Sheldon, Ryan, Deci, & Kasser. The independent effects. 475, 485; Chancellor, J., & Lyubomirsky, S. (2011). Happiness and thrift: When (spending) less is (hedonically) more. *Journal of Consumer Psychology, 21*(2), 135–136. doi.org/10.1016/j.jcps.2011.02.004

[22] Ben-Shahar, T. (2008). *Happier.* McGraw-Hill. 72–73.

[23] Diener, & Biswas-Diener. Will money increase subjective well-being? 161; Angel, J. (2008, November 1). 10 things science says will make you happy. *YES!* yesmagazine.org/issue/sustainable-happiness/2008/11/01/10-things-science-says-will-make-you; Diener, & Biswas-Diener. Will money increase subjective well-being? 124; Ben-Shahar. *Happier.* 72–73; Biswas-Diener, R., & Diener, E. (2009). Making the best of a bad situation: Satisfaction in the slums of Calcutta. *Social Indicators Research Series, 38,* 349. doi.org/10.1007/978-90-481-2352-0_13 ; Aked, J., Marks, N., Cordon, C., & Thompson, S. (2008, October 22). Five ways to wellbeing. *New Economics Foundation.* neweconomics.org/2008/10/five-ways-to-wellbeing

[24] Elias, M. (2002, December 8). Psychologists now know what makes people happy. *USA Today.* usatoday.com/news/health/2002-12-08-happy-main_x.htm

[25] Nickerson, C., Schwarz, N., Diener, E., & Kahneman, D. (2003). Zeroing in on the dark side of the American Dream. *Psychological Science, 14*(6), 531–536.doi.org/10.1046/j.0956-7976.2003.psci_1461.x

[26] Quoidbach, J., Dunn, E. W., Petrides, K. V., & Mikolajczak, M. (2010). Money giveth, money taketh away: The dual effect of wealth on happiness. *Psychol Sci, 21*(6), 759–63. doi: 10.1177/0956797610371963

27 Brickman, P., Coates, D., & Janoff-Bulman, R. (1978). Lottery winners and accident victims: Is happiness relative? *Journal of Personality and Social Psychology*, 36(8), 917–927. doi.org/10.1037/0022-3514.36.8.917
28 Exploring the psychology of wealth, 'pernicious' effects of economic inequality. (2013, June 21). *PBS NewsHour*. pbs.org/newshour/show/pernicious-effects-of-economic-inequality; Lewis, M. (2014, November 12). Extreme wealth is bad for everyone—especially the wealthy. *The New Republic*. newrepublic.com/article/120092/billionaires-book-review-money-cant-buy-happiness; Jilani, Z. (2015, January 2). 5 studies that show how wealth warps your soul. *AlterNet*. alternet.org/2015/01/5-studies-show-how-wealth-warps-your-soul; Miller, L. (2012, June 29). The money-empathy gap. *New York Magazine*. nymag.com/news/features/money-brain-2012-7
29 Shteir, R. (2011). *The steal: A cultural history of shoplifting*. The Penguin Press. 88; Lewis. Extreme wealth is bad for everyone—especially the wealthy; Anwar, Y. (2012, February 28). Affluent people more likely to be scofflaws. *Greater Good Science Center*. greatergood.berkeley.edu/article/item/affluent_people_more_likely_to_be_scofflaws; Exploring the psychology of wealth; Lewis. Extreme wealth is bad for everyone—especially the wealthy; Lopatto, E. (2012, February 28). Self-interest spurs society's 'elite' to lie, cheat, study finds. *Bloomberg*. bloomberg.com/news/2012-02-27/wealthier-people-more-likely-than-poorer-to-lie-or-cheat-researchers-find.html; Grewal, D. (2012, April 10). How wealth reduces compassion. *Scientific American*. scientificamerican.com/article/how-wealth-reduces-compassion; Marsh, J. (2012, September 25). Why inequality is bad for the one percent. *Greater Good Science Center*. greatergood.berkeley.edu/article/item/why_inequality_is_bad_for_the_one_percent; Goldeman, D. (2013, October 5). Rich people just care less. *The New York Times*. opinionator.blogs.nytimes.com/2013/10/05/rich-people-just-care-less; Phillips, L. T., & Lowery, B. S. (2020). I ain't no fortunate one: On the motivated denial of class privilege. *Journal of personality and social psychology*, 119(6), 1403–1422. doi.org/10.1037/pspi0000240 ; Dietze, P., & Knowles, E. D. (2020). Social class predicts emotion perception and perspective-taking performance in adults. *Personality and Social Psychology Bulletin*, 47(1), 42–56. doi.org/10.1177/0146167220914116
30 Exploring the psychology of wealth; Lewis. Extreme wealth is bad for everyone—especially the wealthy.
31 Lewis; Exploring the psychology of wealth.
32 Miller. The money-empathy gap.

16 - The Dopamine Reward System, Hijacked

1 Kinsella, S. (2003). *Confessions of a Shopaholic*. Dell Publishing Company. 27.
2 Kinsella. *Confessions*. 14.
3 Hamilton, J. O'C. (2008, November). This is your brain on bargains. *Stanford Magazine*. alumni.stanford.edu/get/page/magazine/article/?article_id=30924; Harding, A. (2007, January 21). Brain scans predict shoppers' purchasing choices. reuters.com/article/us-brain-scans-idUSCOL06128120070110 ; Layton, J. (n.d.). Is my brain making me buy things I don't need? *HowStuffWorks*. science.howstuffworks.com/life/brain-shopping1.htm; Heitler, S. (2011, November 21). Antidotes to boredom: Why shopping is fun. *Psychology Today*. psychologytoday.com/blog/resolution-not-conflict/201111/antidotes-boredom-why-shopping-is-fun
4 PBS. (2010, April 27). *Mind Over Money*, transcript. pbs.org/wgbh/nova/video/mind-over-money; Diener, E., & Biswas-Diener, R. (2002). Will money increase subjective well-being? *Social Indicators Research*, 57, 161. doi.org/10.1023/A:1014411319119
5 Parker-Pope, T. (2005, December 6). This is your brain at the mall: Why shopping makes you feel so good. *The Wall Street Journal*. wsj.com/articles/SB113382650575214543; Thompson, D. (2013, June 11). Why wanting expensive things makes us so much happier than buying them. *The Atlantic*. theatlantic.com/business/archive/2013/06/why-wanting-expensive-things-makes-us-so-much-happier-than-buying-them/276717; Tierney, J. (2007, November 5). The mystery of buyer's remorse—or, should you look for a money-back guarantee? *The New York Times*.tierneylab.blogs.nytimes.com/2007/11/05/the-mystery-of-buyers-remorse-or-should-you-look-for-a-money-back-guarantee; Woodruffe-Burton, H. (2001). "Retail therapy": An investigation of compensatory consumption and shopping behaviour (doctoral thesis). Lancaster University. EThOS. 269–272; Mertzer, M. (2014, November 24). Survey: 3 in 4 Americans make impulse purchases. *CreditCards.com*. creditcards.com/credit-card-news/impulse-purchase-survey.php#ixzz3SsrAiy38 ; Antanasijevic, I., Markovic, Z., & Ninkovic, B. (2012). The relation to money as a factor of the consumer´s behavior. *Management – Journal for Theory and Practice of Management*, 17(62), 97–105. doi.org/10.7595/management.fon.2011.0004 ; Rosenzweig, E., & Gilovich, T. (2012). Buyer's remorse or missed opportunity? Differential regrets for material and experiential purchases. *Journal of personality and social psychology*, 102(2), 215–223; Shame a key risk factor for compulsive buying, study shows. (2012, September 20). *University of Guelph*. uoguelph.ca/news/2012/09/shameproneness.html; Faber, R. J., & Christenson, G. A. (1996). In the mood to buy: Differences in the mood states experienced by compulsive buyers and other consumers. *Psychology & Marketing*, 13(8), 803–19.
6 PBS. *Mind Over Money*; Ferlaino, J. (n.d.) Retail Therapy. Business in Focus. 94–7; Faber, R.J. (200). The Urge to Buy. In Ratneshwar, S.. (Ed). The Why of Consumption. Rutledge. 185, 192; Lerner, J. S., Li, Y., & Weber, E. U. (2012). The financial costs of sadness. *Psychological Science*, 24(1), 73–78. doi.org/10.1177/0956797612450302 ; Lerner, J. S., Small, D. A., & Loewenstein, G. (2004). Heart strings and purse strings: Carryover effects of emotions on economic decisions. *Psychological science*, 15(5), 338–340; Woodruffe-Burton. "Retail therapy." 216, 220–224; Mertzer. Survey; Twitchell, J. B. (1999). *Lead us into temptation: The triumph of American materialism*. Columbia University Press. 244.
7 Hamilton. This is your brain; Harding. Brain scans; Layton. Is my brain.
8 Dagher, A. (2007). Shopping centers in the brain. *Neuron*, 53(1), 7–8.
9 Dagher. Shopping centers. 7–8; Lea, S. E., & Webley, P. (2006). Money as tool, money as drug: The biological psychology of a strong incentive. *Behavioral and Brain Sciences*, 29(2), 162. doi.org/10.1017/s0140525x06009046
10 Friedman, R. A. (2017, June 30). What cookies and meth have in common. *The New York Times*. nytimes.com/2017/06/30/opinion/sunday/what-cookies-and-meth-have-in-common.html; Center for Disease Control. (2022, May). Adult Obesity Prevalence Maps. cdc.gov/obesity/data/prevalence-maps.html
11 Boyle, J. P., Thompson, T. J., Gregg, E. W., Barker, L. E., & Williamson, D. F. (2010). Projection of the year 2050 burden of diabetes in the US adult population: Dynamic modeling of incidence, mortality, and prediabetes prevalence. *Population Health Metrics*, 8(1). doi.org/10.1186/1478-7954-8-29 ; Armstrong, A. (2006, December 29). One-third of Americans die of heart disease: report. *ABC News*. abcnews.go.com/Health/CardiacHealth/story?id=2759392&page=1
12 Friedman. What cookies and meth.
13 Based on 34oz soda = 8.5 feet of sugar cane. Warinner, C. [TEDx Talks]. (2013, February 12). *Debunking the paleo diet*. [Video]. YouTube. youtube.com/watch?v=BMOjVYgYaG8
14 Whybrow, P. C. (2005). American Mania: When More Is Not Enough. United Kingdom: W.W. Norton. 71.
15 See Chapter 20. Dunn, E. W., Akin, L. B., & Norton, M. I. (2008). Spending money on others promotes happiness. *Science*, 319(5870), 1687.
16 DeLeire, T., & Kalil, A. (2010). Does consumption buy happiness? Evidence from the United States. *International Review of Economics*, 57(2), 163–176. doi.org/10.1007/s12232-010-0093-6 ; Rosenbloom, S. (2010, August 7). But

will it make you happy? *The New York Times.* nytimes.com/2010/08/08/business/08consume.html;   Dunn, E., & Norton, M. (2014). *Happy money: The science of happier spending.* Simon and Schuster. 6, 23–24; Rosenzweig & Gilovich. Buyer's remorse. 215–23.

[17] Lyubomirsky, S. (2013, September 2). How to buy happiness. *Psychology Today.* psychologytoday.com/blog/how-happiness/201309/how-buy-happiness

[18] Note, these studies do not themselves make the distinction I make about being relevant only to people living above the abundance point. I make that distinction because neither of these studies was designed to address the impact of increased spending on basic needs for people living in extreme poverty. DeLeire & Kalil. Does consumption buy happiness? 163–176; Dunn, Aknin, & Norton. Spending money on others. 1687–1688.

[19] Lane, R. E. (2005). *The loss of happiness in market democracies.* Yale University Press. 72. Ed Diener speculates about this as well: Diener, & Biswas-Diener. Will money increase subjective well-being? 160.

[20] Taylor, S. (2012). *Back to sanity: Healing the madness of our minds.* Hay House. 85.

17 - Sex, Status, and Stuff [Conspicuous Consumption]

[1] Sivanathan, N., & Pettit, N. C. (2010). Protecting the self through consumption: Status goods as affirmational commodities. *Journal of Experimental Social Psychology*, 46(3), 564.doi.org/10.1016/j.jesp.2010.01.006 ; Bagwell, L. S., & Bernheim, B. D. (1996). Veblen effects in a theory of conspicuous consumption. *The American economic review*, 349–73; Kasser, T., Cohn, S., Kanner, A. D., & Ryan, R. M. (2007). Some costs of American corporate capitalism: A psychological exploration of value and Goal Conflicts. *Psychological Inquiry*, 18(1), 13. doi.org/10.1080/10478400701386579

[2] Miller, G. (2009). *Spent: Sex, evolution and the secrets of consumerism.* William Heinemann. 1, 2, 22, 83–4, 100–4; Stewart-Williams, S., & Thomas, A. G. (2013). The ape that thought it was a peacock: Does evolutionary psychology exaggerate human sex differences? *Psychological Inquiry*, 24(3), 137–68. doi.org/10.1080/1047840x.2013.804899 ; Smith, R. H. (2013). *The joy of pain: Schadenfreude and the dark side of human nature.* Oxford University Press. 10.

[3] Schor, J. B. (1999). *The overspent American: Why we want what we don't need.* Harper Collins. 90–98. Research shows our concern about status also makes us more likely to ingratiate ourselves to some people, while bullying, embarrassing and blaming others, as well as lying and gossiping about them, and even sabotaging their work. It also leads us to take more risks when others are observing or evaluating us (the better to impress or intimidate them). (Appelbaum, S. H., & Hughes, B. (1998). Ingratiation as a political tactic: Effects within the organization. *Management Decision, 36*(2), 85–95); Wert, S. R., & Salovey, P. (2004). A social comparison account of gossip. *Review of General Psychology, 8*(2), 122–137. doi.org/10.1037/1089-2680.8.2.122 ; White, J. B., Langer, E. J., Yariv, L., & Welch, J. C. (2006). Frequent social comparisons and destructive emotions and behaviors: The dark side of social comparisons. *Journal of Adult Development, 13*(1), 36. doi.org/10.1007/s10804-006-9005-0 ; Smith. *The joy of pain.* 116; Namie, G., & Namie, R. (2009). *Bully at work: What you can do to stop the hurt and reclaim your dignity on the job.* Sourcebooks, Inc. 3–18; Anderson, M. S., Ronning, E. A., De Vries, R., & Martinson, B. C. (2007). The perverse effects of competition on scientists' work and relationships. *Science and Engineering Ethics, 13*(4), 437–61. doi.org/10.1007/s11948-007-9042-5 ; Ermer, E., Cosmides, L., & Tooby, J. (2008). Relative status regulates risky decision making about resources in men: Evidence for the co-evolution of motivation and cognition. *Evolution and Human Behavior, 29*(2), 106–118. doi.org/10.1016/j.evolhumbehav.2007.11.002

[4] Birdal, M., & Ongan, T. H. (2015). Why do we care about having more than others? Socioeconomic determinants of positional concerns in different domains. *Social Indicators Research, 126*(2), 731. doi.org/10.1007/s11205-015-0914-9

[5] Sivanathan & Pettit. (2010). Protecting the self. 564; Bagwell & Bernheim. Veblen effects. 349–373; Miller. *Spent.* 14–5, 71–76, 90–92, 106–111, 114–15.

[6] Solnick, S. J., & Hemenway, D. (2005). Are positional concerns stronger in some domains than in others? *American Economic Review, 95*(2), 147. doi.org/10.1257/000282805774669925 ; Schor. *The Overspent American.* 47.

[7] Miller. *Spent.* 114, 125–26; Koles, B., Wells, V., & Tadajewski, M. (2017). Compensatory consumption and consumer compromises: A state-of-the-art review. *Journal of Marketing Management, 34*(1–2), 110. doi.org/10.1080/0267257x.2017.1373693

[8] Bellezza, S., Paharia, N., & Keinan, A. (2016). Conspicuous consumption of time: When busyness and lack of leisure time become a status symbol. *Journal of Consumer Research.* doi.org/10.1093/jcr/ucw076

[9] Diener, E., Tay, L., & Oishi, S. (2013). Easterlin paradox. *Journal of Personality and Social Psychology, 104*(2), 267–276. doi.org/10.1037/a0030487

[10] Miller. *Spent.* 76–81, 84–85.

[11] Blanchflower, D., & Oswald, A. (2004). Money, sex, and happiness: An empirical study. *NBER Working Paper Series*, 11–12. doi.org/10.3386/w10499

[12] Layard, R., Mayraz, G., & Nickell, S. (2010). Does relative income matter? Are the critics right? *International Differences in Well-Being, 28*, 139–65. doi.org/10.1093/acprof:oso/9780199732739.003.0006 ; Ball, R. J., & Chernova, K. (2005). Absolute income, relative income, and happiness. *SSRN Electronic Journal*, 497–529. doi.org/10.2139/ssrn.724501

[13] Luttmer, E. F. P. (2004). Neighbors as negatives: Relative earnings and well-being. *National Bureau of Economic Research.*doi.org/10.3386/w10667 ; Note, this title and abstract refer to happiness, but the actual measure used was life satisfaction: Firebaugh, G., & Schroeder, M. B. (2009). Does your neighbor's income affect your happiness? *American Journal of Sociology, 115*(3), 805–831. doi.org/10.1086/603534

[14] White, Langer, Yariv, & Welch. Frequent social comparisons. 36.

[15] Emery, L. F., Muise, A., Dix, E. L., & Le, B. (2014). Can you tell that I'm in a relationship? Attachment and relationship visibility on Facebook. *Personality and Social Psychology Bulletin, 40*(11), 1466–1479. doi.org/10.1177/0146167214549944 ; Group bragging betrays insecurity, study finds. (2008, October 28). *Science Blog.* scienceblog.com/17613/group-bragging-betrays-insecurity-study-finds/#ZixX4W8mFQektjjo.97 ; Wu, L., Li, C., & Johnson, D. E. (2011). Role of self-esteem in the relationship between stress and ingratiation. *Psychological Reports, 108*(1), 239–251. doi.org/10.2466/07.09.20.pr0.108.1.239-251 ; Tice, D. M. (1993). The social motivations of people with low self-esteem. *Self-Esteem*, 37–53. doi.org/10.1007/978-1-4684-8956-9_3 ; Fast, N. J., & Chen, S. (2009). When the boss feels inadequate: Power, incompetence, and aggression. *Psychological Science, 20*(11), 1406–1413. doi.org/10.1111/j.1467-9280.2009.02452.x ; Wert & Salovey. A social comparison. 122–37; John, W. (2002, August 24). Sorrow so sweet: a guilty pleasure in another's woe. *The New York Times.* nytimes.com/2002/08/24/arts/sorrow-so-sweet-a-guilty-pleasure-in-another-s-woe.html?pagewanted=2 ; Smith. *The joy of pain.* 8, 9.

[16] Cushman, A. (2007, Aug 28). The Wellspring of Joy. *Yoga Journal.* yogajournal.com/yoga-101/philosophy/the-wellspring-of-joy/

[17] Cushman. The Wellspring of Joy.

18 - Retail Therapy: Taking Our Insecurities ...

[1] Becker, E. (1997). *The denial of death*. Simon and Schuster. 284

[2] Koles, B., Wells, V., & Tadajewski, M. (2017). Compensatory consumption and consumer compromises: A state-of-the-art review. *Journal of Marketing Management, 34*(1–2), 96. doi.org/10.1080/0267257x.2017.1373693

[3] Twitchell, J. B. (1999). *Lead us into temptation: The triumph of American materialism*. Columbia University Press. 271

[4] Lee, J., & Shrum, L. J. (2013). Self-threats and consumption. In *The Routledge companion to identity and consumption*, ed. Ayalla, A., Ruvio, & Russell, W. Belk. Routledge, 216–24. doi.org/10.4324/9780203105337.ch22 ; Rucker, D. D., & Galinsky, A. D (2013). Compensatory consumption. In *The Routledge companion to identity and consumption*, ed. Ayalla, A., Ruvio, & Russell, W. Belk. Routledge, 207–215. doi.org/10.4324/9780203105337.ch21 ; Koles, Wells, & Tadajewski. Compensatory consumption. 96–133; Yi, S. (2012). Shame-proneness as a risk factor of compulsive buying. *Journal of Consumer Policy, 35*(3), 393–410. doi.org/10.1007/s10603-012-9194-9 ; Kasser, T., Ryan, R. M., Couchman, C. E., & Sheldon, K. M. (2004). Materialistic values: Their causes and consequences. In Kasser, T., & Kanner, A. D. (Eds.), *Psychology and consumer culture: The struggle for a good life in a materialistic world*. American Psychological Association. 14–15.doi.org/10.1037/10658-002 ; Lisjak, M., Bonezzi, A., Kim, S., & Rucker, D. D. (2015). Perils of compensatory consumption: Within-domain compensation undermines subsequent self-regulation. *Journal of Consumer Research, 41*(5), 1186–1187. doi.org/10.1086/678902 ; Benson, A. L. (2008). *To buy or not to buy: Why we overshop and how to stop*. Shambhala Publications; Robinson, B. E. (2014). *Chained to the desk*. New York University Press; Sivanathan, N., & Pettit, N. C. (2010). Protecting the self through consumption: Status goods as affirmational commodities. *Journal of Experimental Social Psychology, 46*(3), 564–565, 569. doi.org/10.1016/j.jesp.2010.01.006 ; Diener, E., & Biswas-Diener, R. (2002). Will money increase subjective well-being? *Social Indicators Research, 57*, 160. doi.org/10.1023/A:1014411319119 ; Cryder, C. E., Lerner, J. S., Gross, J. J., & Dahl, R. E. (2008). Misery is not miserly: Sad and self-focused individuals spend more. *Psychological Science, 19*(6), 525–530. doi.org/10.1111/j.1467-9280.2008.02118.x ; Woodruffe-Burton, H. (2001). "Retail therapy": An investigation of compensatory consumption and shopping behaviour (Doctoral thesis). Lancaster University. EThOS. 196–97, 216, 220–24.

[5] Benson. *To buy or not*. 2–4; Hamilton, J. O'C. (2008, November). This is your brain on bargains. *Stanford Magazine*. alumni.stanford.edu/get/page/magazine/article/?article_id=30924

[6] Robinson. *Chained to the desk*. 16–19, 22.

[7] Benson. *To buy or not*. 13.

[8] Robinson. *Chained to the desk*. 16, 48.

[9] Lee & Shrum. Self-threats and consumption. 216–224; Rucker & Galinsky. Compensatory consumption.207–215; Sivanathan & Pettit. Protecting the self. 564–565, 569.

[10] Becker. *The denial of death*. xvii.

[11] Henrich, J., Heine, S., & Norenzayan, A. (2010). The weirdest people in the world? *The Behavioral and Brain Sciences. 33*(2-3), 75.

[12] Landau, M. J., Solomon, S., Pyszczynski, T., & Greenberg, J. (2007). On the compatibility of terror management theory and perspectives on human evolution. *Evolutionary Psychology*. 478–482. doi.org/10.1177/147470490700500303 ; Solomon, S., Greenberg, J. L., & Pyszczynski, T. A. (2004). Lethal consumption: Death-denying materialism. In Tim Kasser and Allen D. Kanner, Eds. *Psychology and Consumer Culture: The Struggle for a Good Life in a Materialistic World*. American Psychological Association. 133. doi.org/10.1037/10658-008

[13] Arndt, J., Solomon, S., Kasser, T., & Sheldon, K. M. (2004). The urge to splurge: A terror management account of materialism and consumer behavior. *Journal of Consumer Psychology, 14*(3), 203–208, 225-29. doi.org/10.1207/s15327663jcp1403_2 ; Solomon, Greenberg, & Pyszczynski. Lethal consumption. 133.

[14] Twitchell. *Lead us into temptation*. 56.

[15] Sifferlin, A. (2017, August 15). 13 percent of Americans take antidepressants. *TIME*. time.com/4900248/antidepressants-depression-more-common

[16] Yalom, I. D. (2009). *Staring at the sun*. John Wiley & Sons. 6.

[17] Most misquoted. (2014, February 26). *Legacy.com*. legacy.com/news/best-quotes/most-misquoted/2069/

[18] Rick, S. I., Pereira, B., & Burson, K. A. (2014). The benefits of retail therapy: Making purchase decisions reduces residual sadness. *Journal of Consumer Psychology, 24*(3), 373–74. doi.org/10.1016/j.jcps.2013.12.004 ; Benson. *To buy or not*; Kasser, Ryan, Couchman, & Sheldon. Materialistic values. 14–15.

[19] Shame a key risk factor for compulsive buying, study shows. (2012, September 20). *University of Guelph*. uoguelph.ca/news/2012/09/shameproneness.html; Lisjak, Bonezzi, Kim, & Rucker. Perils of compensatory consumption. 1186–87; Faber, R. J., & Christenson, G. A. (1996). In the mood to buy: Differences in the mood states experienced by compulsive buyers and other consumers. *Psychology & Marketing, 13*(8), 803–819.

[20] Monbiot, G. (2013, December 9). Materialism: a system that eats us from the inside out. *The Guardian*. theguardian.com/commentisfree/2013/dec/09/materialism-system-eats-us-from-inside-out

[21] Schmich, M. T. (1986, December 12). A Stopwatch on Shopping. *Chicago Tribune*. chicagotribune.com/news/ct-xpm-1986-12-24-8604060073-story.html

[22] Robinson. *Chained to the desk*. 134–35, 153, 157.

[23] Benson. *To buy or not*. 246; Benson, A. L. (2009, April 1). A crisis is a terrible thing to waste: How the recession can help overshoppers. *Psychology Today*. psychologytoday.com/blog/buy-or-not-buy/200904/crisis-is-terrible-thing-waste-how-the-recession-can-help-overshoppers

[24] Yalom. *Staring at the sun*. 7, 204; Landau, Solomon, Pyszczynski & Greenberg. On the Compatibility. 480; Burke, B. L., Martens, A., & Faucher, E. H. (2010). Two decades of terror management theory: a meta-analysis of mortality salience research. *Personality and social psychology review : an official journal of the Society for Personality and Social Psychology, Inc, 14*(2), 186. doi.org/10.1177/1088868309352321 ; Arndt, Solomon, Kasser, & Sheldon. The urge to splurge: 203–208.

[25] Yalom, I. D. (2001). *Existential psychotherapy*. TPB. 33–40; Cozzolino, P. J., Staples, A. D., Meyers, L. S., & Samboceti, J. (2004). Greed, death, and values: From terror management to transcendence management theory. *Personality and Social Psychology Bulletin, 30*(3), 278–292. doi.org/10.1177/0146167203260716 ; Burke, Martens, & Faucher. Two decades. 186-187; Arndt, Solomon, Kasser, & Sheldon. The urge to splurge: 203–208; Yalom. *Existential psychotherapy*. 33–40; Cozzolino, Staples, Meyers, & Samboceti. Greed, death, and values. 278–92.

[26] Sivanathan & Pettit. Protecting the self. 564–70.

[27] Benson. *To buy or not*. 60.

[28] Sivanathan & Pettit. Protecting the self. 564-70; Lisjak, Bonezzi, Kim, & Rucker. Perils of compensatory consumption. 1186–1187; Robinson. *Chained to the desk*. 134–35, 153, 157; Benson. *To buy or not*. 246; Benson. A crisis.

19 - The Great Breath Retention Contest: Security ...

[1] Gilbert, D. (2006, July 2). If only gay sex caused global warming. *Los Angeles Times*.

latimes.com/archives/la-xpm-2006-jul-02-op-gilbert2-story.html

[2] Schneier, B. (2008, January 21). The psychology of security. *Schneier on Security*. schneier.com/essay-155.html

[3] Schneier. The psychology of security.

[4] 2014 US military spending: $581 billion, China: $129 billion, Russia $70 billion (Adam, T. [2015, February 11]. Chart: US defense spending still dwarfs the rest of the world. *The Washington Post*. washingtonpost.com/news/worldviews/wp/2015/02/11/chart-u-s-defense-spending-still-dwarfs-the-rest-of-the-world); Iran: $30 billion (Greenburg, J. [2015, April 9]. Obama: Iran spends $30 billion on defense; US about $600 billion. *Politifact*. politifact.com/truth-o-meter/statements/2015/apr/09/barack-obama/obama-iran-spends-30-billion-defense-us-about-600-); North Korea: $1 billion (Shim, E. [2016, March 31]. North Korea underreporting defense spending, analyst says. *UPI*. upi.com/Top_News/World-News/2016/03/31/North-Korea-underreporting-defense-spending-analyst-says/2811459437466); Estimated cost to eliminate world hunger: $30 billion (Abdallah, L. [2015, February 15]. The cost to end world hunger. *The Borgen Project*. borgenproject.org/the-cost-to-end-world-hunger).

[5] Clark, M. S., Greenberg, A., Hill, E., Lemay, E. P., Clark-Polner, E., & Roosth, D. (2011). Heightened interpersonal security diminishes the monetary value of possessions. *Journal of Experimental Social Psychology*, 47(2), 359–364. doi.org/10.1016/j.jesp.2010.08.001 ; The more secure you feel, the less you value your stuff, UNH research shows. (2011, March 3). *University of New Hampshire*. unh.edu/unhtoday/news/release/2011/03/03/more-secure-you-feel-less-you-value-your-stuff-unh-research-shows

[6] Schneier. The psychology of security; Richards, C. (2013, December 9). Overcoming an aversion to loss. *The New York Times*. nytimes.com/2013/12/09/your-money/overcoming-an-aversion-to-loss.html; Weaver, J. (2015). Study: Monkeys ape humans' economic traits. *Yale Bulletin*. anderson.ucla.edu/faculty/keith.chen/articles/YaleBulletin percent20text percent207_15_05.pdf; Schneier, B. (2010, October). *The security mirage* [Video]. TED Talk. ted.com/talks/bruce_schneier/transcript?language=en

[7] Gilbert, D. (2004, February). *The surprising science of happiness* [Video]. TED Talk.ted.com/talks/dan_gilbert_the_surprising_science_of_happiness#t-3510

[8] The health benefits of strong relationships. (2010, December 1). *Harvard Health Publishing*.health.harvard.edu/newsletter_article/the-health-benefits-of-strong-relationships; Raypole, C. (2020, August 17). 6 ways friendship is good for your health. *Healthline*.healthline.com/health/benefits-of-friendship#resilience

[9] Note, this study does not explicitly compare people with high and low sense of self-worth, but compares people with intrinsic and extrinsic goal orientations, which largely aligns with sense of self-worth, as noted at the end of Chapter 19: Duriez, B., Meeus, J., & Vansteenkiste, M. (2012). Why are some people more susceptible to ingroup threat than others? The importance of a relative extrinsic to intrinsic value orientation. *Journal of Research in Personality*, 46(2), 164–172.

[10] Ramsey, D. (2013). *The total money makeover*. Thomas Nelson. 133; Bach, D. (2005). *The automatic millionaire: A powerful one-step plan to live and finish rich*. Crown Pub. 19; Pagliarini, R. (2008). *The six-day financial makeover: Transform your financial life in less than a week!* St. Martin's Griffin. 111; Opdyke, J. D. (2010). *The wall street journal: Complete personal finance guidebook*. Currency. Chapter 3, Budgeting; Robin, V., Dominguez, J., & Tilford, M. (2008). *Your Money or your life: 9 Steps to transforming your relationship with money and achieving financial independence*. Penguin. 270; On 3–6 months is excessive: Chilton, D. B., & Wiggins, C. (1995). *The wealthy barber*. Audio Élan.

[11] Dixon, A. (2019, July 1). A growing percentage of Americans have no emergency savings whatsoever. *Bankrate*. bankrate.com/banking/savings/financial-security-june-2019/

[12] Dixon. A growing percentage; Marte, J. (2014, August 7). Almost 20 percent of people near retirement age have not saved for it. *The Washington Post*. washingtonpost.com/blogs/wonkblog/wp/2014/08/07/almost-20-percent-of-people-near-retirement-age-have-no-retirement-savings

[13] Schor, J. B. (1999). *The overspent American: Why we want what we don't need*. Harper Perennial. 7. 1995 dollars converted to 2021 using the US Bureau of Labor Statistics CPI Inflation Calculator (data.bls.gov/cgi-bin/cpicalc.pl).

[14] Keller, H. (1917). Excerpts of The Open Door. Helen Keller Archive/American Foundation for the Blind. afb.org/HellenKellerArchive?a=d&d=A-HK02-B228-F05-002&e=-------en-20--1--txt--------1-2-1-undefined-3-----------0-1

[15] Lane, R. E. (2005). *The loss of happiness in market democracies*. Yale University Press. 70.

[16] Kristof, K. M. (2005, January 14). Study: Money can't buy happiness, security either. *Los Angeles Times*.latimes.com/archives/la-xpm-2005-jan-14-fi-richpoll14-story.html; Annual social and economic supplement. (2009). *US Census Bureau*. census.gov/data/datasets/time-series/demo/cps/cps-asec.2009.html; Milanović, B. (2012). *The haves and the have-nots: A brief and idiosyncratic history of global inequality*. Basic Books. 168–169.

[17] Schneier, B. (2009, Nov). Beyond Security Theater. *New Internationalist*. schneier.com/essays/archives/2009/11/beyond_security_thea.html

[18] Schneier. The psychology of security.

[19] Watts, A. (2011). *The Wisdom of Insecurity*. Vintage. 77–78.

[20] Watts, A. (2001). *What is tao?* New World Library. 57–58.

20 – Soulcraft ... Consumption, Culture, and Ideology

[1] Coolidge, C. (1925, January 17). Address to the American society of newspaper editors, Washington, DC. *The American Presidency Project*. presidency.ucsb.edu/documents/address-the-american-society-newspaper-editors-washington-dc

[2] Groves, M. (n.d.). American bison. *Buffalo Groves, Inc.* growingyourfuture.com/civi/sites/default/files/Bison_0.pdf

[3] Isenberg, A. C. (2000). *The destruction of the bison: an environmental history, 1750–1920*. Cambridge University Press.

[4] Gugliotta, G. (2013, February). When did humans come to the Americas? *Smithsonian Magazine*. smithsonianmag.com/science-nature/when-did-humans-come-to-the-americas-4209273/?all

[5] Isenberg. *The destruction of the bison*. 81, 85; The buffalo: Yesterday and today. (n.d.). *PBS*. pbs.org/buffalowar/buffalo.html

[6] Lewy, G. (2004, September). Were American Indians the victims of genocide? *History News Network*. historynewsnetwork.org/article/7302; Isenberg. *The destruction of the bison*. 23–26, 29.

[7] Demographic history of the United States. (n.d). *Wikipedia*. en.wikipedia.org/wiki/Demographic_history_of_the_United_States

[8] US Fish and Wildlife Service. (n.d.). Time Line of the American Bison. fws.gov/bisonrange/timeline.htm

[9] Isenberg. *The destruction of the bison*. 130–132, 160–162; Smits, D. D. (1994). The frontier army and the destruction of the buffalo: 1865–1883. *The Western Historical Quarterly*, 25(3), 312-19, 326. doi.org/10.2307/971110 ; King, G. (2012, July 17). Where the buffalo no longer roamed. *Smithsonian Magazine*. smithsonianmag.com/history/where-the-buffalo-no-longer-roamed-3067904; American buffalo: Spirit of a nation. (1998, November 10). *PBS*. pbs.org/wnet/nature/american-buffalo-spirit-of-a-nation-introduction/2183

[10] Smits. The frontier army. 321.

[11] Lewy. Were American Indians?

[12] Smits. The frontier army. 312–338; King. Where the buffalo; Jawort, A. (2017, September 24). Genocide by other means: US army slaughtered buffalo in Plains Indian wars. *Indian Country Today.* indiancountrytodaymedianetwork.com/2011/05/09/genocide-other-means-us-army-slaughtered-buffalo-plains-indian-wars-30798

[13] Jawort. Genocide by other means.

[14] Isenberg. *The destruction of the bison.* 101–122; Smits. The frontier army. 323-326.

[15] Raygorodtetsky, G. (2018, Nov 16). Indigenous peoples defend Earth's biodiversity—but they're in danger. *National Geographic.* nationalgeographic.com/environment/article/can-indigenous-land-stewardship-protect-biodiversity-

[16] Global Footprint Network. (2012). Ecological footprint and biocapacity in 2011. *National Footprint Accounts 2015 Edition.* footprintnetwork.org/content/images/uploads/NFA_2011_Edition.pdf

[17] McLellan, R., Iyengar, L., Jeffries, B., & Oerlemans, N. (2014). *Living planet report 2014: species and spaces, people and places.* WWF International. 37.

[18] Yunghans, R. (2011, July 20). Average home sizes around the world. *Apartment Therapy.* apartmenttherapy.com/average-home-sizes-around-the-151738; Wilson, L. (n.d.). How big is a house? Average house size by country. *Shrink That Footprint.* shrinkthatfootprint.com/how-big-is-a-house

[19] Lowrey, A. (2011, June 27). Your big car is killing me. *Slate.* slate.com/articles/business/moneybox/2011/06/your_big_car_is_killing_me.html; "Vehicle mass in running order in the EU is defined as mass of the empty vehicle plus 75 kg of weight for the driver," plus luggage and fluids, so I used the average mass in running order for the EU-27 minus 75kg: Campestrini, M., & Mock, P. (2011). European vehicle market statistics. *International Council on Clean Transportation.* 40. theicct.org/sites/default/files/publications/Pocketbook_LowRes_withNotes-1.pdf; Chitravanshi, R. (2014, December 30). Government agrees not to increase the limit for average weight of vehicles. *The Economic Times.* economictimes.indiatimes.com/industry/auto/news/passenger-vehicle/cars/government-agrees-not-to-increase-the-limit-for-average-weight-of-vehicles/articleshow/45682562.cms

[20] DOT releases new NHTS showing vehicles in households outnumber drivers. (2003, May 26). *Bureau of Transportation Statistics.* bts.gov/newsroom/2003-statistical-releases; Hu, P. S., & Reuscher, T. R. (2004). Summary of travel trends: 2001 national household travel survey. *Federal Highway Administration.* 11. nhts.ornl.gov/2001/pub/STT.pdf; Owen, D. (2010). The efficiency dilemma. *The New Yorker.* 78. https://www.davidowen.net/files/the-efficiency-dilemma.pdf; What is it? (n.d.). *PBS.* pbs.org/kcts/affluenza/diag/what.html

[21] Tilford, D. (n.d.). Why Consumption Matters. The Sierra Club. Retrieved 9/10: sierraclub.org/sustainable_consumption/tilford.asp; All about affluenza. (n.d.). *PBS.* pbs.org/kcts/affluenza/show/about.html

[22] Bocco, D. (n.d.). How much garbage does a person create in one year? *Infobloom.* wisegeek.com/how-much-garbage-does-a-person-create-in-one-year.htm

[23] List of countries by energy consumption per capita (for 2003) Wikipedia [Energy use (kg of oil equivalent per capita). The World Bank; "Total Primary Energy Consumption per Capita (Million Btu per Person). US Energy Information Administration]. en.wikipedia.org/wiki/List_of_countries_by_energy_consumption_per_capita; "List of countries by total primary energy consumption and production," (for 2007) Wikipedia [Total Primary Energy Consumption (Quadrillion Btu). US Energy Information Administration]. en.wikipedia.org/wiki/List_of_countries_by_total_primary_energy_consumption_and_production

[24] US material use factsheet. (2020). *Center for Sustainable Systems, University of Michigan.* css.umich.edu/factsheets/us-material-use-factsheet

[25] Ipsos. (2010. Feb 23). Majority (65%) of global citizens agree money is more important to them nowadays than previously. ipsos.com/en-us/majority-65-global-citizens-agree-money-more-important-them-nowadays-previously

[26] Wuthnow, R. (1993). Pious materialism: How Americans view faith and money. *The Christian Century, 3,* 240.

[27] Schor, J. B. (1999). *The overspent American: Why we want what we don't need.* Harper Collins. 13.

[28] Wuthnow, R. (1998). *God and mammon in America.* Simon and Schuster. 181.

[29] Eagan, K., Stolzenberg, E. B., Ramirez, J. J., Aragon, M. C., Suchard, M. R., & Hurtado, S. (2014). The American freshman: National norms fall 2014. *Los Angeles: Higher Education Research Institute, UCLA,* 44; American Council on Education. Office of Research. (1970). *National norms for entering college freshmen—fall 1970.* Staff of the Office of Research, American Council on Education. 50.

[30] Hacker J. D. (2020). From '20. and odd' to 10 million: The growth of the slave population in the United States. *Slavery & abolition, 41*(4), 840–855. doi.org/10.1080/0144039x.2020.1755502

[31] Biello, D. (2009, January 28). Spent nuclear fuel: A trash heap deadly for 250,000 years or a renewable energy source? *Scientific American.* scientificamerican.com/article/nuclear-waste-lethal-trash-or-renewable-energy-source

[32] This is my own estimate based on Mexico in 1846–48, Cuba in 1898–1902, the Philippines in 1898–1946, Cuba 1906–09, Colombia/Panama in 1903, the Dominican Republic in 1905, Nicaragua in 1909, Haiti in 1915–34, Iran in 1953, Guatemala in 1954, Congo in the early 1960s, the Dominican Republic in 1961, Chile in 1970–73, Argentina in 1976, Grenada in 1983, and Iraq in 2003. American motives in each of these situations varied, but financial interests can be documented as playing anything from a meaningful to a primary role in each of them—and likely others as well, but which I am either not familiar with or am not clear meet these criteria.

[33] Solotaroff, P. (2013, December 10). In the belly of the beast. *Rolling Stone.* rollingstone.com/feature/belly-beast-meat-factory-farms-animal-activists

[34] EarthTalk. (2009, Jan 22). Are Old-Growth Forests Protected in the U.S.? *Scientific American.* scientificamerican.com/article/are-old-growth-forests

[35] Van Noppen, T. (2013, April 11). Dirty water: Can US clean up its act? *LiveScience.* livescience.com/28669-dirty-water-report.html

[36] Wilcove, D. S., & Master, L. L. (2005). How many endangered species are there in the United States? *Frontiers in Ecology and the Environment, 3*(8), 426. doi.org/10.1890/1540-9295(2005)003[0414:hmesat]2.0.co;2

[37] Schwartz, B. (2007). There must be an alternative. *Psychological Inquiry, 18*(1), 48. doi.org/10.1080/10478400701389086

[38] Diener, E., & Biswas-Diener, R. (2002). Will money increase subjective well-being? *Social Indicators Research, 57,* 160. doi.org/10.1023/A:1014411319119

[39] Kasser, T., Cohn, S., Kanner, A. D., & Ryan, R. M. (2007). Some costs of American corporate capitalism: A psychological exploration of value and Goal Conflicts. *Psychological Inquiry, 18*(1), 6. doi.org/10.1080/10478400701386579 ; Kasser, T. (2014). What psychology says about materialism and the holidays. *American Psychological Association.* apa.org/news/press/releases/2014/12/materialism-holidays.aspx

[40] Schor. *The Overspent American.* 24.

[41] West, C. (2011, August 25). Dr. King weeps from his grave. *The New York Times.* nytimes.com/2011/08/26/opinion/martin-luther-king-jr-would-want-a-revolution-not-a-memorial.html

[42] Rudmin, W., Materialism and Militarism. (1992). De Tocqueville on America's Hopeless Hurry to Happiness. In F.W. Rudmin & M. Richins (Eds.), *Meaning, Measure, and Morality of Materialism*. 110–112. Association for Consumer Research.
[43] Guttmann, A. (2021, March 1). Global advertising revenue 2012–2024. *Statista*. statista.com/statistics/236943/global-advertising-spending; Guttmann, A. (2019, March 28). Advertising spending in the US 2015–2022. *Statista*. statista.com/statistics/272314/advertising-spending-in-the-us
[44] How many advertisements is a person exposed to in a day? (2007). ams.aaaa.org/eweb/upload/faqs/adexposures.pdf
[45] Focus on the Family. (n.d.) Entertainment Statistics. focusonthefamily.com/entertainment/mediawise/advertising-and-kids/statistics.aspx
[46] Miller, G. (2009). *Spent: Sex, evolution and the secrets of consumerism*. William Heinemann. 37, 46–47.
[47] Kasser, Cohn, Kanner, & Ryan. (2007). Some costs of American corporate capitalism. 12–13.
[48] Benson, A. L. (2009, February, 7). Are there more shopaholics who want to confess? *Psychology Today*. psychologytoday.com/blog/buy-or-not-buy/200902/are-there-more-shopaholics-who-want-confess
[49] Kasser, Cohn, Kanner, & Ryan. (2007). Some costs of American corporate capitalism. 6–12
[50] Schor. The overspent American. 81–82.
[51] Unsworth, B. (2012). *Sacred Hunger*. United Kingdom: Knopf Doubleday Publishing Group. 325.
[52] Speth, J. G. (2008). *The bridge at the edge of the world*. Yale University Press. 47–48.
[53] Speth. *The bridge*. 47.
[54] Bernanke, B. (2010, May 08). The economics of happiness. *Board of Governors of the Federal Reserve System*. federalreserve.gov/newsevents/speech/bernanke20100508a.htm
[55] Costanza, R., Hart, M., Talberth, J., & Posner, S. (2009). Beyond GDP: The need for new measures of progress. *The Pardee Papers*. 3. bu.edu/pardee/files/documents/PP-004-GDP.pdf
[56] Costanza, Hart, Talberth, & Posner. Beyond GDP. 7.
[57] Costanza, R., d'Arge, R., de Groot, R. *et al*. (1997). The value of the world's ecosystem services and natural capital. *Nature*. 253–60. 387. doi.org/10.1038/387253a0
[58] Speth. *The bridge*. 9.
[59] Tabuchi, H. (2016, December 3). How big banks are putting rain forests at peril. *The New York Times*. nytimes.com/2016/12/03/business/energy-environment/how-big-banks-are-putting-rain-forests-in-peril.html; Thirty-eight percent of world's surface in danger of desertification. (2010, February 10). *ScienceDaily*. sciencedaily.com/releases/2010/02/100209183133.htm; Douglass, S. (2012, Oct 1). Southwest forests face grim future as climate warms. Examiner.com. examiner.com/article/southwest-forests-face-grim-future-as-climate-warms; Minard, A.
(2007, April 5). U.S. Southwest drought could be start of new dust bowl. *National Geographic*. news.nationalgeographic.com/news/2007/04/070405-us-drought.html; for much more expansive information, see DeBuys, W. (2012). *A great aridness: Climate change and the future of the American Southwest*. Oxford University Press; Carrington, D. (2014, September 30). Earth has lost half of its wildlife in the past 40 years, says WWF. *The Guardian*. theguardian.com/environment/2014/sep/29/earth-lost-50-wildlife-in-40-years-wwf?CMP=share_btn_fb
[60] Roberts, J. (2012, June 18). Climate change is simple: We do something or we're screwed. *Grist*.grist.org/climate-change/climate-change-is-simple-we-do-something-or-were-screwed
[61] Nath, B. (2009). Environmental Education and Awareness - Volume I. EOLSS Publications. 302.

Money Coda

[1] Smith, A., & Stewart, D. (1892). *The Theory of Moral Sentiments*. United Kingdom: G. Bell & Sons. Chapter III: Of the Influence and Authority of Conscience. oll.libertyfund.org/title/smith-the-theory-of-moral-sentiments-and-on-the-origins-of-languages-stewart-ed; Brown, V. (1997). 'Mere inventions of the imagination': A survey of recent literature on Adam Smith. *Economics and Philosophy*, *13*(2), 281–312. doi.org/10.1017/s0266267100004521; Berry, C. J. (2018). Adam Smith: A very short introduction. Oxford University Press. 101; Sharma, R. (2020, February 16). Adam Smith: The father of economics. *Investopedia*. investopedia.com/updates/adam-smith-economics/
[2] Smith & Stewart. *The Theory of Moral Sentiments*; Brown. 'Mere inventions.' 281–312; Berry. Adam Smith. 101; Sharma. Adam Smith.
[1] Freire, P., & Ramos, B. M. (2014). *Pedagogy of the oppressed: 30th anniversary edition*. Bloomsbury Academic. 51, 58.

21 - Abundance Denied: An Illustrative Tale

[2] Reliving that night. (n.d.). International Campaign for Justice in Bhopal. bhopal.net/what-happened/that-night-december-3-1984/reliving-that-night; Union Carbide's Disaster. (n.d.). The Bhopal Medical Appeal. bhopal.org/what-happened/union-carbides-disaster; Saira. (n.d.). The Bhopal Medical Appeal. bhopal.org/what-happened/peoples-stories/saira
[3] The death toll. (n.d.). *International Campaign for Justice in Bhopal*. bhopal.net/what-happened/that-night-december-3-1984/the-death-toll; Yes Men hoax on BBC reminds world of Dow Chemical's refusal to take responsibility for Bhopal disaster. (2004, December 06). *Democracy Now!* democracynow.org/2004/12/6/yes_men_hoax_on_bbc_reminds
[4] International Campaign for Justice in Bhopal. (2021, December 2). Dow Discriminates – 37th Anniversary of the World's Worst Industrial disaster in Bhopal. (E-mail newsletter).
[5] 5 worst industrial disasters. (2013, May 10). *Salon*. salon.com/2013/05/10/5_worst_industrial_disasters_ap; Sim, D. (2014, December 01). Bhopal gas tragedy: 30 years on from the world's worst industrial disaster [photo report]. *International Business Times*. ibtimes.co.uk/bhopal-gas-tragedy-30-years-worlds-worst-industrial-disaster-photo-report-1477510
[6] Yes Men hoax on BBC.
[7] Dow 'Help' Announcement Is Elaborate Hoax. (n.d.). DowEthics.com. dowethics.com/r/about/corp/bbc.htm
[8] Yes Men hoax on BBC.
[9] Carbide cripples Bhopal,. (n.d.). International Campaign for Justice in Bhopal. bhopal.net/old_bhopal_net/UCC25.html
[10] Carbide cripples Bhopal; Fortun, K. (2009). *Advocacy after Bhopal: Environmentalism, disaster, New Global Orders*. The University of Chicago Press. 118.
[11] Personality spotlight; NEWLN: Warren Anderson: Union carbide chief arrested in India. (n.d.). *United Press International* (UPI). upi.com/Archives/1984/12/07/Personality-SpotlightNEWLNWarren-Anderson-Union-Carbide-chief-arrested-in-India/9883471243600

[12] Broughton, E. (2005). The Bhopal disaster and its aftermath: a review. *Environmental Health*. 6. ncbi.nlm.nih.gov/pmc/articles/PMC1142333; How Dow Chemical can end the Bhopal tragedy. (2012, July 27). *The Motley Fool*. fool.com/investing/general/2012/07/27/how-dow-chemical-can-end-the-bhopal-tragedy.aspx
[13] Dougal, A.S. (n.d.). Union Carbide Corporation. Encyclopedia of Business, 2nd ed. referenceforbusiness.com/history/Ul-Vi/Union-Carbide-Corporation.html; Union carbide corporation. (2018, June 27). *Encyclopedia.com*. encyclopedia.com/social-sciences-and-law/economics-business-and-labor/businesses-and-occupations/union-carbide-corp
[14] Carbide cripples Bhopal.
[15] Broughton. The Bhopal disaster; Yes Men hoax on BBC; Hymans, L. (2004, December 04). Brilliant bit of Bhopal activism. *Grist*. grist.org/article/brilliant-bit-of-bhopal-activism
[16] Deb, S. (2014, October). The worst industrial disaster in the history of the world. *The Baffler*. thebaffler.com/salvos/worst-industrial-disaster-history-world
[17] Carbide cripples Bhopal.
[18] How Dow Chemical can end; Why did India accept a mere $500 for every Bhopal life? (2010. June 11). *NDTV*. ndtv.com/india-news/why-did-india-accept-a-mere-500-for-every-bhopal-life-420494
[19] Greenpeace International. (2002, Dec 10). Dow fights for its image, but not the victims in Bhopal. greenpeace.org/international/en/news/features/dow-fights-for-its-image-but
[20] How Dow Chemical can end.
[21] Fernández, L. (2021, July 6). Net income of Dow from 2005 to 2020. (n.d.). *Statista*. statista.com/statistics/267331/net-income-of-dow-chemical-since-2005
[22] Form DEF 14A. (2015). *SECDatabase.com*. 21.pdf.secdatabase.com/1096/0001193125-15-109360.pdf
[23] Hanh, T. N. (2006). *True love: A practice for awakening the heart*. Shambhala Publications. 15, 16.

22 - Fourth Sister ... Bottlenecks, Toxic Waste, Sorcery

[1] Hettinger, N. (2002). The problem of finding a positive role for humans in the natural world. *Ethics and the Environment, 7*(1), 109–23.
[2] Kimmerer, R. W. (2013). *Braiding sweetgrass: Indigenous wisdom, scientific knowledge and the teachings of plants*. Milkweed Editions. 6, 124, 140, 150; Baraniuk, C. (2017, August 01). The animals thriving in the Anthropocene. *BBC Future*.bbc.com/future/article/20170801-the-animals-thriving-in-the-anthropocene
[3] Kimmerer. *Braiding sweetgrass*. 139, 140.
[4] Stromberg, J. (2013, January). What is the anthropocene and are we in it? *Smithsonian Magazine*.smithsonianmag.com/science-nature/what-is-the-anthropocene-and-are-we-in-it-164801414/
[5] Kolbert, E. (2009, May 25). The sixth extinction? *The New Yorker*. newyorker.com/magazine/2009/05/25/the-sixth-extinction; Miller, G. (2009). *Spent: Sex, evolution, and the secrets of consumerism*. William Heinemann. 45–46.
[6] Carrington, D. (2014, September 30). Earth has lost half of its wildlife in the past 40 years, says WWF. *The Guardian*. theguardian.com/environment/2014/sep/29/earth-lost-50-wildlife-in-40-years-wwf; Tabuchi, H. (2016, December 3). How big banks are putting rain forests at peril. *The New York Times*. nytimes.com/2016/12/03/business/energy-environment/how-big-banks-are-putting-rain-forests-in-peril.html
[7] Tran, L. (2022, April 27). UN report: People have wrecked 40% of all the land on Earth. Grist. grist.org/international/un-report-land-use-damage; Mulvihill, M., Goldenman, G., & Blum, A. (Aug. 27, 2021). The proliferation of plastics and toxic chemicals must end. *The New York Times*. nytimes.com/2021/08/27/opinion/plastics-fossil-fuels.html
[8] Roberts, D. (2012, June 18). Climate change is simple: We do something or we're screwed. *Grist*. grist.org/climate-change/climate-change-is-simple-we-do-something-or-were-screwed; Scott, M. (2014, August 12). What's the hottest Earth's ever been? *Climate*. gov.climate.gov/news-features/climate-qa/whats-hottest-earths-ever-been
[9] Gillis, J. & Popovich, N. (2017, June 1). The U.S. Is the Biggest Carbon Polluter in History. It Just Walked Away From the Paris Climate Deal. *New York Times*. nytimes.com/interactive/2017/06/01/climate/us-biggest-carbon-polluter-in-history-will-it-walk-away-from-the-paris-climate-deal.html; CarbonBrief. (2021, Oct 5). Analysis: Which countries are historically responsible for climate change? carbonbrief.org/analysis-which-countries-are-historically-responsible-for-climate-change; Mooney, C. (2015, Jan 22). The U.S. has caused more global warming than any other country. Here's how the Earth will get its revenge. *Washington Post*. washingtonpost.com/news/energy-environment/wp/2015/01/22/the-u-s-has-contributed-more-to-global-warming-than-any-other-country-heres-how-the-earth-will-get-its-revenge/
[10] Roberts. Climate change is simple.
[11] Sjödin, P., E. Sjöstrand, A., Jakobsson, M., & Blum, M. G. B. (2012). Resequencing data provide no evidence for a human bottleneck in Africa during the penultimate glacial period. *Molecular Biology and Evolution*, 29(7), 1851–1860. doi.org/10.1093/molbev/mss061
[12] Estimates of historical world population. (n.d.). *Wikipedia*. en.wikipedia.org/wiki/World_population_estimates
[13] Historical perspectives of energy consumption. (n.d.). *Western Oregon University*. people.wou.edu/~courtna/GS361/electricity percent20generation/HistoricalPerspectives.htm
[14] Taylor, B. & Tilford, D. (2011). Why Consumption Matters. In Schor, J.B. & Holt, D.B. (Eds.) *The consumer society reader*. New Press. 463.
[15] Leonard, A. (n.d.). Story of stuff, referenced and annotated script. 10. storyofstuff.org/wp-content/uploads/2020/01/StoryofStuff_AnnotatedScript.pdf; Taylor & Tilford. Why Consumption Matters; Rogich, D., Cassara, A., Wernick, I., & Miranda, M. (2008). Material flows in the United States. *World Resources Institute*. 1, 2. pdf.wri.org/material_flows_in_the_united_states.pdf; Households and NPISHs final consumption expenditure per capita (constant 2010 US$). (n.d.). *The World Bank*. data.worldbank.org/indicator/NE.CON.PRVT.PC.KD
[16] Townsend, M., & Burke, J., (2002, July 07). Earth "will expire by 2050." *The Guardian*.guardian.co.uk/uk/2002/jul/07/research.waste; *Population in the world is currently growing at a rate of around 1.14 percent per year. World population (2020 and historical)*. (n.d.). *Worldometers*.worldometers.info/world-population/
[17] Harrabin, R. (2020, March 16). Climate change: The rich are to blame, international study finds. BBC News. bbc.com/news/business-51906530
[18] Gomory, R. E., & E. Sylla, R. E. (2013). The American corporation. *Dædalus*. amacad.org/publication/american-corporation; Our hidden history of corporations in the US (n.d.). *Reclaim Democracy!* reclaimdemocracy.org/corporate-accountability-history-corporations-us
[19] Speth, J. G. (2008). *The Bridge at the Edge of the World*. Yale University Press. 62.
[20] Speth. *The bridge*. 166.
[21] Speth. 62.
[22] Speth. 62.
[23] Speth. 168.

[24] Top spenders. (n.d.). *OpenSecrets.* opensecrets.org/lobby/top.php?indexType=s&showYear=a

[25] Gilens, M., & Page, B. I. (2014). Testing theories of American Politics: Elites, interest groups, and average citizens. *Perspectives on Politics, 12*(3), 564–81. doi.org/10.1017/s1537592714001595

[26] Andrzejewski, A. (2019, May 14). How the fortune 100 turned $2 billion in lobbying spend into $400 billion of taxpayer cash. *Forbes.* forbes.com/sites/adamandrzejewski/2019/05/14/how-the-fortune-100-turned-2-billion-in-lobbying-spend-into-400-billion-of-taxpayer-cash

[27] Speth. *The bridge.* 166.

[28] Speth. 55, 169, 170–71.

[29] Winkler, A. (2018, March 05). "Corporations are people" is built on an incredible 19th-century lie. *The Atlantic.* theatlantic.com/business/archive/2018/03/corporations-people-adam-winkler/554852/

[30] Stout, L. A. (2007). Why we should stop teaching Dodge v. Ford. *SSRN Electronic Journal*, 166–167. doi.org/10.2139/ssrn.1013744 ; Lattman, P. (2012, November 09). In unusual move, Delaware supreme court rebukes a judge. *The New York Times.* dealbook.nytimes.com/2012/11/09/in-unusual-move-the-delaware-supreme-court-rebukes-a-judge/

[31] Strine Jr, L. E. (2012). Our continuing struggle with the idea that for-profit corporations seek profit. *SSRN Electronic Journal, 47.* 135, 155. ssrn.com/abstract=2523977

[32] Kasser, T., Cohn, S., Kanner, A. D., & Ryan, R. M. (2007). Some costs of American corporate capitalism: A psychological exploration of value and goal conflicts. *Psychological Inquiry, 18*(1), 5.

[33] Bainbridge, S. (2015, May 05). Case law on the fiduciary duty of directors to maximize the wealth of corporate shareholders. *ProfessorBainbridge.com.* professorbainbridge.com/professorbainbridgecom/2012/05/case-law-on-the-fiduciary-duty-of-directors-to-maximize-the-wealth-of-corporate-shareholders.html; Stout. Why we should stop. 166–67.

[34] Bainbridge. Case law; Bakan, J. (2019). *The Corporation: The Pathological Pursuit of Profit and Power.* United States: Free Press. 36.

[35] Bainbridge. Case law.

[36] Strine. Our continuing struggle. 135, 155; Stout. Why we should stop. 166–67; Bainbridge. Case law; Henderson, T. (2010, July 27). The shareholder wealth maximization myth. *Truth On the Market.* truthonthemarket.com/2010/07/27/the-shareholder-wealth-maximization-myth/

[37] Strine. Our continuing struggle. 135, 155.

[38] Bakan. *The Corporation.* 80.

[39] Hanh, T. N. (2006). *True love: A practice for awakening the heart.* Shambhala Publications. 74.

23 - For Otters, Redwoods, Bhopal: The Reality ...

[1] Williams, T. T. (2019). *Erosion: Essays of undoing.* Farrar, Straus and Giroux. 119.

[2] Powers, R. (2018). *The overstory: A novel.* W. W. Norton. 386.

[3] How much oil is consumed in the United States? (n.d.). *US energy information administration (EIA).* eia.gov/tools/faqs/faq.php?id=33&t=6 ; Oil and petroleum products explained. (n.d.). *US energy information administration (EIA).* eia.gov/energyexplained/oil-and-petroleum-products/use-of-oil.php

[4] Plastics. (n.d.). *American Chemistry Council.* americanchemistry.com/chemistry-in-america/chemistry-in-everyday-products/plastics; Winter, D. (2015, December 4). The violent afterlife of a recycled plastic bottle. *The Atlantic.* theatlantic.com/technology/archive/2015/12/what-actually-happens-to-a-recycled-plastic-bottle/418326; Reuseit (n.d.). The 7 Most Common Plastics and How They are Typically Used. reuseit.com/product-materials/learn-more-the-7-most-common-plastics-and-how-they-are-typically-used.htm; Sullivan, J. (2012, December 23). 10 everyday things that started life as oil. *Listverse.* listverse.com/2012/12/23/10-everyday-things-that-started-life-as-oil

[5] Manning, R. (2004, Feb). The oil we eat. *Harpers.* harpers.org/archive/2004/02/the-oil-we-eat; Lott, M. C. (2011, August 11). 10 calories in, 1 calorie out – the energy we spend on food. *Scientific American.* blogs.scientificamerican.com/plugged-in/10-calories-in-1-calorie-out-the-energy-we-spend-on-food

[6] Manning. The oil we eat.

[7] Manning.

[8] Manning.

[9] Church, N. J. (2005, April 01). Why our food is so dependent on oil. *Resilience.* resilience.org/stories/2005-04-01/why-our-food-so-dependent-oil

[10] Hess, A., & Frohlich, T. C. (2014, October 21). The 20 most profitable companies in the world. *247 Wall St.* 247wallst.com/special-report/2014/10/21/the-20-most-profitable-companies-in-the-world

[11] Somewhere between 2 million and 6 million barrels, roughly 84 million gallons and 252 million gallons, were released into the Gulf: Gulf war oil spill. (n.d.). *Wikipedia.* en.wikipedia.org/wiki/Gulf_War_oil_spill; But 660 million barrels were released on land and sea combined: Carlisle, T. (2010, December 21). Kuwait moves to clean up oil lakes. *The National.* thenational.ae/business/energy/kuwait-moves-to-clean-up-oil-lakes; Kuwaiti oil fires. (n.d.). *Wikipedia.* en.wikipedia.org/wiki/Kuwaiti_oil_fires#Oil_spills

[12] Tutton, M. (2010, June 4). Lessons learned from the largest oil spill in history. *CNN.* cnn.com/2010/WORLD/meast/06/04/kuwait.oil.spill; Yant, M. D. (2011). *Desert mirage.* Prometheus. 52.

[13] Between 3.19 million barrels and 4.9 million, or 130 million gallons and 220 million: Griffin, D., Black, N., & Devine, C. (2015, April 20). 5 years after the Gulf oil spill: What we do (and don't) know. *CNN.* cnn.com/2015/04/14/us/gulf-oil-spill-unknowns/

[14] Deepwater horizon oil spill. (n.d.). *Wikipedia.* en.wikipedia.org/wiki/Deepwater_Horizon_oil_spill; Biello, D. (2015, May 20). BP oil spill responsible for Gulf of Mexico dolphin deaths. *Scientific American.* scientificamerican.com/article/bp-oil-spill-responsible-for-gulf-of-mexico-dolphin-deaths; Harbison, M. (2014, May 06). More than one million birds died during Deepwater horizon disaster. *Audubon.* audubon.org/news/more-one-million-birds-died-during-deepwater-horizon-disaster

[15] Deepwater horizon oil spill.

[16] Mufson, S. (2014, September 04). BP's "gross negligence" caused Gulf oil spill, federal judge rules. *The Washington Post.* washingtonpost.com/business/economy/bps-gross-negligence-caused-gulf-oil-spill-federal-judge-rules/2014/09/04/3e2b9452-3445-11e4-9e92-0899b306bbea_story.html; Judge Carl J Barbier's Ruling On 2010 Gulf Oil Spill. (2014, Sept). documentcloud.org/documents/1283700-judge-carl-j-barbiers-ruling-on-2010-gulf-oil.html p. 115; Robertson, C., & Krauss, C. (2014, September 4). BP may be fined up to $18 billion for spill in gulf. *The New York Times.* nytimes.com/2014/09/05/business/bp-negligent-in-2010-oil-spill-us-judge-rules.html

[17] BP net income 2006–2021 | BP. (n.d.). *Macrotrends.* macrotrends.net/stocks/charts/BP/bp/net-income; BP: 2021 global 500. (n.d.). *Fortune.* fortune.com/company/bp/global500/

[18] Hess & Frohlich. The 20 most.

[19] Nossiter, A. (2010, June 16). Far from gulf, a spill scourge 5 decades old. *The New York Times.* nytimes.com/2010/06/17/world/africa/17nigeria.html

[20] Royal Dutch shell net income 2006–2021 | RDS.B. (n.d.). *Macrotrends.*

macrotrends.net/stocks/charts/RDS.B/royal-dutch-shell/net-income; Royal Dutch shell. (n.d.). *Fortune.* fortune.com/global500/2019/royal-dutch-shell/

[21] Forero, J. (2003, October 23). Texaco goes on trial in Ecuador pollution case. *The New York Times.* nytimes.com/2003/10/23/business/texaco-goes-on-trial-in-ecuador-pollution-case.html

[22] Exxon net income 2006–2021 | XOM. (n.d.). *Macrotrends.* macrotrends.net/stocks/charts/XOM/exxon/net-income; Exxon Mobil. (n.d.). *Fortune.* fortune.com/global500/2019/exxon-mobil/

[23] Forero. Texaco goes on trial.

[24] Carroll, C., & Gullo, K. (2011, February 15). Chevron's Ecuador award "unenforceable," analysts say. Bloomberg. bloomberg.com/news/articles/2011-02-14/chevron-to-appeal-adverse-judgment-in-ecuador-pollution-case

[25] Chevron using 60 law firms and 2,000 legal personnel to evade Ecuador environmental liability, company reports. (2013, March 03). *CSRwire.* csrwire.com/press_releases/35294-Chevron-Using-60-Law-Firms-and-2-000-Legal-Personnel-To-Evade-Ecuador-Environmental-Liability-Company-Reports

[26] Carroll & Gullo. Chevron's Ecuador award.

[27] Carroll & Gullo. Chevron's Ecuador award; Reuters. (2012, Oct 9). Chevron fails to block $18 billion Ecuador judgment. reuters.com/article/us-usa-court-ecuador/chevron-fails-to-block-18-billion-ecuador-judgment-idUSBRE8980UQ20121009

[28] Chevron net income 2006–2021 | CVX. (n.d.). *Macrotrends.* macrotrends.net/stocks/charts/CVX/chevron/net-income; Chevron. (n.d.). *Fortune.* fortune.com/fortune500/2019/chevron/

[29] Winkler, A. (2018, March 05). "Corporations are people" is built on an incredible 19th-century lie. *The Atlantic.* theatlantic.com/business/archive/2018/03/corporations-people-adam-winkler/554852/

[30] Kerschberg, B. (2010, December 02). Corporate executives: Get ready for a billion-dollar lawsuit. *HuffPost.* huffingtonpost.com/ben-kerschberg/corporate-executives-get-_b_791292.html; Kiobel v. Royal Dutch Petroleum Co. (amicus). (n.d.). *Center for Constitutional Rights.* ccrjustice.org/home/what-we-do/our-cases/kiobel-v-royal-dutch-petroleum-co-amicus

[31] Chevron reaches end of torture liability suit. (2012, April 23). *Courthouse News Service.* courthousenews.com/chevron-reaches-end-of-torture-liability-suit; Bowoto v. Chevron. (n.d.). *EarthRights International.* earthrights.org/legal/plaintiffs-bowoto-v-chevron-ask-supreme-court-hear-case; Chevron lawsuit (re Nigeria). (2005, August 01). *Business & Human Rights Resource Centre.* business-humanrights.org/en/chevron-lawsuit-re-nigeria

[32] I have not been able to find the original source for the quote, however I am including because it seems consistent with other things Kundera has written.

[33] Corporate control in agriculture. (n.d.). *About Farm Aid.* farmaid.org/issues/corporate-power/corporate-power-in-ag; McGreal, C. (2019, March 9). How America's food giants swallowed the family farms. *The Guardian.* theguardian.com/environment/2019/mar/09/american-food-giants-swallow-the-family-farms-iowa

[34] The Humane Society of the United States. (2009). The welfare of animals in the meat, egg, and dairy industries. animalstudiesrepository.org/cgi/viewcontent.cgi?article=1004&context=hsus_reps_impacts_on_animals; Hessler, K. & Balaban, T. (2009, July/Aug). Agricultural animals and the law. americanbar.org/newsletter/publications/gp_solo_magazine_home/gp_solo_magazine_index/agriculturalanimals.html; Foer, J. S. (2010). *Eating Animals.* United States: Hachette Book Group USA. 154-55.

[35] History World. (n.d.) History of the domestication of animals. historyworld.net/wrldhis/PlainTextHistories.asp?historyid=ab57

[36] Pew Commission on Industrial Farm Animal Production. (2008, April 29). Putting meat on the table: Industrial farm animal production in America. vii. pewtrusts.org/en/research-and-analysis/reports/0001/01/01/putting-meat-on-the-table

[37] Leonard, C. (2014). *The meat racket: The secret takeover of America's food business.* United Kingdom: Simon & Schuster. 3; Food and Water Watch. (2010, Nov). Factory Farm Nation. 7, 9, 11, 18. documents.foodandwaterwatch.org/doc/FactoryFarmNation-web.pdf

[38] Pew Commission on Industrial Farm Animal Production. Putting meat. 6; Hessler & Balaban. Agricultural animals and the law.

[39] Food and Water Watch. Factory Farm Nation. 6.

[40] Farm Animal Rights Movement. (n.d.). Report: Number of animals killed In US increases in 2010. farmusa.org/statistics11.html; Headlines from the front lines. (n.d.). *The Humane Society.* humanesociety.org/news/resources/research/stats_slaughter_totals.html; Foer. *Eating Animals.* 34; Farm Forward; (n.d.). Ending factory farming. farmforward.com/ending-factory-farming; The percentage of animals raised on factory farms varies by animal, but is extremely high for all of them: 99.9 percent of broiler chickens, 97 percent of egg-laying hens, 99 percent of turkeys, 95 percent of pigs and 78 percent of cattle. (Foer. *Eating Animals.* 109.) Between 90 and 95 percent of the 10 billion land animals killed each year in the US for food are poultry, and with over 99 percent of them raised on factory farms: Hessler & Balaban. Agricultural animals and the law; Garcés, L. (2013, January 24). Why we haven't seen inside a broiler chicken factory farm in a decade. *Food Safety News.* foodsafetynews.com/2013/01/why-we-havent-seen-inside-a-broiler-chicken-factory-farm-in-a-decade. The result is that 99 percent or more of all animals raised in the US for food are factory farmed. Some experts, such as the Union of Concerned Scientists, indicate that only 50 percent of animals raised for meat in the US come from factory farms (Union of Concerned Scientists. [2008, April]. CAFOs uncovered. 2. ucsusa.org/resources/confined-animal-feeding-operations-uncovered), but it seems that they are referring to exclusively Concentrated Animal Feeding Operations, which account for only 15 percent of all Animal Feeding Operations. (EPA region 7 [Midwest]. [n.d.]. *EPA.* epa.gov/region07/water/cafo). AFOs operate largely the same as CAFOs, just on a smaller scale. For example, to automatically qualify as a CAFO, a factory must have a minimum of 37,500 chickens, 16,500 turkeys, 750 pigs, and 300 cows (Environmental Protection Agency. [n.d.]. Regulatory Definitions of Large CAFOs, Medium CAFO, and Small CAFOs. epa.gov/npdes/pubs/sector_table.pdf).

[41] Foer. *Eating Animals.* 79.; The Issues. (n.d.). *PETA.* peta2.com/issue/factory-farming; Solotaroff. In the belly; Philpott, T. (2013, July/August). You won't believe what pork producers do to pregnant pigs. *Mother Jones.* motherjones.com/environment/2013/06/pregnant-sows-gestation-crates-abuse; Specter, M. (2003, April 14). The extremist. 64. *The New Yorker.* newyorker.com/magazine/2003/04/14/the-extremist

[42] Specter. The extremist. 63.

[43] Solotaroff. In the belly.

[44] Murray, L. (n.d.). Factory-farmed chickens: Their difficult lives and deaths. *Encyclopædia Britannica.* advocacy.britannica.com/blog/advocacy/2007/05/the-difficult-lives-and-deaths-of-factory-farmed-chickens/

[45] Craig, J. V.(n.d.). Debeaking. *United Poultry Concerns, Inc.* upc-online.org/merchandise/debeak_factsheet.html; Foer. *Eating Animals.* 309.

[46] Hessler & Balaban. Agricultural animals and the law.

[47] McKenna, M. (2010, December 24). Update: Farm animals get 80 percent of antibiotics sold in US *WIRED.* wired.com/2010/12/news-update-farm-animals-get-80-of-antibiotics-sold-in-us

[48] Beck, J. (2015, June 27). Antibiotic resistance is everyone's problem. *The Guardian*. theatlantic.com/health/archive/2015/06/how-to-stop-prevent-antibiotic-resistance-resistant-bacteria/397058/
[49] Foer. *Eating Animals*. 130, 188; Gerzilov, V., Datkova, V., Mihaylova, S., & Bozakova, N. (2012). Effect of poultry housing systems on egg production. *Bulgarian Journal of Agricultural Science*. agrojournal.org/18/06-18-12.pdf
[50] Welfare issues for meat chickens. (n.d.). *Compassion in World Farming*. ciwf.com/farm-animals/chickens/meat-chickens/welfare-issues; Park, M. (2010). *Gristle: From factory farms to food safety*. The New Press. 49. "20 to 30 percent," "0.19 to 0.46" 8.76 million: The Humane Society of the United States. The welfare of animals.
[51] Foer. *Eating Animals*. 111; Foer, J. S. (2009, October 7). Against meat. *The New York Times Magazine*. nytimes.com/2009/10/11/magazine/11foer-t.html
[52] Foer. *Eating Animals*. 48; The chicken industry. (n.d.). *PETA*. peta.org/issues/animals-used-for-food/factory-farming/chickens/chicken-industry; Specter. The extremist. 63.
[53] Specter. The extremist. 64; Foer. *Eating Animals*. 114; The Humane Society of the United States. The welfare of animals.
[54] Lohan, T. (2010, January 26). Got milk? A disturbing look at the dairy industry. *AlterNet*. alternet.org/2010/01/got_milk_a_disturbing_look_at_the_dairy_industry
[55] Foer. *Eating Animals*. 48, 60.
[56] A history of chickens: then (1900) vs now (2020). (n.d.). *The Happy Chicken Coop*. thehappychickencoop.com/a-history-of-chickens
[57] Foer. *Eating Animals*. 48-9; Animal Planet. (2009. Sept 7). Egg industry grinds millions of baby chicks alive. blogs.discovery.com/animal_news/2009/09/horrific-egg-industry-grinds-millions-of-baby-chicks-alive.html
[58] Foer. *Eating Animals*. 93.
[59] Consumer perceptions of farm animal welfare. (n.d.). *Animal Welfare Institute*. awionline.org/sites/default/files/uploads/documents/fa-consumer_perceptionsoffarmwelfare_-112511.pdf
[60] Garcés. Why we haven't seen; Foer. *Eating Animals*. 84; Leonard. *The meat racket*. 2, 4, 9; Pew Commission on Industrial Farm Animal Production. Putting meat. 8.
[61] Garcés. Why we haven't seen; Solotaroff. In the belly.
[62] Genoways, T. (2013, July/August). Gagged by big ag. *Mother Jones*. motherjones.com/environment/2013/06/ag-gag-laws-mowmar-farms; Hirsch, J. (2013, April 10). Ag-gag laws: State of the states. *Modern Farmer*. modernfarmer.com/2013/04/ag-gag-laws-state-of-the-states; Anti-whistleblower ag-gag bills hide factory farming abuses from the public. (n.d.). *The Humane Society of the United States*. humanesociety.org/issues/campaigns/factory_farming/fact-sheets/ag_gag.html
[63] Bottemiller, H. (2012, January 24). Supreme court blocks California's downer livestock law. *Food Safety News*. foodsafetynews.com/2012/01/supreme-court-blocks-californias-downer-livestock-law; Solotaroff. In the belly.
[64] Foer. *Eating Animals*. 134.
[65] Brulliard, (2016, August 6). How eggs became a victory for the animal welfare movement. *The Washington Post*. washingtonpost.com/news/animalia/wp/2016/08/06/how-eggs-became-a-victory-for-the-animal-welfare-movement-if-not-necessarily-for-hens; California court affirms proposition 2's constitutionality. (2013, August 30). *The Humane Society of the United States*. humanesociety.org/news/press_releases/2013/08/california-court-affirms-prop2-083013.html
[66] World's most admired companies. (n.d.). *Fortune*. fortune.com/worlds-most-admired-companies
[67] Duhigg, C., & Kocieniewski, D. (2012, April 28). How apple sidesteps billions in taxes. *The New York. Times*. nytimes.com/2012/04/29/business/apples-tax-strategy-aims-at-low-tax-states-and-nations.html
[68] Duhigg & Kocieniewski. How apple sidesteps billions.
[69] Merle, R. (2016, August 30). Apple owes $14.5 billion in back taxes, European authorities say. *The Washington Post*. washingtonpost.com/amphtml/business/economy/apple-owes-145-billion-in-back-taxes-european-authorities/2016/08/30/e7f6ed80-6ea2-11e6-9705-23e51a2f424d_story.html
[70] Subsidy tracker parent company summary. (n.d.). *Good Jobs First*. subsidytracker.goodjobsfirst.org/parent/apple-inc
[71] Rushe, D. (2018, July 2). US cities and states give big tech $9.3bn in subsidies in five years. *The Guardian*. theguardian.com/cities/2018/jul/02/us-cities-and-states-give-big-tech-93bn-in-subsidies-in-five-years-tax-breaks
[72] Marx, P. (2017, September 12). Love your iPhone? Don't thank apple. Thank the US government. *Medium*. theboldtalic.com/love-your-iphone-dont-thank-apple-thank-the-us-government-4f702dd7117e
[73] WP Company. (2020, March 16). Apple fined a record $1.2 billion by French antitrust regulators; company plans to appeal. *The Washington Post*. washingtonpost.com/business/economy/apple-fined-a-record-12-billion-by-french-antitrust-regulators-company-plans-to-appeal/2020/03/16/d673f604-6780-11ea-9923-57073adce27c_story.html; Holpuch, A. (2014, January 15). Apple to pay $32.5m over practice that let children make in-app purchases. *The Guardian*. theguardian.com/technology/2014/jan/15/apple-practice-children-make-in-app-purchases; Kim S. (2020, November 30). Apple hit with $12 million fine over iPhone waterproof claim as European penalties top $1.36 billion in 2020. *Newsweek*. newsweek.com/apple-fined-12-million-dollars-waterproof-claim-iphone-warranty-european-penalties-1551138; Apple fined for misleading iPad adverts. (n.d.). *The Financial Times*. ft.com/cms/s/0/cc167316-bb72-11e1-9436-00144feabdc0.html; Apple fined over misleading product guarantees. (n.d.). *International Charter* .icharter.org/news/apple_fined_over_misleading_product_guarantees.html; Apple fined by China court for copyright violation. (2012, December 28). *BBC News*. bbc.com/news/business-20856199; Khanna, M. (2021, March 22). Apple to pay $308 million in lawsuit, fined $2 million by Brazil govt. *IndiaTimes*. indiatimes.com/technology/news/apple-pay-308-million-dollars-in-lawsuit-fined-2-million-dollars-brazil-govt-536767.html; Apple fined $500m for iTunes patent infringement. (2015, February 25). *Sky News*. news.sky.com/story/apple-fined-500m-for-itunes-patent-infringement-10370093; Worstall, T. (2013, December 25). Apple fined $670,000 in Taiwan for price fixing. *Forbes*. forbes.com/sites/timworstall/2013/12/25/apple-fined-670000-in-taiwan-for-price-fixing
[74] Kara, S. (2018, October 12). Is your phone tainted by the misery of the 35,000 children in Congo's mines? *The Guardian*. theguardian.com/global-development/2018/oct/12/phone-misery-children-congo-cobalt-mines-drc
[75] Apple Inc. (n.d.). *Ethical Consumer*. ethicalconsumer.org/company-profile/apple-inc
[76] Bilton, R. (2014, December 18). Apple "failing to protect Chinese factory workers." BBC News. bbc.com/news/business-30532463
[77] Albergotti, R. (2020, November 20). Apple is lobbying against a bill aimed at stopping forced labor in China. *The Washington Post*. washingtonpost.com/technology/2020/11/20/apple-uighur; Canales, K. (2021, May 10). 7 Apple suppliers in China have links to forced labor programs, including the use of Uyghur Muslims from Xinjiang, according to a new report. *Insider*. businessinsider.com/apple-china-suppliers-uyghur-muslims-forced-labor-report-2021-5
[78] The stark reality of iPod's Chinese factories. (2006, August 18). *Mail Online*. mailonsunday.co.uk/news/article-401234/The-stark-reality-iPods-Chinese-factories.html; Morphy, E. (2008, Jan 31). Apple, IT and the specter of sweatshop labor. TechNewsWorld. technewsworld.com/story/61454.html; Staff, R. (2010, November 5). Foxconn worker plunges to death at China plant: report. *Reuters*. reuters.com/article/us-china-foxconn-death-idUSTRE6A41M920101105; Moore, M. (2011, February 15). Apple's child labour issues worsen. *Telegraph*. telegraph.co.uk/technology/apple/8324867/Apples-child-labour-issues-worsen.html; Caulfield, B. (2012, March 29).

Apple-Foxconn investigation finds "serious" violations of Chinese labor laws. *Forbes.* forbes.com/sites/briancaulfield/2012/03/29/apple-foxconn-investigation-finds-sense-of-unsafe-working-conditions-among-workers; Jim Armitage, J. (2013, July 30). 'Even worse than Foxconn': Apple rocked by child labour claims. *Independent.* independent.co.uk/life-style/gadgets-and-tech/even-worse-than-foxconn-apple-rocked-by-child-labour-claims-8736504.html; Luk, L. (2013, October 10). Foxconn admits to labor violations at factory. *The Wall Street Journal.* wsj.com/articles/SB10001424052702303382004579127393929164558; Bilton. Apple "failing to protect."

[79] The stark reality; Morphy. Apple, IT and the specter.

[80] Staff. Foxconn worker plunges.

[81] Moore. Apple's child labour.

[82] Duffy, K. (2020, December 10). Former Apple employees have accused the company of turning a blind eye to suppliers that were violating Chinese labor laws. *Insider.* businessinsider.com/former-apple-employees-apple-ignored-suppliers-violated-china-labor-laws-2020-12

[83] Price, D. (2017, January 03). Why Apple was bad for the environment (and why that's changing). Macworld. macworld.co.uk/news/apple/why-apple-was-bad-environment-why-changing-green-3450263

[84] Passy, J. (2018, October 08). Can Apple make an iPhone that isn't bad for the environment? *MarketWatch.* marketwatch.com/story/why-apples-new-iphone-is-bad-for-the-environment-2017-09-12; Price. Why Apple was bad; Ong, J. (2011, October 25). Apple supplier Pegatron facing pollution concerns in China. *AppleInsider.* appleinsider.com/articles/11/10/25/apple_supplier_pegatron_facing_pollution_concerns_in_china; Lane, S. (2011, January 2020). Chinese environmental groups accuse Apple of ignoring health concerns. *AppleInsider.* appleinsider.com/articles/11/01/20/chinese_environmental_groups_accuse_apple_of_ignoring_health_concerns.html

[85] Bergen, M. (2021, May 20). Microsoft and Apple wage war on gadget right-to-repair laws. *Bloomberg.* bloomberg.com/news/articles/2021-05-20/microsoft-and-apple-wage-war-on-gadget-right-to-repair-laws

[86] Stempel, J. (2020, March 2). Apple to pay up to $500 million to settle US lawsuit over slow iPhones. *Reuters.* reuters.com/article/us-apple-iphones-settlement/apple-to-pay-up-to-500-million-to-settle-u-s-lawsuit-over-slow-iphones-idUSKBN20P2E7; Allyn, B. (2020, November 18). Apple agrees to pay $113 million to settle "batterygate" case over iPhone slowdowns. *NPR.* npr.org/2020/11/18/936268845/apple-agrees-to-pay-113-million-to-settle-batterygate-case-over-iphone-slowdowns; Amaro, S. (2020, March 16). Apple fined a record $1.2 billion by French antitrust authorities. *CNBC.* cnbc.com/2020/03/16/apple-fined-1point2-billion-by-french-competition-authorities.html; Apple to pay $3.4 million over programmed iPhone obsolescence in Chile. (2021, April 08). *Gadgets.* 360.gadgets.ndtv.com/mobiles/news/apple-lawsuit-chile-iphone-obsolescence-programmed-limited-lifespan-fine-usd-3-4-million-2408793

[87] Are iPhones bad for the environment? (2020, March, 03). *Compare and Recycle.* compareandrecycle.co.uk/blog/are-iphones-bad-for-the-environment; Minter, A. (2020, October 17). Maybe Apple isn't as green as it claims. *Bloomberg.* bloomberg.com/opinion/articles/2020-10-17/maybe-apple-isn-t-as-green-as-it-claims; Mikolajczak, C. (2020, June 04). Apple crushes one-man repair shop in Norway's Supreme Court, after three-year battle. *Right to Repair.* repair.eu/news/apple-crushes-one-man-repair-shop/; Bergen. Microsoft and Apple wage war.

[88] Urry, A. (2016, October 11). Apple's recycling robot wants your old iPhone. Don't give it to him. *Grist.* grist.org/business-technology/apples-recycling-robot-wants-your-old-iphone-dont-give-it-to-him

[89] Minter. Maybe Apple isn't; Urry. Apple's recycling robot.

[90] Evans, J. (2020, July 21). The planet can't wait, says Tim cook as Apple goes carbon neutral. *Computerworld.* computerworld.com/article/3567654/the-planet-cant-wait-says-tim-cook-as-apple-goes-carbon-neutral.html

[91] Evans. The planet can't wait; Are iPhones bad?

[92] Veksler, D. L. (2017, October 15). Apple is not as green as it seems. *FEE.* fee.org/articles/apples-environmental-claims-are-misleading

[93] Bakan, J. (2019). The Corporation: The Pathological Pursuit of Profit and Power. United States: Free Press. 351.

[94] Bakan. *The Corporation.* 50.

[95] Strine Jr, L. E. (2012). Our continuing struggle with the idea that for-profit corporations seek profit. *SSRN Electronic Journal, 47,* 135, 155. ssrn.com/abstract=2523977

[96] Bottled water. (n.d.). *Container Recycling Institute.* container-recycling.org/index.php/issues/bottled-water; This is an estimate based on an estimated 18 million barrels of oil required in 2006, and consumption in 2020 was roughly double what it was in 2010: Peter Gleick with more reasons to stop drinking bottled water. (2010, August 10). *Fast Company.* fastcompany.com/1678595/peter-gleick-more-reasons-stop-drinking-bottled-water; Sales volume of bottled water in the United States from 2010 to 2020 (in billion gallons). (n.d.). *Statista.* statista.com/statistics/237832/volume-of-bottled-water-in-the-us

[97] Winter. The violent afterlife; Plastic breakdown. (n.d.). *Plastic Soup Foundation.* plasticsoupfoundation.org/en/plastic-problem/plastic-environment/break-down

[98] Carrington, D. (2020, May 22). Microplastic pollution in oceans vastly underestimated—study. *The Guardian.* theguardian.com/environment/2020/may/22/microplastic-pollution-in-oceans-vastly-underestimated-study; Readfearn, G. (2018, March 15). WHO launches health review after microplastics found in 90 percent of bottled water. *The Guardian.* theguardian.com/environment/2018/mar/15/microplastics-found-in-more-than-90-of-bottled-water-study-says

[99] Lenzer, A. (2009, August). Fiji Water: Spin the bottle. *Mother Jones.* motherjones.com/politics/2009/08/fiji-spin-bottle; Naidu, N. (2020, May 14). Fiji 33 years after the first fateful coup – a failed democracy? stuff.co.nz/world/south-pacific/121508384/thirtythree-years-after-the-first-military-coup-fiji-is-still-a-failed-democracy

[100] Organic market summary and trends. (2021, February 12). *Economic Research Service US Department of Agriculture.* ers.usda.gov/topics/natural-resources-environment/organic-agriculture/organic-market-summary-and-trends

[101] Rabin, R. C. (2018, October 23). Can eating organic food lower your cancer risk? *The New York Times.* nytimes.com/2018/10/23/well/eat/can-eating-organic-food-lower-your-cancer-risk.html

[102] Exposed and ignored: How pesticides are endangering our nation's farmworkers. (n.d.). *Farmworker Justice.* kresge.org/sites/default/files/Exposed-and-ignored-Farmworker-Justice-KF.pdf

[103] Vanbergen, A. J. (2021, August 04). A cocktail of pesticides, parasites and hunger leaves bees down and out. *Nature.* nature.com/articles/d41586-021-02079-4; Pesticides and the loss of biodiversity. (n.d.). *PAN Europe.* pan-europe.info/issues/pesticides-and-loss-biodiversity

[104] Pandey, K. (2018, July 18). Fashion industry may use quarter of world's carbon budget by 2050. *Down To Earth* .downtoearth.org.in/news/environment/fashion-industry-may-use-quarter-of-world-s-carbon-budget-by-2050-61183

[105] Pandey. Fashion industry may.

[106] Young, S. (2021, May 28). The fabrics with the worst environmental impact revealed, from polyester to fur. *Independent.* independent.co.uk/climate-change/sustainable-living/fast-fashion-sustainable-worst-fabrics-b1855935.html

[107] Polyvinyl chloride production in the United States from 1990 to 2019. (n.d.). *Statista.* statista.com/statistics/975603/us-polyvinyl-chloride-production-volume

[108] Building, construction & design. (n.d.). *Vinyl Institute.* vinylinfo.org/uses/building-and-construction

[109] Hites, R. A. (2011). Dioxins: An overview and history. *Environmental Science & Technology*, 45(1), 16–20. doi.org/10.1021/es1013664 ; Ehrlich, B. (2014, February 3). The PVC debate: A fresh look. *Building Green*. buildinggreen.com/feature/pvc-debate-fresh-look; PVC: The poison plastic. (2003, August 18). *Greenpeace.org*. greenpeace.org/usa/wp-content/uploads/legacy/Global/usa/report/2009/4/pvc-the-poison-plastic.html; Dioxins and their effects on human health. (2016, October 04). World Health Organization .who.int/news-room/fact-sheets/detail/dioxins-and-their-effects-on-human-health; MacKie, D., Liu, J., Loh, Y.-S., & Thomas, V. (2003). No evidence of dioxin cancer threshold. Environmental Health Perspectives, 111(9), 1145–47. jstor.org/stable/3435500; PVC: The poison plastic.

[110] Car emissions and global warming. (2014, July 18). *Union of Concerned Scientists*. ucsusa.org/resources/car-emissions-global-warming

[111] Berners-Lee, M., & Clark, D. (2010, September 23). What's the carbon footprint of … a new car? *The Guardian*. theguardian.com/environment/green-living-blog/2010/sep/23/carbon-footprint-new-car

[112] Laville, S. (2019, October 10). Exclusive: carmakers among key opponents of climate action. *The Guardian*. theguardian.com/environment/2019/oct/10/exclusive-carmakers-opponents-climate-action-us-europe-emissions; Zlata Rodionova, Z. (2016, April 11). Oil companies including Exxon and Shell spent £81m 'obstructing' climate laws in 2015, NGO says. *Independent*. independent.co.uk/news/business/news/royal-dutch-shell-exxon-american-petroleum-institute-western-states-petroleum-association-lobbying-climate-change-money-investors-a6978336.html; Meyer, R. (2020, February 20). The oil industry is quietly winning local climate fights. *The Atlantic*. theatlantic.com/science/archive/2020/02/oil-industry-fighting-climate-policy-states/606640; Tabuchi, H. (2021, August 05). Oil producers used Facebook to counter President Biden's clean energy message, a study shows. *The New York Times*. nytimes.com/2021/08/05/climate/oil-facebook-ads-biden.html; Franta, B. (2018, September 19). Shell and Exxon's secret 1980s climate change warnings. *The Guardian*. theguardian.com/environment/climate-consensus-97-per-cent/2018/sep/19/shell-and-exxons-secret-1980s-climate-change-warnings; Hiltzik, M. (2017, August 22). Column: A new study shows how Exxon Mobil downplayed climate change when it knew the problem was real. *Los Angeles Times*. latimes.com/business/hiltzik/la-fi-hiltzik-exxonmobil-20170822-story.html; McGreal, C. (2021, Jun 30). Big oil and gas kept a dirty secret for decades. Now they may pay the price. *The Guardian*. theguardian.com/environment/2021/jun/30/climate-crimes-oil-and-gas-environment

[113] Aronoff, K, (2020, October 28). Car companies have been knowingly screwing the planet for half a century. *The New Republic*. newrepublic.com/article/159969/car-companies-knowingly-screwing-planet-half-century

[114] Collins, M. (2015, July 14). The big bank bailout. *Forbes*. forbes.com/sites/mikecollins/2015/07/14/the-big-bank-bailout

[115] Biggest bank settlements. (n.d.). *WSJ*. graphics.wsj.com/lists/SEC0918

[116] Taibbi, M. (2012, December 13). Outrageous HSBC settlement proves the drug war is a joke. *Rolling Stone*. rollingstone.com/politics/politics-news/outrageous-hsbc-settlement-proves-the-drug-war-is-a-joke-230696; Tabuchi, H. (2016, December 03). How big banks are putting rain forests in peril. *The New York Times*. nytimes.com/2016/12/03/business/energy-environment/how-big-banks-are-putting-rain-forests-in-peril.html; McCoy, K. (2016, September 08). Wells Fargo fined $185M for fake accounts; 5,300 were fired. *USA Today*. usatoday.com/story/money/2016/09/08/wells-fargo-fined-185m-over-unauthorized-accounts/90003212; Calamur, K. (2015, July 08). JPMorgan Chase fined $136m over how it collects debt. *NPR*. npr.org/sections/thetwo-way/2015/07/08/421277881/jpmorgan-chase-fined-136m-over-how-it-collects-debt; Taibbi, M. (2012, March 14). Bank of America: Too crooked to fail. *Rolling Stone*. rollingstone.com/politics/news/bank-of-america-too-crooked-to-fail-20120314

[117] Dayen, D. (2017, October 5). Special investigation: How America's biggest bank paid its fine for the 2008 mortgage crisis—with phony mortgages! *The Nation*. thenation.com/article/archive/how-americas-biggest-bank-paid-its-fine-for-the-2008-mortgage-crisis-with-phony-mortgages

[118] Collins. The big bank bailout.

[119] Taibbi. Outrageous HSBC settlement.

[120] Groeger, L. (2014, February 24). Big pharma's big fines. *Pro Publica*. projects.propublica.org/graphics/bigpharma

[121] Angell, M. (2009, January 15). Drug companies & doctors: A story of corruption. *The New York Review*. nybooks.com/articles/2009/01/15/drug-companies-doctorsa-story-of-corruption

[122] Rosenberg, P., & Jerricks, T. (2016, November 02). Project censored. *Salt Lake City Weekly*. cityweekly.net/utah/project-censored/Content?oid=3514564

[123] Hawksley, H. (2001, April 12). Mali's children in chocolate slavery. *BBC*. news.bbc.co.uk/2/hi/africa/1272522.stm; Blunt, L. (2000, September 28). The bitter taste of slavery. *BBC*. news.bbc.co.uk/2/hi/africa/946952.stm; Fountain, A. C., & Huetz-Adams, F. (2020). Cocoa barometer. *Voice Network*. 4.voicenetwork.eu/wp-content/uploads/2021/03/2020-Cocoa-Barometer-EN.pdf; International programme on the elimination of child labour (IPEC). (n.d.). *ILO*. ilo.org/public//english/standards/ipec/themes/cocoa/download/2005_02_cl_cocoa.pdf; Harkin–Engel Protocol. (n.d.). *Wikipedia*. en.wikipedia.org/wiki/Harkin%E2%80%93Engel_Protocol

[124] On Halloween, Nestlé claims no responsiblity for child labor. (2006, October 30). *International Labor Rights Forum*. laborrights.org/stop-child-labor/cocoa-campaign/news/10993; Doe v. Nestle. (2011, November 04). *EarthRights International*. earthrights.org/publication/amicus-brief-doe-v-nestle

[125] Rice, D. (2020, February 14). One-third of all plant and animal species could be extinct in 50 years, study warns. *USA Today*. usatoday.com/story/news/nation/2020/02/14/climate-change-study-plant-animal-extinction/4760646002

[126] Hanh, T. N. Guided meditations. *Thich Nhat Hanh Foundation*. thichnhathanhfoundation.org/covid-resources-meditations

24 - We Are the Machine, Blossoms: Exit Strategies

[1] McKibben, B. (2019, September 12). Hello from the year 2050. We avoided the worst of climate change—but everything is different. *TIME*. time.com/5669022/climate-change-2050

[2] Korten, D. (n.d.). *Origin of the term*. Living Economies Forum. davidkorten.org/great-turning/origin-of-the-term; van Gelder, S. (2000, April 1). The Great Turning. *Yes Magazine*. yesmagazine.org/issue/issues-new-stories/2000/04/01/the-great-turning

[3] I have not been able to find the original source for this quote, though it is quite common and is consistently attributed to Thomas Friedman.

[4] Lockhart, P.R. (2019, Aug 16). *How slavery became America's first big business*. Vox. vox.com/identities/2019/8/16/20806069/slavery-economy-capitalism-violence-cotton-edward-baptist; Zickuhr, K. (2021, June 24). *New research shows slavery's central role in U.S. economic growth leading up to the Civil War*. Washington Center for Equitable Growth. equitablegrowth.org/new-research-shows-slaverys-central-role-in-u-s-economic-growth-leading-up-to-the-civil-war

[5] 50,000 British soldiers fought against the American Revolution (Brooks, R.B. [2017, Nov 27]. *British soldiers in the revolutionary war*. History of Massachusetts Blog. historyofmassachusetts.org/british-soldiers-revolutionary-war). I couldn't find totals for how many British and French troops fought against the revolution in Haiti, but I did find that

100,000 British troops died or were injured (45,000 died, 55,000 injured), suggesting that far more than 100,000 actually fought. Some 75,000 French died, about 50% more than the British number, suggesting that far more than 150,000 fought (50% more than the British). To be conservative, I assumed a total of 250,000 troops total. (Perry, James (2005). *Arrogant Armies: Great Military Disasters and the Generals Behind Them.* Edison: CastleBooks. 75–76; Estimated number of deaths in the Haitian Revolution from 1790 to 1804. [n.d.] Statista. statista.com/statistics/1069645/estimated-death-toll-haitian-revolution-by-race)

[6] Graham, B. (2015, Oct). *Labor law highlights, 1915–2015.* Bureau of Labor Statistics, Monthly Labor Review. bls.gov/opub/mlr/2015/article/pdf/labor-law-highlights-1915-2015.pdf; Whaple, R. (n.d.). Hours of work in U.S. history. EH.net. eh.net/encyclopedia/hours-of-work-in-u-s-history/; Lisa, A. (2019, May 21). *Major laws that changed the workplace over the last 100 years.* Stacker. stacker.com/stories/3093/major-laws-changed-workplace-over-last-100-years

[7] Cillizza, C. (2015, June 26). How unbelievably quickly public opinion changed on gay marriage, in 5 charts. *Washington Post.* washingtonpost.com/news/the-fix/wp/2015/06/26/how-unbelievably-quickly-public-opinion-changed-on-gay-marriage-in-6-charts

[8] Iron Eyes, C. (2011, Sept). thelastrealindian.blogspot.com/2011/09/last-real-indian.html

[9] Nicholas, K. A. (2017). The climate mitigation gap: Education and government recommendations miss the most effective individual actions. *Environmental Research Letters, 12*(7).doi.org/10.1088/1748-9326/aa7541

[10] Arguedas, D. (2018, November 4). Ten simple ways to act on climate change. *BBC Future.* bbc.com/future/article/20181102-what-can-i-do-about-climate-change

[11] Key facts and findings. (n.d.). *Food and Agriculture Organization of the United Nations.* fao.org/news/story/en/item/197623/icode; Dutkiewicz, J. (2020, August 31). The climate activists who dismiss meat consumption are wrong. *The New Republic* .newrepublic.com/article/159153/climate-change-dismiss-meat-emissions-wrong; Maisto, M. (2012, April 28). Eating less meat is world's best chance for timely climate change, say experts. *Forbes.* forbes.com/sites/michellemaisto/2012/04/28/eating-less-meat-is-worlds-best-chance-for-timely-climate-change-say-experts; Jacquet, J. (2021, May 14). The meat industry is doing exactly what Big Oil does to fight climate action. *The Washington Post.* washingtonpost.com/outlook/the-meat-industry-is-doing-exactly-what-big-oil-does-to-fight-climate-action/2021/05/14/831e14be-b3fe-11eb-ab43-bebddc5a0f65_story.html; Key facts and findings.

[12] Yeo, S. (2018, August 14). Can the psychological technique of "pre-conformity" help change our harmful behaviors? *Pacific Standard.* psmag.com/environment/peer-pressure-can-make-americans-eat-less-meat

[13] Your questions about food and climate change, answered. (2019, April 30). *The New York Times.* nytimes.com/interactive/2019/04/30/dining/climate-change-food-eating-habits.html

[14] Goldstein, B., Gounaridis, D., & Newell, J. P. (2020). The carbon footprint of household energy use in the United States. *PNAS, 117*(32), 19122-19130. doi.org/10.1073/pnas.1922205117; Feeling powerless? Switch to green power. (2020, April 11). *Earthday.org.*earthday.org/feeling-powerless-switch-to-green-power

[15] Green electricity – how much does it really cost? (n.d.). *Worcester Energy.* worcesterenergy.org/take-action/energy-and-cost-saving-tips/purchasing-green-electricity

[16] Coffey, H. (2020, Jan 10). Flygskam: What is the Flight Shaming Environmental Movement that's Sweepng Europe? *Independent.* independent.co.uk/travel/news-and-advice/flygskam-anti-flying-flight-shaming-sweden-greta-thornberg-environment-air-travel-train-brag-tagskryt-a8945196.html

[17] Broom, D. (2019, August 30). 1 in 7 people would choose not to fly because of climate change. *World Economic Forum.* weforum.org/agenda/2019/08/1-in-7-people-would-choose-not-to-fly-because-of-climate-change

[18] Carrington, D. (2020, November 17). 1% of people cause half of global aviation emissions – study. *The Guardian.* theguardian.com/business/2020/nov/17/people-cause-global-aviation-emissions-study-covid-19; Mitloehner, F. (2019, November 13). It's time to stop comparing meat emissions to flying. GHGGuru Blog.ghgguru.faculty.ucdavis.edu/2019/11/13/its-time-to-stop-comparing-meat-emissions-to-flying

[19] Timperley, J. (2020, February 18). Should we give up flying for the sake of the climate? *BBC.* bbc.com/future/article/20200218-climate-change-how-to-cut-your-carbon-emissions-when-flying

[20] Nicholas. The climate mitigation gap.

[21] Sales of electrified vehicles jump up 81% in the first quarter of 2021. (2021, April 19). *Cox Automotive.* coxautoinc.com/news/sales-of-electrified-vehicles-jump-up-81-in-the-first-quarter-of-2021

[22] Berners-Lee & Clark. What's the carbon footprint?

[23] Diamond, J. (2008, January 2). What's your consumption factor? *The New York Times.* nytimes.com/2008/01/02/opinion/02diamond.html; 20% of the current global population of roughly eight billion people live in industrialized countries: PRB. [n.d.]. Has the world's population distribution changed much over time? When could world population stop growing? prb.org/resources/human-population

[24] Savage, N. (2020, Sept 15). *The fires this time: public goods, the Jewish community, different time horizons.* Hazon. hazon.salsalabs.org/newsletter-sept-15?wvpId=cd3e2d5e-5d16-4f9b-a3f0-c46c4d1d81af

[25] People are more likely to make changes if they know other people are making them as well, and they are more likely to respond to calls for change if the calls come from people who have already made those changes. Patrick, K. (2018, March 4). Vegans, you're doing it wrong: How to get people to eat less meat, with Gregg Sparkman. *Medium* .medium.com/how-to-save-the-world-by-katie-patrick/vegans-youre-doing-it-wrong-with-gregg-sparkman-social-psychologist-ed55f0c9f3a9 ; Yeo. Can the psychological?; Rowlatt, J. (2019, September 20). Climate change action: We can't all be Greta, but your choices have a ripple effect. BBC News. bbc.com/news/science-environment-49756280; Dutkiewicz. The climate activists.

[26] Rowlatt. Climate change action.

[27] Rose, T. (2014, Oct 1). Personal Communication from Tiffany Rose, Senior Action Team Coordinator with PETA.

[28] Average cost of a slave worldwide today: $90. Roughly 6 percent of the cost of a candy bar goes to the people growing the cocoa. Let's say a bar costs $1.50. Eighteen cents a month, adds up to $90 in 41.6 years. Slavery is everywhere. (n.d.) *Free the Slaves.*freetheslaves.net/about-slavery/slavery-today; NIeburg, O. (2014, July 09). Paying the price of chocolate: Breaking cocoa farming's cycle of poverty. *Confectionery.* confectionerynews.com/Commodities/Price-of-Chocolate-Breaking-poverty-cycle-in-cocoa-farming

[29] Average miles driven: 13,474, average mileage: 25.5, cost of gas: $3/gallon. DiLallo, M. (2018, October 3). The average American drives this much each year -- how do you compare? *The Motley Fool.* fool.com/investing/general/2015/01/25/the-average-american-drives-this-much-each-year-ho.asp; Autmotive News. (2015, June 4). Average US mpg edges up to 25.5 in May. autonews.com/article/20150604/OEM05/150609925/average-US-mpg-edges-up-to-25.5-in-may

[30] Jensen, D. (2009, July/August). Forget shorter showers. *Orion Magazine.* orionmagazine.org/article/forget-shorter-showers

[31] McGrath, M. (March 11). Climate change: 'Default effect' sees massive green energy switch. BBC News. bbc.com/news/science-environment-56361970

[32] McKibben, B. (n.d.). A moral atmosphere. Orion Magazine. orionmagazine.org/article/a-moral-atmosphere

[33] Schendler, A. (2021, August 31). Worrying about your carbon footprint is exactly what big oil wants you to do. *The New York Times.* nytimes.com/2021/08/31/opinion/climate-change-carbon-neutral.html

[34] Urpelainen, J., & George, E. (2021, July 14). Reforming global fossil fuel subsidies: How the United States can restart international cooperation. Brookings. brookings.edu/research/reforming-global-fossil-fuel-subsidies-how-the-united-states-can-restart-international-cooperation

[35] Hein, J. (2015, June 16). Oil companies are drilling on public land for the price of a cup of coffee. Here's why that should change. *The Washington Post*. washingtonpost.com/posteverything/wp/2015/06/16/oil-companies-are-drilling-on-public-land-for-the-price-of-a-cup-of-coffee-heres-why-that-should-change

[36] McKibben, B. (2020, May 1). "A bomb in the center of the climate movement": Michael Moore damages our most important goal. *Rolling Stone*. rollingstone.com/politics/political-commentary/bill-mckibben-climate-movement-michael-moore-993073

[37] Climate Leadership Council. (2019, Jan 17). Economists' statement on carbon dividends. clcouncil.org/economists-statement

[38] McKibben. Hello from the year 2050.

[39] Sewell, C. (2020, February 11). Removing the meat subsidy: Our cognitive dissonance around animal agriculture. *Journal of International Affairs*. jia.sipa.columbia.edu/removing-meat-subsidy-our-cognitive-dissonance-around-animal-agriculture

[40] Smedley, T. (2019, August 6). How shorter workweeks could save Earth. *BBC*. bbc.com/worklife/article/20190802-how-shorter-workweeks-could-save-earth

[41] This quote is all over the internet, but I couldn't find a primary source.

[42] White, C. (2007, April). The ecology of work. *Orion Magazine*. orionmagazine.org/article/the-ecology-of-work

[43] Jevons, W. S. (2007). *The coal question*. Encyclopedia of Earth. 123.

[44] Owen, D. (2010, December 13). The efficiency dilemma. *The New Yorker*. newyorker.com/magazine/2010/12/20/the-efficiency-dilemma

[45] Saunders, H. D. (1992). The Khazzoom-Brookes postulate and neoclassical growth. *The Energy Journal, 13*(4), 131–148. doi.org/10.5547/issn0195-6574-ej-vol13-no4-7 ; Owen. The efficiency dilemma.

[46] Brockway, P., & Sorrell, S. (2021, March 2). Guest post: Why 'rebound effects' may cut energy savings in half. Carbon Brief. carbonbrief.org/guest-post-why-rebound-effects-may-cut-energy-savings-in-half

[47] One summary of studies suggests that the rebound effect might overall be around 50 percent: Roberts, D. (2012, January 31). Does the rebound effect matter for policy? *Grist*. grist.org/energy-efficiency/does-the-rebound-effect-matter-for-policy

[48] Owen. The efficiency dilemma.

[49] Owen.

[50] Motor vehicle fuel consumption and travel in the U.S., 1960–2006. (n.d.). *Infoplease*. infoplease.com/ipa/A0004727.html; Owen. The efficiency dilemma; FAQ: Google fusion tables. (2019, December 3). *Google*. fusiontables.google.com/DataSource?dsrcid=225439#rows:id=1

[51] This statistic, like many in this section, is a bit tricky because the issues are so complex. For example, during that time period the US also experienced significant fuel switching, resulting in efficiency gains. There was also significant outsourcing, as well as "financialization" of the economy. For more on this: Heinberg, R. (2011). *The end of growth: Adapting to our new economic reality*. New Society Publishers, Part II, Chapter 4; Owen. The efficiency dilemma; US population in 1984: 235,824,902, and in 2005: 296,000,000: US population from 1900. (n.d.). *Demographia*. demographia.com/db-uspop1900.htm; Passel, J. S., & Cohn, D. (2008, February 11). US population projections: 2005–2050. *Pew Research Center*. pewsocialtrends.org/2008/02/11/us-population-projections-2005-2050

[52] Leonard, A. (n.d.). Story of stuff, referenced and annotated script.storyofstuff.org/wp-content/uploads/2020/01/StoryofStuff_AnnotatedScript.pdf; Tilford, D. Why Consumption Matters. The Sierra Club.sierraclub.org/sustainable_consumption/tilford.asp; Rogich, D., Cassara, A., Wernick, I., & Miranda, M. (2008). Material flows in the United States. *World Resources Institute*. 1, 2. pdf.wri.org/material_flows_in_the_united_states.pdf; Households and NPISHs final consumption expenditure per capita (constant 2010 US$). (n.d.). *The World Bank*. data.worldbank.org/indicator/NE.CON.PRVT.PC.KD

[53] Townsend, M., & Burke, J., (2002, July 07). Earth 'will expire by 2050'. *The Guardian*. guardian.co.uk/uk/2002/jul/07/research.waste; "Population in the world is currently growing at a rate of around 1.14 percent per year." World population (2020 and historical). (n.d.). *Worldometers*. worldometers.info/world-population

[54] Tverberg, G. (2012, March 12). World energy consumption since 1820 in charts. *Our Finite World*. ourfiniteworld.com/2012/03/12/world-energy-consumption-since-1820-in-charts

[55] This is extrapolated from calculations based on the period 1984–2004. See Chapter 8.

[56] Hodges, J. (2020, October 19). Wind, solar are cheapest power source in most places, BNEF says. *Bloomberg*. bloomberg.com/news/articles/2020-10-19/wind-solar-are-cheapest-power-source-in-most-places-bnef-says; Chrobak, U. (2021, October 8). Solar power got cheap. So why aren't we using it more? *Popular Science*.popsci.com/story/environment/cheap-renewable-energy-vs-fossil-fuels

[57] Lovins, A. B. (2018). How big is the energy efficiency resource? *Environmental Research Letters, 13*(9). doi.org/10.1088/1748-9326/aad965

[58] Transcript: Greta Thunberg's speech at the U.N. climate action summit. (2019, September 23). *NPR*. npr.org/2019/09/23/763452863/transcript-greta-thunbergs-speech-at-the-u-n-climate-action-summit

[59] Dutkiewicz, J. (2020, August 31). The climate activists who dismiss meat consumption are wrong. *The New Republic*. newrepublic.com/article/159153/climate-change-dismiss-meat-emissions-wrong

[60] Barry, W. (2017, August 30). In distrust of movements. *Orion Magazine*. orionmagazine.org/article/in-distrust-of-movements/

[61] Cohen, J. E. (1995). Presidential rhetoric and the public agenda. *American Journal of Political Science, 39*(1), 87–107. doi.org/10.2307/2111759

[62] Mann, M. E. (2019, September 12). Lifestyle changes aren't enough to save the planet. Here's what could. *TIME*. time.com/5669071/lifestyle-changes-climate-change/

[63] Jensen, D. (2013, Jan/Feb). Culture of Plunder. *Orion*. 12. derrickjensen.org/2013/01/culture-of-plunder

[64] Joanna Macy and the Great Turning. (n.d.). Joanna Macy Film. joannamacyfilm.org/about

[65] I couldn't find a primary source for this quote, but offer this as one example of the many websites/people/organizations that have published it: St. Teresa of Avila (n.d.). *Quotes from Teresa of Avila*. Our Lady of Mercy Lay Carmelite. olmlaycarmelites.org/quote/teresa-avila

[66] Hanh, T. N. (2006). *True love: A practice for awakening the heart*. Shambhala Publications. 15, 16.

[67] Korten. *Origin of the Term*.

[1] I could not identify the original source of this Mooji quote, but it is very much in the spirit of his teachings so I have chosen to include it.

25 - Another World Is Possible

[2] Holden, R. (2010). *Be happy!* Hay House, Inc. 37.

3 Posth, C., Renaud, G., Mittnik, A., Drucker, D. G., Rougier, H., Cupillard, C., et al. (2016). Pleistocene mitochondrial genomes suggest a single major dispersal of Non-Africans and a late glacial population turnover in Europe. *Current Biology*, 26(6), 827–833. doi.org/10.1016/j.cub.2016.02.022; Karmin, M., Saag, L., Vicente, M., Sayres, M. A. W., Järve, M., et al. (2015). A recent bottleneck of Y chromosome diversity coincides with a global change in culture. *Genome research*, 25(4), 459–466. doi.org/10.1101/gr.186684.114; Haber, M., Jones, A. L., Connell, B. A., Asan, Arciero, E., Yang, H., Thomas, M. G., Xue, Y., & Tyler-Smith, C. (2019). A rare deep-rooting D0 African Y-chromosomal haplogroup and its implications for the expansion of modern humans out of Africa. *Genetics*, 212(4), 1421–1428. doi.org/10.1534/genetics.119.302368

4 Hignett, K. (2018, January 26). Out of Africa: When did prehistoric humans actually leave—and where did they go? *Newsweek*. newsweek.com/out-africa-prehistoric-humans-791822

5 Reséndez, A. (2016). *The other slavery: The uncovered story of Indian enslavement in America*. United States: Houghton Mifflin Harcourt. 5,6, 324.

6 Franklin, B. (n.d.). The autobiography of Benjamin Franklin. Independence Hall Association. ushistory.org/franklin/autobiography/page57.htm

7 Jackson, A. (1833, Dec 3). *Fifth Annual Message*. The American Presidency Project. presidency.ucsb.edu/ws/?pid=29475; Roosevelt, T. (1889). *The winning of the West*. Best Books. 90.

8 Junger, S. (2016). *Tribe: On homecoming and belonging*. Twelve. 3–11.

9 Brooks, D. (2016, August 9). The great affluence fallacy. *The New York Times*. nytimes.com/2016/08/09/opinion/the-great-affluence-fallacy.html

10 Dawkins, R., & Wong, Y. (2005). *The Ancestor's Tale: A Pilgrimage to the Dawn of Evolution*. United States: Houghton Mifflin.

11 Bishop, M. (2001). *The Middle Ages*. Houghton Mifflin Harcourt. 7.

12 Caradonna, J. (2014, September 9). Is "progress" good for humanity? *The Atlantic*. theatlantic.com/business/archive/2014/09/the-industrial-revolution-and-its-discontents/379781

13 Sahlins, M. (1998). The original affluent society. *Limited wants, unlimited means: A reader on hunter-gatherer economics and the environment*. Island Press. 5-42. For an overview of critiques and commentary since that essay was published, see Solway, J. S. (2006). "The original affluent society": Four decades on. In Solway, J. (Ed) *The Politics of Egalitarianism: Theory and Practice*. United Kingdom: Berghahn Books. 69.

14 Solway. "The original affluent society." 75.

15 Sahlins. The original affluent society. 5-42.

16 Diamond, J. M. (2010). *The worst mistake in the history of the human race*. Oplopanax Publishing. 64–66; Sahlins. The original affluent society. 5-42; Hunter-gatherers: Insights from a golden affluent age. (2009). *Pacific Ecologist*. pacificecologist.org/archive/18/pe18-hunter-gatherers.pdf

17 Junger. *Tribe*. 11.

18 Isidore, C., & Luhby, T. (2015, July 9). Turns out Americans work really hard … but some want to work harder. *CNN Business*. money.cnn.com/2015/07/09/news/economy/americans-work-bush/index.html; Bureau of Labor Satistics. (2011, June 22). American Time Use Survey—2010 Results, News Release. 2. bls.gov/news.release/pdf/atus.pdf; Moore, D. W. (2002, February 12). Eyes wide open: Americans, sleep and stress. *Gallup*. gallup.com/poll/5314/eyes-wide-open-americans-sleep-stress.aspx; Rampell, C. (2011, September 22). Commuter nation. *The New York Times*. economix.blogs.nytimes.com/2011/09/22/commuter-nation

19 Solway. "The original affluent society." 69.

20 Junger. *Tribe*. 15–23.

21 Harter, J., & Arora, R. (2008, June 5). Social time crucial to daily emotional wellbeing in US *Gallup*. gallup.com/poll/107692/social-time-crucial-daily-emotional-wellbeing.aspx; Bureau of Labor Satistics. American Time Use Survey. 2.

22 Junger. *Tribe*. 14–15.

23 Cordain, L., Gotshall, R., Eaton, S., & Eaton, S. (1998). Physical activity, energy expenditure and fitness: An evolutionary perspective. *International Journal of Sports Medicine*, 19(05), 328–35. doi.org/10.1055/s-2007-971926; Solway. "The original affluent society." 70.

24 Jaslow, R. (2013, May 3). CDC: 80 percent of American adults don't get recommended exercise. *CBS News*. cbsnews.com/news/cdc-80-percent-of-american-adults-dont-get-recommended-exercise; Hendrick, B. (2010, February 10). Percentage of overweight, obese Americans swells. *WebMD*. webmd.com/diet/news/20100210/percentage-of-overweight-obese-americans-swells#1

25 Wolchover. (2011, February 16). Busting the 8-hour sleep myth; Warren, J. (2008, February 19). How to sleep like a hunter-gatherer. *Discover Magazine*. discovermagazine.com/2007/dec/sleeping-like-a-hunter-gatherer

26 See Chapter 7.

27 Stress in America: Our health at risk. (2012). *American Psychological Association*. 15, 17. apa.org/news/press/releases/stress/2011/final-2011.pdf

28 Hobbes, T. (1651). *Leviathan: Or the matter, form, & power of a common-wealth ecclesiastical and civil. McMaster University*. 77–78. socserv2.socsci.mcmaster.ca/econ/ugcm/3ll3/hobbes/Leviathan.pdf

29 Koerth-Baker, M. (2013, March 19). Who lives longest? *The New York Times Magazine* .nytimes.com/2013/03/24/magazine/who-lives-longest.html; Gurven, M., & Kaplan, H. (2007). Longevity among hunter-gatherers: A cross-cultural examination. *Population and Development Review*, 33(2), 349. doi.org/10.1111/j.1728-4457.2007.00171.x ; Kanazawa, S. (2008, November 20). Common misconceptions about science II: Life expectancy. *Psychology Today*. psychologytoday.com/blog/the-scientific-fundamentalist/200811/common-misconceptions-about-science-ii-life-expectancy

30 Gurven & Kaplan. Longevity among hunter-gatherers. 321; World Health Organization. (2016, Sept). Children: reducing mortality. who.int/mediacentre/factsheets/fs178/en

31 Wilford, J. N. (2002, October 29). Don't blame Columbus for all the Indians' ills. *The New York Times*. nytimes.com/2002/10/29/science/don-t-blame-columbus-for-all-the-indians-ills.html

32 Diamond. *The worst mistake*. 64–66; Wilford. Don't blame.

33 Wilford. Don't blame.

34 Diamond. *The worst mistake*. 64–66

35 Finkel, M. (2009, December). The hadza. *National Geographic*. nationalgeographic.com/magazine/article/hadza

36 Diamond. *The worst mistake*. 64–66

37 McMillan, T. (n.d.). The new face of hunger. *National Geographic*. nationalgeographic.com/foodfeatures/hunger

38 Junger. *Tribe*. 23; Hidaka B. H. (2012). Depression as a disease of modernity: explanations for increasing prevalence. *Journal of Affective Disorders*, 140(3), 205–214. doi.org/10.1016/j.jad.2011.12.036

39 Chacon, R.J. & Mendoza, R.G. (Eds.). (2007). *Latin American indigenous warfare and ritual violence*. United States: University of Arizona Press.

40 O'Hehir, (2009, Aug 6). Sacrificial virgins of the Mississippi. *Salon*. salon.com/2009/08/06/cahokia

[41] I cannot find my reference notes for this information. I remember the article clearly and feel comfortable including this fact, but I apologize that I cannot cite the original source to allow you to confirm this or pursue it further.
[42] Sahlins. The original affluent society. 5.
[43] Scharping, N. (2020, May 15). When did Homo sapiens first appear? *Discover Magazine.* discovermagazine.com/planet-earth/when-did-homo-sapiens-first-appear; Solway. "The original affluent society." 69; Gurven & Kaplan. Longevity among hunter-gatherers. 321.

26 - From Owning to Belonging

[1] Kimmerer, R. W. (2013). *Braiding sweetgrass: Indigenous wisdom, scientific knowledge and the teachings of plants.* Milkweed Editions. 208.
[2] Blackstone, W. (1763). Blackstone on property. *Online Library of Liberty.* oll.libertyfund.org/page/blackstone-on-property-1753
[3] George, A. (2014, March 26). Stuff: Humans as hunters and mega-gatherers. *NewScientist.* newscientist.com/article/mg22129620-700-stuff-humans-as-hunters-and-mega-gatherers
[4] History of land as private property. (n.d.). *Henry George Institute.* henrygeorge.org/pchp29.htm
[5] Alden Wily, L. (2018). Collective land ownership in the 21st Century: Overview of global trends. *Land, 7*(2), 2–3. dpi.com/2073-445X/7/2/68/pdf
[6] Warren, L. (2014). Owning nature: Toward an environmental history of private property. *The Oxford Handbook of Environmental History,* 399, 401. doi.org/10.1093/oxfordhb/9780195324907.013.0015
[7] Von Mises, L. (2016). *Human action.* Lulu Press, Inc. 679.
[8] Murphy, L., & Nagel, T. (2002). *The myth of ownership: Taxes and justice.* Oxford University Press. 8.
[9] Emphasis mine. Mill, J. S. (1878). *Principles of political economy with some of their applications to social philosophy.* London: Longmans, Green, Reader and Dyer. 123.
[10] Roy, E. A. (2017, March 16). New Zealand river granted same legal rights as human being. *The Guardian.* theguardian.com/world/2017/mar/16/new-zealand-river-granted-same-legal-rights-as-human-being
[11] Torres-Spelliscy, C. (2014, April 8). The history of corporate personhood. *Brennan Center for Justice.* brennancenter.org/blog/hobby-lobby-argument
[12] Noble, B. (2008). Owning as belonging, owning as property: The crisis of power and respect in First Nations heritage transactions with Canada. *First Nations Cultural Heritage and Law: Case Studies, Voices and Perspectives.* 465–73; Noisecat, J.B. (2017, March 27). The western idea of private property is flawed. Indigenous peoples have it right. *The Guardian.* theguardian.com/commentisfree/2017/mar/27/western-idea-private-property-flawed-indigenous-peoples-have-it-right; Gilman, R. (n.d.). The idea of owning land. Context Institute. context.org/iclib/ic08/gilman1
[13] To offer just one example in addition to the one that follows from Natalie Diaz, writer and climate activist Julian Brave NoiseCat notes that in his own Indigenous language, Secwepemcstin, the word for the people and the word for the community both in part derive from a word meaning world, dirt, nature, earth, land and spirit. "Ingrained in each Salish community then is the idea—even older than our indigenous languages—that the people are of the land and the land is of the people. These kindred spirits are alive and inseparable." (Noisecat. The western idea.)
[14] Diaz, N. (n.d.). The first water is the body. *Orion Magazine.* orionmagazine.org/article/women-standing-rock
[15] Kimmerer. *Braiding sweetgrass.* 55.
[16] Keeney, B. (2010). *The bushman way of tracking god: The original spirituality of the kalahari people.* Simon and Schuster. 40.
[17] Harper, D. (n.d.). *Own.* Etymonline: Online Etymology Dictionary. etymonline.com/search?q=own; Harper, D. (n.d.). *Possess.* Etymonline: Online Etymology Dictionary. etymonline.com/search?q=possess
[18] Wall, R. (2015, March 30). Nature needs a new pronoun: To stop the age of extinction, let's start by ditching "it". *YES!* yesmagazine.org/issue/together-earth/2015/03/30/alternative-grammar-a-new-language-of-kinship
[19] Rousseau, J. J. (2008). *The social contract and the first and second discourses.* Yale University Press. 207.
[20] Rousseau. *The social contract.* 207.

27 - Infinite Gratitude and the Gift of Everything

[1] Sheldon, W. L. (1893). What justifies private property? *The International Journal of Ethics, 4*(1), 30–31. jstor.org/stable/pdf/2375709.pdf
[2] Findings from a national survey & focus groups on economic mobility. (2009). *Economic Mobility Project, The Pew Charitable Trusts Greenberg Quinlan Rosner Research Public Opinion Strategies.* 14, 15. pewtrusts.org/-/media/legacy/uploadedfiles/pcs_assets/2009/survey_on_economic_mobility_findings(1).pdf
[3] Manne, A. (2014, July 9). "The a**hole effect": What wealth does to the brain. *AlterNet.* alternet.org/2014/07/ahole-effect-what-wealth-does-brain; Siegelbaum, D. (2014, March 26). American dream breeds shame and blame for job seekers. *BBC News.* bbc.com/news/magazine-26669971
[4] Public Broadcasting Service. (2013, June 21). *Exploring the psychology of wealth, 'pernicious' effects of economic inequality.* [Video]. PBS.pbs.org/newshour/show/pernicious-effects-of-economic-inequality
[5] Grewal, D. (2012, April 10). How Wealth Reduces Compassion. *Scientific American.* scientificamerican.com/article/how-wealth-reduces-compassion; Marsh, J. (2012, September 25). Why inequality is bad for the one percent. *Greater Good Science Center.* greatergood.berkeley.edu/article/item/why_inequality_is_bad_for_the_one_percent.
[6] Siegelbaum. American dream.
[7] Milanović, B. (2011). *The haves and the have-nots: A brief and idiosyncratic history of global inequality.* Basic Books. 115–23.
[8] Chang, H. J. (2012). *23 things they don't tell you about capitalism.* Bloomsbury Publishing USA. 30.
[9] Buffett, W. (n.d.). My philanthropic pledge. The Giving Pledge. givingpledge.org/Pledger.aspx?id=177
[10] Milanović. *The haves and the have-nots.* 115–23.
[11] Milanović. 115–123; Hertz, T. (2006). Understanding mobility in America. *Center for American Progress.* americanprogress.org/wp-content/uploads/kf/hertz_mobility_analysis.pdf; Corak, M. (n.d.). Inequality from generation to generation: The United States in comparison. Graduate School of Public and International Affairs University of Ottawa. 10.milescorak.files.wordpress.com/2012/01/inequality-from-generation-to-generation-the-united-states-in-comparison-v3.pdf
[12] Edler, M. (2007). Moving on up? Kent State Magazine. s3-live.kent.edu/s3fs-root/s3fs-public/file/Winter2007-Volume7-Issue2.pdf
[13] Hertz. Understanding mobility in America.
[14] Pizzigati, S. (2012, September 23). The "self-made" myth: our hallucinating rich. *Inequality.org.* inequality.org/selfmade-myth-hallucinating-rich

[15] Cassidy, J. (2013, June 17). Why America still needs affirmative action. *The New Yorker*. newyorker.com/news/john-cassidy/why-america-still-needs-affirmative-action

[16] Milanović. *The haves and the have-nots*. 144.

[17] The inverse of the statistics documented here: Hertz. Understanding mobility in America.

[18] A translation of the statistic "21 percent less." Robertson, R. E. (2003). Women's earnings: Work patterns partially explain difference between men's and women's earnings. *United States General Accounting Office*. 29.gao.gov/new.items/d0435.pdf

[19] Milanović. *The haves and the have-nots*. 115–123.

[20] Black, R. (2011, August 23). Species count put at 8.7 million. *BBC News*. bbc.com/news/science-environment-14616161

[21] Alexander, M. (2018, December 21). None of us deserve citizenship. *The New York Times*. nytimes.com/2018/12/21/opinion/sunday/immigration-border-policy-citizenship.html

[22] Noah, T. (2010, Sept 3).The United States of inequality. Slate. slate.com/articles/news_and_politics/the_great_divergence/features/2010/the_united_states_of_inequality/introducing_the_great_divergence.html; Ingraham, C. (2017, Dec 6). The richest 1 percent now owns more of the country's wealth than at any time in the past 50 years. *Washington Post*. washingtonpost.com/news/wonk/wp/2017/12/06/the-richest-1-percent-now-owns-more-of-the-countrys-wealth-than-at-any-time-in-the-past-50-years; Hanauer, N. & Rolf, D.M. (2020, Sept 14). The Top 1% of Americans Have Taken $50 Trillion From the Bottom 90%—And That's Made the U.S. Less Secure. *Time*. time.com/5888024/50-trillion-income-inequality-america

[23] Merton, T. (2011). *Thoughts in solitude*. Farrar, Straus and Giroux. 33.

28 - Feel the Wind, Drink the Stars, and Take Only What You Need

[1] Kimmerer, R. W. (2013). *Braiding sweetgrass: Indigenous wisdom, scientific knowledge and the teachings of plants*. Milkweed Editions. 177.

[2] Kimmerer. *Braiding sweetgrass*. 189,190.

[3] Turner, G. & Alexander, C. (2014, Sept 1). Limits to Growth was right. New research shows we're nearing collapse. *The Guardian*. theguardian.com/commentisfree/2014/sep/02/limits-to-growth-was-right-new-research-shows-were-nearing-collapse

[4] Du Pisani, J.A. (2006). Sustainable development – historical roots of the concept. *Environmental Sciences*. 3:2, 83–96. doi.org/10.1080/15693430600688831; Castro, C. J. (2004). Sustainable development: Mainstream and critical perspectives. *Organization & Environment*, 17(2), 195–225.jstor.org/stable/26162866; Kaur, M. (2020, Nov 19). Shifting paradigms: Embedding the economy in nature. bu.edu/eci/2020/11/19/shifting-paradigms-embedding-the-economy; History of SD. (n.d.). Sustainable Development Commission. sd-commission.org.uk/pages/history_sd.html

[5] Castro. Sustainable development.

[6] Kimmerer. *Braiding sweetgrass*. 177.

[7] Kimmerer. 180.

[8] Kimmerer. 183.

[9] Kimmerer. 181.

[10] Foer, J. S. (2009). *Eating animals*. United States: Little, Brown.

[11] Kimmerer. 185.

[12] Kimmerer. 185.

[13] Biswas-Diener, R. (2008). Material wealth and subjective well-being. *The Science of Subjective Well-Being*, 312; Brown, K. W., & Kasser, T. (2005). Are psychological and ecological well-being compatible? The role of values, mindfulness, and lifestyle. *Social Indicators Research*, 74(2), 349–368; Rich, S. A., Hanna, S., Wright, B. J., & Bennett, P. C. (2017). Fact or fable: Increased wellbeing in Voluntary Simplicity. *International Journal of Wellbeing*, 7(2), 64–77.doi.org/10.5502/ijw.v7i2.589

29 - The Foundation of the Universe: Being Abundance

[1] Buckminster fuller's holistic view. (1991, Nov 27). The Chronicle of Higher Education. chronicle.com/article/buckminster-fullers-holistic-view

[2] Sieden, L. S. (2012). *A fuller view*. Divine Arts; Twist, L., & Barker, T. (2003). *The soul of money: Transforming your relationship with money and life*. WW Norton & Company. 58–60.

[3] New International Version. Leviticus 25:8–22. biblehub.com/leviticus/25-23.htm; Sider, R. J. (2005). Rich Christians in an age of hunger: Moving from affluence to generosity. Thomas Nelson. 66–74.

[4] Sider. Rich Christians. 75.

[5] Sawhill, I. V., & Pulliam, P. (2019, June 25). Six facts about wealth in the United States. *Brookings*. brookings.edu/blog/up-front/2019/06/25/six-facts-about-wealth-in-the-united-states

[6] Silver, C. (2020, December 24). The top 25 economies in the world. *Investopedia*. investopedia.com/insights/worlds-top-economies

[7] Sacks, J. (n.d.). Teshuvah, tefilla and tzedakah. *Chabad.org*. chabad.org/holidays/JewishNewYear/template_cdo/aid/4453/jewish/Teshuvah-Tefilla-and-Tzedakah.htm; Spitzer, J. (n.d.). Pe'ah: The corners of our fields. *My Jewish Learning*.myjewishlearning.com/article/peah-the-corners-of-our-fields; Dosick, W. D. (2009). *Living Judaism: The complete guide to Jewish belief, tradition, and practice*. Harper Collins. 249–452; New International Version. Corinthians 8:1–15. https://biblehub.com/1_corinthians/8-1.htm; Zakât Foundation of America. (2008). *The Zakat Handbook: A Practical Guide for Muslims in the West*. AuthorHouse. xiii, 2–18; Qazi, M. (June 12, 2018). Zakat reawakens spirit of camaraderie. *Daily Sabah*. dailysabah.com/feature/2018/06/12/zakat-reawakens-spirit-of-camaraderie; Al-Dhâriyât, 51:15–19; Findly, E. B. (2003). *Dāna: giving and getting in Pali Buddhism* (No. 52). Motilal Banarsidass Publishing. xiv; Dictionary of Pāli Proper Names.(2016, 23 December). What Buddha Said. what-buddha-said.net/library/DPPN/wtb/b_f/daana.htm; Bodhi, & Buddha. (2015). *In the Buddha's words: An anthology of discourses from the Pāli canon*. Wisdom Publications. 125–26 (cāga = generosity); Buddhavaṃsa, Chapter 2, as cited in Pāramitā. (n.d.). Wikipedia. en.wikipedia.org/wiki/Paramita; Long, J.D. (2008) Charity and service in the Hindu tradition: Reflections and ideas for a. ONE Seva. hafsite.org/pdf/ONE percent20Seva percent20Guide.pdf

[8] Kidwell, C.S. (1990, May-June). True Indian giving. Foundations News, Council on Foundations. 27–29; Berry, M., & Adamson, R. (n.d.). The wisdom of the giveaway: A guide to native American philanthropy, (Extension Course). *First Nations Development Institute*. 3–6. philanthropy.org/documents/FG_TheWisdomoftheGiveawayAGuidetoNativeAmericanPhilanthropy_000.pdf; Hanson,

J.H. (1997, Feb 1). Power, philanthropy and potlatch: what tribal exchange rituals can tell us about giving, Fund Raising Management. allbusiness.com/specialty-businesses/non-profit-businesses/608912-1.html

[9] Swimme, B. (1996). *The Hidden heart of the cosmos: humanity and the new story*. Orbis Books. 39–44.

[10] Calculate the impact you can have. (n.n.). *The Life You Can Save*. thelifeyoucansave.org/impact-calculator

[11] Singer, P. (2010). *The life you can save: How to do your part to end world poverty*. Random House Incorporated. 100.

[12] Singer. *The life you can save*. 100.

[13] Singer. 85–89; Wenar, L. (2011). Poverty is no pond: Challenges for the affluent. In Wenar, L., Illingworth, P., & Pogge, T. (Eds.), *Giving well: The ethics of philanthropy*. United Kingdom: Oxford University Press, USA. 124.

[14] Angelou, M. (1994). *Wouldn't take nothing for my journey now*. Bantam Books. 13–15.

[15] Singer. *The life you can save*. 15–18.

[16] Jerri. [@LikeBureau]. (2020, November 17). Let's be clear: Dolly Parton is a millionaire and not a billionaire because she *keeps giving money away*. Being a [Tweet]. Twitter. twitter.com/LikeBureau/status/1328738250152845312

[17] Collins, C., Rogers, P., & Garner, J. P. (2001). *Robin Hood was right: A guide to giving your money for social change*. WW Norton & Company. 16.

[18] Singer. *The life you can save*. 24; The Center on Philanthropy at Indiana University. Giving USA 2011, executive summary. (2011). *Giving USA Foundation*. 6, 7. americansforthearts.org/sites/default/files/GivingUSA_2011_ExecSummary1.pdf; The Center for Global Prosperity (CGP). (2011). Index of global philanthropy and remittances 2011. *Hudson Institute*. 30. hudson.org/content/researchattachments/attachment/977/2011_index_of_global_philanthropy_and_remittances_downloadable_version.pdf. Only 9 percent of our sharing goes to domestic human services, and I assume most or all of that goes to meeting basic needs. About 5 percent goes to health programs for "underserved populations," though what percentage of that goes to meeting basic needs is unclear. I unscientifically assume half, so 2.5 percent. That brings the total to roughly 11.5 percent. 4.3 percent of all giving goes to international programs, which includes things like exchange programs, peace and security, etc. Assuming rather unscientifically that half of that goes to aid to help meet basic needs, that yields an additional 2, bringing the total to roughly 13.5 percent. Then, Singer notes that of the 1/3 of charity that goes to religious institutions, less than 10 percent is passed on as aid for developing nations (a number that according to Giving USA excludes missionary activities, meaning it's fair to assume the whole amount is aimed at meeting basic needs). That would mean less than 3 percent of all charitable giving. I assume roughly 2 percent. That brings our total to roughly 15.5 percent. (Singer notes that some additional small amount of donations to education goes to helping students from developing countries and to research on poverty, which I assume is not significant enough to impact our final figure. Swagel, P.L. (2009, June 11). Broad benefits: Health-related giving by private and community foundations. philanthropycollaborative.org/BroadBenefits061109.pdf. Swagel found that roughly 68 percent of health-related funding by private and community foundations goes to "underserved populations." The numbers used here assume that the portion of *all* health-related giving in the US that went to underserved populations is roughly the same.

[19] Buffett, W. (n.d.). My philanthropic pledge. The Giving Pledge. givingpledge.org/Pledger.aspx?id=177

[20] Buffett. My philanthropic pledge.

[21] The Giving Pledge. (n.d.). givingpledge.org

[22] Business insider India. (2022, June 16). Top 10 richest people in the world. businessinsider.in/business/news/top-10-richest-people-in-the-world/articleshow/74415117.cms; Gose, B. (2022, Feb 16). Elon Musk Gave $5.7 Billion to Charity in 2021, SEC Filing Reveals. *The Chronicle of Philanthropy*. philanthropy.com/article/elon-musk-gave-5-7-billion-to-charity-in-2021-sec-filing-reveals

[23] Schleifer, T. (2018, September 13). $2 billion in charity is not enough for Jeff Bezos to slink out of the public limelight. *Vox*. vox.com/2018/9/13/17855950/jeff-bezos-donation-2-billion-philanthropy-charity-amazon-ceo; Wamsley, L. (June 15, 2021). MacKenzie Scott is giving away another $2.7 billion to 286 organizations. NPR. npr.org/2021/06/15/1006829212/mackenzie-scott-is-giving-away-another-2-7-billion-to-286-organizations

[24] This following report is from 2017 (Anderson, S. & Wakamo, B. [2021, Dec 15]. The year in inequality in 10 charts. Inequality.org. inequality.org/great-divide/year-in-inequality-10-charts), but the amount of wealth that America's richest people control has only grown since then (Kirsch, N. [2017, November 9]. The 3 richest Americans hold more wealth than bottom 50 percent of the country, study finds. *Forbes*.f orbes.com/sites/noahkirsch/2017/11/09/the-3-richest-americans-hold-more-wealth-than-bottom-50-of-country-study-finds

[25] Singer. *The life you can save*. 151–52.

[26] Singer. 163–65.

[27] Singer. 100. 173.

[28] *Reading Adrienne Rich: Reviews and re-visions*, 1951-81. (1984). United States: University of Michigan Press. 106.

[29] Tierney, T. J., & Fleishman, J. L. (2011). *Give smart: Philanthropy that gets results*. PublicAffairs. 13–14.

[30] The Life You Can Save. (2010, June 6). *The Life You Can Save in 3 Minutes* [Video]. YouTube. youtube.com/watch?v=onsIdBanynY

[31] Hübl, T. (2014, July 31). Pure giving is participating in creation [Speech audio recording]. Celebrate Life Festival. soundcloud.com/thomashuebl/pure-giving-is-participating-in-creation

31 - Awakening from a Dream Into the Universe

[1] Kimmerer, R. W. (2013). *Braiding sweetgrass: Indigenous wisdom, scientific knowledge and the teachings of plants*. Milkweed Editions. 308.

32 - A Love Song to You and the World

[1] Hafiz. (1999). The Gift: Poems by the Great Sufi Master. United Kingdom: Arkana. 160–61.

[2] hooks, b. (2018). *All about love: New visions*. HarperCollins Publishers. 67–68.

[3] Brown, B. (2012). *Daring greatly: How the courage to be vulnerable transforms the way we live, love, parent, and lead*. Penguin. 146.

[4] Brown, B. (2017). *Braving the wilderness: The quest for true belonging and the courage to stand alone*. Random House. 14, 40.

[5] Turner, T-P. (2017). *Belonging: Remembering ourselves home*. Her Own Room Press. 145.

[6] Goodall, J., & Berman, P. (1999). *Reason for hope: A spiritual journey*. Grand Central Publishing. 173–174.

[7] Kimmerer, R. W. (2013). *Braiding sweetgrass: Indigenous wisdom, scientific knowledge and the teachings of plants*. Milkweed Editions. 47.

[8] Buhner, S. H. (2014). *Plant intelligence and the imaginal realm: Beyond the doors of perception into the dreaming of Earth*. Inner Traditions/Bear. 33.

[9] Buhner. *Plant intelligence*. 32.

[10] Buhner. 34–36.

[11] Buhner. 32, 62, 63.

[12] Buhner. 61.

[13] Pomeroy, R. (2012, Jan 23). Scientists have learned from cases of animal cruelty. Real Clear Science. realclearscience.com/blog/2012/01/scientists-can-be-cruel.html

[14] Crazy like a loon. (2020, September 23). Go Wild Institute. gowildinstitute.org/crazy-like-a-loon

[15] Lorde, A. (2004). Conversations with Audre Lorde. University Press of Mississippi. 91.

[16] Buhner. Plant intelligence. 18.

[17] Buhner. 19.

[18] Buhner. 19.

[19] Buhner. 17.

[20] Twitchell, J. B. (1999). Lead us into temptation: The triumph of American materialism. Columbia University Press. 56.

[21] See Chapter 12; Homo sapiens emerged roughly 300,000 years ago, the planet formed bout 4.5 billion years ago. Scharping, N. (2020, May 15). When did Homo sapiens first appear? Discover Magazine.discovermagazine.com/planet-earth/when-did-homo-sapiens-first-appear; Redd, N. T. (2016, October 31). How was earth formed? Space. space.com/19175-how-was-earth-formed.html; Gibbons, A. (2012, June 13). Bonobos join chimps as closest human relatives. Science. sciencemag.org/news/2012/06/bonobos-join-chimps-closest-human-relatives; Bryner, J. (2006, November 9). Surprise! Your cousin's a sea urchin. LiveScience. livescience.com/1103-surprise-cousin-sea-urchin.html; Sagan, C. (2006). The varieties of scientific experience: A personal view of the search for God. Penguin. 38–39; Johnson, G. (2014, July 21). Beyond energy, matter, time and space. The New York Times .nytimes.com/2014/07/22/science/beyond-energy-matter-time-and-space.html

[22] Sagan. The varieties. 64.

[23] Johnston, L. (2018, March 14). "I'm not afraid": What Stephen Hawking said about God, his atheism and his own death. The Washington Post. washingtonpost.com/news/acts-of-faith/wp/2018/03/14/im-not-afraid-what-stephen-hawking-said-about-god-his-atheism-and-his-own-death/?utm_term=.ce90f4a8e7f7 ; Also Carl Sagan (Varieties) and Friedrich Nietzsche among the most prominent.

[24] Ferris, T. (1998). The shole Shebang: A state of the universe report. United States: Simon & Schuster. 291.

[25] Strawson, G. (2016, May 16). Consciousness isn't a mystery. It's matter. The New York Times. nytimes.com/2016/05/16/opinion/consciousness-isnt-a-mystery-its-matter.html

[26] Wilber, K. (2001). Quantum questions: Mystical writings of the world's great physicists. Shambhala Publications. 3–9.

[27] Wilber. Quantum questions. 135.

[28] Wilber. 6.

[29] Schrödinger gave the example of delighting in music and having a song make us cry. "Science, we believe, can, in principle, describe in full detail all that happens ... in our sensorium and 'motorium' from the moment the waves of compression and dilation reach our ear to the moment when certain glands secrete a salty fluid that emerges from our eyes. But of the feelings of delight and sorrow that accompany the process science is completely ignorant." (Wilber. Quantum questions. 84, 5.)

[30] Wilber. 83.

[31] Philosopher Thomas Nagel writes that "the scientific image," which many today believe is the only way to accurately represent reality, consists of the language of physics, chemistry, biology, and neurophysiology. He notes that it is not unscientific to acknowledge that there is more to reality than this scientific image can account for. "The spectacular progress of the physical sciences since the seventeenth century was made possible by the exclusion of the mental from their purview ... Science will have to expand to accommodate facts of a kind fundamentally different from those that physics is designed to explain." Nagel, T. (2017, March 9). Is consciousness an illusion? The New York Review. 32, 34

[32] Wilber. Quantum questions. 84.

[33] Buhner. Plant intelligence. 64.

[34] Buhner. 64.

[35] Houston, J. (1996). A mythic life: Learning to live our greater story. HarperSanFrancisco. 65.

[36] Ladinsky, D. (2002). Love poems from God: Twelve sacred voices from the East and West. Penguin. 87.

[37] Lipka, M., & Gecewicz, C. (2017, September 6). More Americans now say they're spiritual but not religious. Pew Research Center. pewresearch.org/fact-tank/2017/09/06/more-americans-now-say-theyre-spiritual-but-not-religious

[38] Lipka, M. (2016, August 24). Why America's "nones" left religion behind. Pew Research Center. pewresearch.org/fact-tank/2016/08/24/why-americas-nones-left-religion-behind

[39] The On Being Project. (2016, Japril 7). David Whyte: Seeking language large enough. onbeing.org/programs/david-whyte-seeking-language-large-enough

[40] Wilber. Quantum questions. 107–13.

[41] Jammer, M. (2011). Einstein and religion: Physics and theology. Princeton University Press. 75; Calaprice, A. (2000). The expanded quotable Einstein. Princeton University Press. 217; Der Einstein-Gutkind brief. (2013, Nov 13). Richard Dawkins Foundation. de.richarddawkins.net/articles/der-einstein-gutkind-brief-mit-transkript-und-englischer-ubersetzung

[42] Raner, G. H. (2004, November 9). Einstein on his personal religious views: Guy H. Raner. Freedom from Religion Foundation. ffrf.org/publications/item/13319-einstein-on-his-personal-religious-views

[43] Raner. Einstein on his personal.

[44] Jammer. Einstein and religion. 97.

[45] Jammer. Einstein and religion. 51, 149.

[46] Einstein, A. (1931). Living Philosophies. Simon and Schuster. 6.

[47] Ladinsky. Love poems from God. 115.

[48] Ladinsky. 125, 148.

[49] Ladinsky. 169.

[50] Ladinsky. 199.

[51] Schachter-Shalomi, Z., & Segel, J. (2013). Jewish with feeling: A guide to meaningful Jewish practice. Jewish Lights Publishing. 5, 7.

[52] Ladinsky. Love poems from God. 277.

[53] Ladinsky. 303.

[54] Ladinsky. 350.

[55] Schachter-Shalomi & Segel, Jewish with feeling. 6.

[56] Phillips, J. B. (2004). Your God is too small: A guide for believers and skeptics alike. Simon and Schuster. 26.

[57] Phillips. Your God is too small. 7, 8.

[58] Forrest church interview. (2009, February 27). PBS. pbs.org/wnet/religionandethics/2009/02/27/february-27-2009-forrest-church-interview/860

[59] I first heard this quote from Rabbi Shir Yaakov Feit, who I am deeply grateful to for this and so much more. Elber, M. (2017). Sacred now: Cultivating Jewish spiritual consciousness. Wipf and Stock Publishers. 24. These words have been

echoed by many others, including Anglican priest Giles Fraser, Rabbi Zalman Schachter-Shalomi, and theologian David Bentley Hart: Fraser, G. (2015, February 2). I don't believe in the God that Stephen Fry doesn't believe in either. *The Guardian*. theguardian.com/commentisfree/2015/feb/02/stephen-fry-god-christianity-evil-maniac; Burkeman, O. (2014, January 14). The one theology book all atheists really should read. *The Guardian*. theguardian.com/news/oliver-burkeman-s-blog/2014/jan/14/the-theology-book-atheists-should-read; Schachter-Shalomi & Segel, *Jewish with feeling*. 8.

[60] McNulty, T. (2007). The wisdom of trees. In Heinemann, B.W. (Ed.), *The nature of wisdom: Inspirations from the natural world*. Barnes & Noble, Inc. 8

[61] Taylor, B. B. (2000). *The luminous web: Essays on science and religion*. Cowley Publications. 54–55.

[62] Ladinsky. *Love poems from God*. 97.

[63] Keeney, B. (2010). *The bushman way of tracking God: the original spirituality of the Kalahari people*. Simon and Schuster. 18.

[64] Kabdebó, L. The "gaps" of the western mind and modern poetics. *Neohelicon* 34, 145–194 (2007). https://doi.org/10.1007/s11059-007-1014-x

[65] Ladinsky. *Love poems from God*. 141.

[66] Ladinsky. 147.

[67] Ladinsky. 132.

[68] Ladinsky. 205.

[69] Ladinsky. 202.

[70] Ladinsky. 179.

[71] Ladinsky. 162.

[72] Hafiz. The Gift. 222.

[73] Ladinsky. *Love poems from God*. 112.

[74] Ladinsky. 90.

[75] Ladinsky. 105.

[76] Ladinsky. 306.

[77] Brown, J. E. (1953). *The sacred pipe; Black Elk's account of the seven rites of the Oglala Sioux*. Norman: University of Oklahoma Press. 115.

[78] Ladinsky. *Love poems from God*. 272.

[79] Ladinsky. 276.

[80] Ladinsky. 295.

[81] Ellsberg, R. (2016). *Blessed among us: Day by day with saintly witnesses*. Liturgical Press. 292.

[82] Ladinsky. *Love poems from God*. 29.

[83] Ladinsky. 35.

[84] Ladinsky. 76.

[85] Ladinsky. 77.

[86] Ladinsky. 19.

[87] I have not been able to find this exact quote, which Carl Sagan cites in *Pale Blue Dot* (Sagan, C., & Druyan, A. (2011). *Pale blue dot*. Ballantine Books. 332). However, I did find related sentiments (James, W. (1961). *The varieties of religious experience*. Amazon Distribution GmbH. 47–48, 51). Simiilar ideas from others: Smith, W. C. (1987). *Faith and belief*. Princeton University Press.12; Clebsch, W. A. (1973). *American religious thought: A history*. The University of Chicago Press. 187.

[88] Gilmore, M. (2003, February 11). Feeling at home in a scientific world. *Los Angeles Times*. latimes.com/archives/la-xpm-2003-feb-11-oe-gilmore11-story.html

[89] Buhner. *Plant intelligence*. 20, 21.

[90] Waldron, J. L. (1998). The life impact of transcendent experiences with a pronounced quality of noesis. *Journal of Transpersonal Psychology, 30*(2), 115.

[91] Yaden, D. B., Haidt, J., Hood, R. W., Vago, D. R., & Newberg, A. B. (2017). The varieties of self-transcendent experience. *Review of General Psychology*, *21*(2), 143–160. doi.org/10.1037/gpr0000102; Levin, J. & Steele, L. (2005). The transcendent experience. *Explore, 1*, 89–101. baylorisr.org/wp-content/uploads/levin_transcendent.pdf; Neher, A. (2013). *Paranormal and transcendental experience: A psychological examination*. Courier Corporation. 122–31.

[92] Davidson, S. (2014). *The December project: An extraordinary rabbi and a skeptical seeker confront life's greatest mystery*. HarperOne. 93.

[93] Lewis, C. S. (2001). A grief observed. United Kingdom: HarperCollins. 56.

[94] Treu, A. (2012, March 10). The voice from the burning bush. The Jewish Theological Seminary. jtsa.edu/the-voice-from-the-burning-bush

[95] Martin, J. (2014, December 18). Finding god in all things. *On Being*. onbeing.org/programs/james-martin-finding-god-in-all-things

[96] Ladinsky. *Love poems from God*. 1, 12.

[97] Schachter-Shalomi & Segel, *Jewish with feeling*. 18, 19.

[98] Keeney. *The bushman way*.

[99] Schachter-Shalomi & Segel, *Jewish with feeling*. 5, 7.

[100] Wilber. *Quantum questions*. 3.

[101] Wilber. 194-195.

[102] Buhner. *Plant intelligence*. 243.

[103] Smith, E. E. (2017, June 30). What a "transcendent experience" really means. *The Cut*. thecut.com/article/what-a-transcendent-experience-really-means.html

[104] Neher. *Paranormal and transcendental experience*. 123.

[105] Wiederkehr, M. (2008). Seven sacred pauses: Living mindfully through the hours of the day. United States: Ave Maria Press. 178

[106] Ehrmann, M. (1927). Desiderata. All Poetry. allpoetry.com/desiderata---words-for-life.

[107] Merton, T. (2009). *Conjectures of a guilty bystander*. Image. 140–42.

[108] Rohde, D. L., Olson, S., & Chang, J. T. (2004). Modelling the recent common ancestry of all living humans. *Nature, 431*(7008), 562–566. doi.org/10.1038/nature02842

[109] Scharping. When did *Homo sapiens*?

[110] Human evolution interactive timeline. (n.d.). Human Origins Initiative. humanorigins.si.edu/evidence/human-evolution-timeline-interactive

[111] MacGregor, N. (2011). *A history of the world in 100 objects*. United Kingdom: Penguin Publishing Group. 18; How many words should a child know? (2009, February 2). *The Language Fix*. languagefix.wordpress.com/2009/02/02/how-many-words-should-a-child-know

[112] Human evolution interactive timeline.

[113] Wayman. E. (2012, October 31). Five early primates you should know. *Smithsonian Magazine*.

smithsonianmag.com/science-nature/five-early-primates-you-should-know-102122862; The rise of mammals. (n.d.). *PBS*. pbs.org/wgbh/evolution/library/03/1/l_031_01.html; Pappas, S. (2017, March 1). What was the first life on earth? *LiveScience*. livescience.com/57942-what-was-first-life-on-earth.html
[114] Swimme, B. (1984). *The universe is a green dragon: A cosmic creation story*. Bear & Company.
[115] Early human migrations. (n.d.). *Wikipedia*. en.wikipedia.org/wiki/Early_human_migrations; Settlement of the Americas. (n.d.). *Wikipedia*. en.wikipedia.org/wiki/Models_of_migration_to_the_New_World
[116] Gates, H. L. (n.d.). How many slaves landed in the US? *PBS* .pbs.org/wnet/african-americans-many-rivers-to-cross/history/how-many-slaves-landed-in-the-us
[117] Courtois, F. (1986). *An experience of enlightenment*. Theosophical Pub. House. 24–25.
[118] Oxford essential quotations. (2017). *Oxford University Press, (5)*. doi.org/10.1093/acref/9780191843730.001.0001
[119] Ladinsky. *Love poems from God*. 91.
[120] Timmons, J. (2018, January 4). When can a fetus hear? *Healthline*. healthline.com/health/pregnancy/when-can-a-fetus-hear#Will-my-baby-to-be-recognize-my-voice?; Mosher, D. (2020, Oct 23). Babies cry in the womb and 18 other surprising facts I learned when I became a dad. *The Independent*. independent.co.uk/life-style/health-and-families/babies-cry-in-the-womb-and-18-other-surprising-facts-i-learned-when-i-became-a-dad-a7315136.html
[121] Angier, N. (2008, Oct 20). The wonders of blood. *New York Times*. nytimes.com/2008/10/21/science/21angi.html
[122] Biological half-life. (n.d.). *Wikipedia*. en.wikipedia.org/wiki/Biological_half-life
[123] Sweetlove, L. (2011). Number of species on Earth tagged at 8.7 million. *Nature*. doi.org/10.1038/news.2011.498
[124] Füleky, G. (n.d.). Cultivated plants, primarily as food sources. *Szent István University, Gödöllő, Hungary*. eolss.net/sample-chapters/C10/E5-02.pdf
[125] The body's ecosystem. (2014, August 1). *The Scientist*.the-scientist.com/features/the-bodys-ecosystem-37085; Human microbiome: Your body is an ecosystem. (2016). *American Museum of Natural History*. amnh.org/content/download/131241/2201972/file/human_microbiome_your_body_is_an_ecosystem_stepread1.pdf; Laura Sanders, L. (2016, March 23). Microbes can play games with the mind. *Science News*. sciencenews.org/article/microbes-can-play-games-mind; Gallagher, J. (2018, April 10). More than half your body is not human. *BBC News*.bbc.com/news/health-43674270
[126] Roger, A. J., Muñoz-Gómez, S. A., & Kamikawa, R. (2017). The origin and diversification of mitochondria. *Current Biology, 27*(21). doi.org/10.1016/j.cub.2017.09.015
[127] Kasprak, A. (2018, April 30). Does the human body replace itself every seven years? *Snopes*. snopes.com/fact-check/does-body-replace-itself-seven-years; Radford, B. (2011, April 4). Does the human body really replace itself every 7 years? *Live Science*. livescience.com/33179-does-human-body-replace-cells-seven-years.html; Kestenbaum, D. (2007, July 14). Atomic tune-up: how the body rejuvenates itself. *NPR* .npr.org/templates/story/story.php?storyId=11893583
[128] Macy, J. (2007). *World as lover, world as self a guide to living fully in turbulent times*. Parallax Press. 30.
[129] Zaehner, R. C. (1969). *Mysticism sacred and profane: An inquiry into some varieties of praeternatural experience*. Oxford University Press. 41–42.
[130] Hay, L. L., Holden, R. (2015). *Life loves you: 7 spiritual practices to heal your life*. United States: Hay House, Incorporated. 25.
[131] Hay & Holden. *Life loves you*. 1–30.
[132] Kimmerer. *Braiding sweetgrass*. 122.
[133] Buhner. *Plant intelligence*. 16.
[134] Alberts, B., Johnson, A., Lewis, J., et al. (2002). Molecular biology of the cell. 4th edition. *National Center for Biotechnology Information*. ncbi.nlm.nih.gov/books/NBK26866/
[135] Balter, M. (2005, May). The seeds of civilization. *Smithsonian Magazine*. smithsonianmag.com/history/the-seeds-of-civilization-78015429
[136] June, L. (n.d.). Reclaiming our Indigenous European roots. *The Moon Magazine*. moonmagazine.org/lyla-june-reclaiming-our-indigenous-european-roots-2018-12-02
[137] Diener, E., & Oishi, S. (2000). Money and happiness: Income and subjective well-being across nations. In Diener E., & Suhg, E. M. (Eds.), *Culture and subjective well-being*. The MIT Press. 207.
[138] Ponce De Leon, J., Subcomandante Marcos., Carrigan, A. (2011). *Our word is our weapon: Selected writings*. Seven Stories Press. 18.
[139] Transcript: Greta Thunberg's speech at the U.N. climate action summit. (2019, September 23). *NPR*. npr.org/2019/09/23/763452863/transcript-greta-thunbergs-speech-at-the-u-n-climate-action-summit
[140] Buhner. *Plant intelligence*. 243.
[141] King Jr, M. L. (1963). Letter from Birmingham jail. csuchico.edu/iege/_assets/documents/susi-letter-from-birmingham-jail.pdf
[142] (1857) Frederick Douglass, "If there is no struggle, there is no progress." (2007, January 25). *BlackPast.org*. blackpast.org/african-american-history/1857-frederick-douglass-if-there-no-struggle-there-no-progress
[143] This quote is all over the Internet, and seems very much in character, but I couldn't find a primary source. derrickjensen.org
[144] Kimmerer. *Braiding sweetgrass*. 124
[145] Nouwen, H. J. M. (1992). *Life of the beloved*. United States: Crossroad. 106.
[146] Kimmerer. 340.
[147] Roy, A. (2003, February 20). Confronting empire. *The Nation*. thenation.com/article/archive/confronting-empire
[148] Roy, A. (2006). *Ordinary person's guide to empire*. Penguin Books India. 44.
[149] I am grateful to Rabbi Shir Yaakov, who first told me this story, reportedly involving journalist Dan Rather. I cannot find any transcript or clear citations for this quote. Dan Rather did definitely interview Mother Teresa; his references to it are frequent. Still, the best I can offer is that poet and academic Scott Cairns writes about the life-changing effect it had on him watching the interview and hearing Mother Teresa say these words. Cairns, S. (2017). Vocation, poetry, and prayer. In Bezzerides, A. M., & Prodromou, E. H. (Eds.). (2017). *Eastern Orthodox Christianity and American higher education: Theological, historical, and contemporary reflections*. University of Notre Dame Press. 381. doi.org/10.2307/j.ctvpj76z9
[150] Hāfiz. The Gift. 160-61.
[151] Brach, T. (2020). *Radical compassion: Learning to love yourself and your world with the practice of RAIN*. United Kingdom: Penguin Books, Limited. 169.
[152] Hāfiz. The Gift. 280.

Printed in Great Britain
by Amazon

38904950R00208